SCHAUM'S OUTLINE OF

THEORY AND PROBLEMS

OF

LOGIC
Second Edition

•

JOHN NOLT, Ph.D.
Associate Professor of Philosophy
University of Tennessee

DENNIS ROHATYN, Ph.D.
Professor of Philosophy
University of San Diego

ACHILLE VARZI, Ph.D.
Assistant Professor of Philosophy
Columbia University

•

SCHAUM'S OUTLINE SERIES
McGRAW-HILL

New York   San Francisco   Washington, D.C.   Auckland   Bogotá   Caracas   Lisbon
London   Madrid   Mexico City   Milan   Montreal   New Delhi
San Juan   Singapore   Sydney   Tokyo   Toronto

JOHN NOLT is Associate Professor of Philosophy at the University of Tennessee, Knoxville, where he has taught since receiving his doctorate from Ohio State University in 1978. He is the author of *Informal Logic: Possible Worlds and Imagination* and numerous articles on logic, metaphysics, and the philosophy of mathematics.

DENNIS ROHATYN is Professor of Philosophy at the University of San Diego, where he has taught since 1977. He is the author of *Two Dogmas of Philosophy*, *The Reluctant Naturalist*, and many other works. He is a regular symposiast on critical thinking at national and regional conferences. In 1987 he founded the Society for Orwellian Studies.

ACHILLE VARZI is Assistant Professor of Philosophy at Columbia University, New York. His works include *Holes and Other Superficialities* and *Fifty Years of Events: An Annotated Bibliography* (both with Roberto Casati) and numerous articles on logic, formal semantics, and analytic metaphysics.

Schaum's Outline of Theory and Problems of
LOGIC

Copyright © 1998, 1988 by The McGraw-Hill Companies, Inc. All rights reserved. Printed in the United States of America. Except as permitted under the Copyright Act of 1976, no part of this publication may be reproduced or distributed in any forms or by any means, or stored in a data base or retrieval system, without the prior written permission of the publisher.

14 15 16 17 18 19 20 CUS CUS 0

ISBN 0-07-046649-1

Sponsoring Editor: Barbara Gilson
Production Supervisor: Pamela Pelton
Editing Supervisor: Maureen B. Walker
Project Supervision: Keyword Publishing Services Ltd.

Library of Congress Cataloging-in-Publication Data

Nolt, John Eric
    Schaum's outline of theory and problems of logic / John Nolt ,
  Dennis Rohatyn, — 2nd ed.
      p. cm. — (Schaum's outline series)
    Includes bibliographical references and index.
    ISBN 0-07-046649-1
    1. Logic--Outlines, syllabi , etc.   I. Rohatyn. Dennis A.
  II. Title.
  BC108.N65 1998
  160 '. 2 ' 02—dc21   98-28482
                            CIP

McGraw-Hill

A Division of The McGraw-Hill Companies

Preface

The roots of logic may be traced to Aristotle, who systematized and codified the subject in a way that was not significantly surpassed for over two millennia. Modern logic, however, stems largely from the work of the German philosopher Gottlob Frege in the late nineteenth century, and has developed tremendously during the twentieth century. Today logic has applications in many areas besides philosophy, including mathematics, linguistics, engineering, and computer science. The aim of this book is to serve as an introduction and reference text to students in all these related fields.

We begin by examining reasoning as it occurs informally in writing and conversation. In so doing, we have occasion to introduce some of the central concepts of logic (such as argument, validity, truth, evidence) while avoiding technicalities. Chapter 1 concerns structural (syntactic) matters, and Chapter 2 presents some fundamental semantic concepts along with basic criteria for argument evaluation.

Chapters 3 and 4 introduce the most elementary system of formal logic, propositional logic, from the semantic point of view (truth tables and refutation trees) and from the syntactic or deductive point of view (the propositional calculus), respectively. Chapter 5 covers the logic of categorical statements, the modern descendant of Aristotle's logical theory. Predicate logic, which is Frege's brainchild, forms the subject of Chapters 6 and 7. This is an overarching system which unifies and extends the systems of the three previous chapters, and is the core of all modern logic. Again we consider it first from a semantic and then from a deductive point of view (the predicate calculus).

In Chapters 8 and 9 we return to an informal standpoint to consider common fallacies in reasoning and some important forms of inductive (probabilistic) argument. Chapter 10 treats probability more rigorously, laying out the axioms and major theorems of the probability calculus.

Finally, Chapter 11 sketches some ways in which predicate logic itself can be strengthened or generalized. We consider, among other things, its expressive limitations, its extensions to stronger systems (higher-order and modal logics), and its application to arithmetic and to the theory of definitions.

The book presupposes no previous acquaintance with the subject and may be used as a text for an introductory course, as a problem supplement to other texts, or as a guide for self-study. As a textbook, there is more material here than can be covered in a single course, and some omission will generally be necessary. All later chapters presuppose the concepts introduced in Chapters 1 and 2, so that these two chapters are indispensable. Thereafter, however, a good bit of flexibility is possible. The following table indicates dependencies that should be taken into account in planning a course:

Chapter	Presupposes Chapter(s)
2	1
3	1, 2
4	1, 2, 3
5	1, 2

6	1, 2, 3, 4, 5
7	1, 2, 3, 4, 5, 6
8	1, 2
9	1, 2
10	1, 2, 3, 4
11	1, 2, 3, 4, 5, 6, 7

This second edition is based on the first edition by John Nolt and Dennis Rohatyn, which appeared in 1988. Every chapter has been revised and updated, and many problems and examples have been added. The most substantial changes occur in Chapters 3 and 4, which have been entirely reorganized and partly rewritten, and in Chapters 6 and 7, which are the result of rearranging and expanding the original Chapter 6. Overall, however, the pedagogical features and style of presentation of the first edition have been preserved.

JOHN NOLT
DENNIS ROHATYN
ACHILLE VARZI

Contents

Chapter 1

Argument Structure

1.1 WHAT IS AN ARGUMENT?

Logic is the study of arguments. An *argument* is a sequence of statements of which one is intended as a *conclusion* and the others, the *premises*, are intended to prove or at least provide some evidence for the conclusion. Here are two simple examples:

> All humans are mortal. Socrates is human. Therefore, Socrates is mortal.
>
> Albert was not at the party, so he cannot have stolen your bag.

In the first argument, the first two statements are premises intended to prove the conclusion that Socrates is mortal. In the second argument, the premise that Albert was not at the party is offered as evidence for the conclusion that he cannot have stolen the bag.

The premises and conclusion of an argument are always *statements* or *propositions*,[1] as opposed to questions, commands, or exclamations. A statement is an assertion that is either true or false (as the case may be) and is typically expressed by a declarative sentence.[2] Here are some more examples:

> Dogs do not fly.
>
> Robert Musil wrote *The Man Without Qualities*.
>
> Brussels is either in Belgium or in Holland.
>
> Snow is red.
>
> My brother is an entomologist.

The first three sentences express statements that are in fact true. The fourth sentence expresses a false statement. And the last sentence can be used to express different statements in different contexts, and will be true or false depending on whether or not the brother of the speaker is in fact an entomologist. By contrast, the following sentences do not express any statements:

> Who is the author of *The Man Without Qualities*?
>
> Please do not call after 11pm.
>
> Come on!

Nonstatements, such as questions, commands, or exclamations, are neither true nor false. They may sometimes suggest premises or conclusions, but they are never themselves premises or conclusions.

SOLVED PROBLEM

1.1   Some of the following are arguments. Identify their premises and conclusions.

(*a*)   He's a Leo, since he was born in the first week of August.

(*b*)   How can the economy be improving? The trade deficit is rising every day.

---

[1]Philosophers sometimes draw a distinction between statements and propositions, but it is not necessary to make that distinction here.

[2]The distinction between a statement or proposition and the sentence used to express it is important. A sentence can be ambiguous or context-dependent, and can therefore express any of two or more statements—even statements that disagree in their being true or false. (Our fifth example below is a case in point.) However, where there is no danger of confusion we shall avoid prolixity by suppressing the distinction. For example, we shall often use the term 'argument' to denote sequences of statements (as in our definition) as well as the sequences of sentences which express them.

(c)   I can't go to bed, Mom. The movie's not over yet.

(d)   The building was a shabby, soot-covered brownstone in a decaying neighbor-hood. The scurrying of rats echoed in the empty halls.

(e)   Everyone who is as talented as you are should receive a higher education. Go to college!

(f)   We were vastly outnumbered and outgunned by the enemy, and their troops were constantly being reinforced while our forces were dwindling. Thus a direct frontal assault would have been suicidal.

(g)   He was breathing and therefore alive.

(h)   Is there anyone here who understands this document?

(i)   Many in the U.S. do not know whether their country supports or opposes an international ban on the use of land mines.

(j)   Triangle *ABC* is equiangular. Therefore each of its interior angles measures 60 degrees.

Solution

(a)   Premise: He was born in the first week of August.
Conclusion: He's a Leo.

(b)   Technically this is not an argument, because the first sentence is a question; but the question is merely rhetorical, suggesting the following argument:
Premise: The trade deficit is rising every day.
Conclusion: The economy cannot be improving.

(c)   Premise: The movie's not over yet.
Conclusion: I can't go to bed.

(d)   Not an argument; there is no attempt here to provide evidence for a conclusion.

(e)   Not an argument; 'Go to college!' expresses a command, not a statement. Yet the following argument is suggested:
Premise: Everyone who is as talented as you are should receive a higher education.
Conclusion: You should go to college.

(f)   Premise: We were vastly outnumbered and outgunned by the enemy.
Premise: Their troops were constantly being reinforced while our forces were dwindling.
Conclusion: A direct frontal assault would have been suicidal.

(g)   Though grammatically this is a single sentence, it makes two distinct statements, which together constitute the following argument:
Premise: He was breathing.
Conclusion: He was alive.

(h)   Not an argument.

(i)   Not an argument.

(j)   Premise: Triangle *ABC* is equiangular.
Conclusion: Each of its interior angles measures 60 degrees.

Though the premises of an argument must be *intended* to prove or provide evidence for the conclusion, they need not *actually* do so. There are bad arguments as well as good ones. Argument 1.1(c), for example, may be none too convincing; yet still it qualifies as an argument. The purpose of logic is precisely to develop methods and techniques to tell good arguments from bad ones.[3]

---

[3]For evaluative purposes, it may be useful to regard the argument in 1.1(c) as incomplete, requiring for its completion the implicit premise 'I can't go to bed until the movie is over'. (Implicit statements will be discussed in Section 1.6.) Even so, in most contexts this premise would itself be dubious enough to deprive the argument of any rationally compelling persuasive force.

Since we are concerned in this chapter with argument structure, not argument evaluation, we shall usually not comment on the quality of arguments used as examples in this chapter. In no case does this lack of comment constitute a tacit endorsement.

Notice also that whereas the conclusion occurs at the end of the arguments in our initial examples and in most of the arguments in Problem 1.1, in argument 1.1(c) it occurs at the beginning. The conclusion may in fact occur anywhere in the argument, but the beginning and end are the most common positions. For purposes of analysis, however, it is customary to list the premises first, each on a separate line, and then to give the conclusion. The conclusion is often marked by the symbol '∴', which means "therefore." This format is called *standard form*. Thus the standard form of our initial example is:

> All humans are mortal.
> Socrates is human.
> ∴ Socrates is mortal.

1.2 IDENTIFYING ARGUMENTS

Argument occurs only when someone intends a set of premises to support or prove a conclusion. This intention is often expressed by the use of *inference indicators*. Inference indicators are words or phrases used to signal the presence of an argument. They are of two kinds: *conclusion indicators*, which signal that the sentence which contains them or to which they are prefixed is a conclusion from previously stated premises, and *premise indicators*, which signal that the sentence to which they are prefixed is a premise. Here are some typical examples of each (these lists are by no means exhaustive):

Conclusion Indicators	*Premise Indicators*
Therefore	For
Thus	Since
Hence	Because
So	Assuming that
For this reason	Seeing that
Accordingly	Granted that
Consequently	This is true because
This being so	The reason is that
It follows that	For the reason that
The moral is	In view of the fact that
Which proves that	It is a fact that
Which means that	As shown by the fact that
From which we can infer that	Given that
As a result	Inasmuch as
In conclusion	One cannot doubt that

Premise and conclusion indicators are the main clues in identifying arguments and analyzing their structure. When placed between two sentences to form a compound sentence, a conclusion indicator signals that the first expresses a premise and the second a conclusion from that premise (possibly along with others). In the same context, a premise indicator signals just the reverse. Thus, in the compound sentence

> He is not at home, so he has gone to the movie.

the conclusion indicator 'so' signals that 'He has gone to the movie' is a conclusion supported by the premise 'He is not at home'. But in the compound sentence

> He is not at home, since he has gone to the movie.

the premise indicator 'since' indicates that 'He has gone to the movie' is a premise supporting the conclusion 'He is not at home'. We can also have

> Since he is not at home, he has gone to the movie.

In this case, the premise indicator at the beginning of the sentence signals that the first subsentence is a premise supporting the second. Premise indicators seldom occur at the beginning of a noncompound sentence (except for 'It is a fact that', 'One cannot doubt that', and the like); but when they do, they indicate that the sentence is a premise supporting a previously stated conclusion.

SOLVED PROBLEMS

1.2 Use the inference indicators of the argument below to determine its inferential structure. Then write it in standard form.

> ①[Gold-argon compounds are not likely to be produced even in the laboratory, much less in nature,] (since) ②[it is difficult to make argon react with anything,] and (since) ③[gold, too, forms few compounds.]

Solution

We have circled the inference indicators for emphasis and bracketed and numbered each statement for ease of reference. Grammatically, the argument consists of a compound sentence whose three component sentences are linked together by two occurrences of the premise indicator 'since'. Each occurrence of 'since' introduces a premise. Statement 1, which is linked to statements 2 and 3 by the occurrences of 'since', is the conclusion. In standard form the argument is:

> It is difficult to make argon react with anything.
>
> Gold, too, forms few compounds.
>
> ∴ Gold-argon compounds are not likely to be produced even in the laboratory, much less in nature.

1.3 Use the inference indicators of the argument below to determine its inferential structure. Then write it in standard form.

> ①[Inflation has dropped considerably, while interest rates have remained high.] (Therefore) ②[in real terms borrowing has become more expensive,] (since) ③[under these conditions borrowed money cannot (as it could when inflation was higher) be paid back in highly inflated dollars.]

Solution

The conclusion indicator 'therefore' typically introduces a conclusion. But what follows it here is the compound sentence consisting of sentences 2 and 3 linked by the premise indicator 'since'. 'Since' signals that statement 2 is a conclusion from statement 3. And 'therefore', since it is prefixed to statement 2, indicates that 2 is a conclusion from 1. Thus the argument consists of two premises, statements 1 and 3, supporting a single conclusion, statement 2. In standard form it is:

> Inflation has dropped considerably, while interest rates have remained high.
>
> Under these conditions borrowed money cannot (as it could when inflation was higher) be paid back in highly inflated dollars.
>
> ∴ In real terms borrowing has become more expensive.

Expressions which function in some contexts as inference indicators generally have other functions in other contexts. Thus, the notion of an inference indicator should not be taken too rigidly. For example, the word 'since' in

It has been six years since we went to France.

indicates duration, not an inference, and hence it is not functioning as a premise indicator. Likewise, the word 'thus' in

He was angry, and he remained thus for several days.

is not a conclusion indicator. It means "in this condition," not "therefore." Explanations of motives or causes which use words like 'since' or 'because' are especially difficult to distinguish from arguments. If, for example, you ask a man the question

Why did you sell your bike?

and he replies

I sold it because I needed the money.

this reply is merely an explanation. In this context 'because' is not a premise indicator. No argument is present; the respondent is merely explaining why he sold his bicycle, not trying to prove that he did so.

Some arguments have no indicators at all (we have seen an example in 1.1(*c*)). In such cases, we must rely on contextual clues or our understanding of the author's intentions in order to differentiate premises from conclusions.

SOLVED PROBLEMS

1.4   Rewrite the argument below in standard form.

①[Al Capone was not all that clever.] ②[Had he been cleverer, the IRS would never have gotten him convicted on income tax evasion charges.]

Solution

It is clear that sentence 2 is *intended* to serve as evidence for sentence 1, even though the argument contains no indicators, and even though the evidence that sentence 2 actually provides is flimsy. The argument is:

Had Al Capone been cleverer, the IRS would never have gotten him convicted on income tax evasion charges.

∴ Al Capone was not all that clever.[4]

1.5   Rewrite the argument below in standard form.

①[Some politicians are hypocrites.] ②[They say we should pay more taxes if the national deficit is to be kept under control.] But then ③[they waste huge amounts of money on their election campaigns.]

Solution

The author's intention is to establish that some politicians are hypocrites. In standard form:

Some politicians say we should pay more taxes if the national deficit is to be kept under control.

They waste huge amounts of money on their election campaigns.

∴ They are hypocrites.

---

[4]Like argument (*c*) of Problem 1.1, this argument may be regarded as incomplete, for the author obviously also assumes that the IRS did get Capone convicted on an income tax dodge, though this assumption is not explicitly stated.

1.3 COMPLEX ARGUMENTS

Some arguments proceed in stages. First a conclusion is drawn from a set of premises; then that conclusion (perhaps in conjunction with some other statements) is used as a premise to draw a further conclusion, which may in turn function as a premise for yet another conclusion, and so on. Such a structure is called a *complex argument*. Those premises which are intended as conclusions from previous premises are called *nonbasic premises* or *intermediate conclusions* (the two names reflect their dual role as conclusions of one step and premises of the next). Those which are not conclusions from previous premises are called *basic premises* or *assumptions*. For example, the following argument is complex:

> All rational numbers are expressible as a ratio of integers. But pi is not expressible as a ratio of integers. Therefore pi is not a rational number. Yet clearly pi is a number. Thus there exists at least one nonrational number.

The conclusion is that there exists at least one nonrational number (namely, pi). This is supported directly by the premises 'pi is not a rational number' and 'pi is a number'. But the first of these premises is in turn an intermediate conclusion from the premises 'all rational numbers are expressible as a ratio of integers' and 'pi is not expressible as a ratio of integers'. These further premises, together with the statement 'pi is a number', are the basic premises (assumptions) of the argument. Thus the standard form of the argument above is:

> All rational numbers are expressible as a ratio of integers.
> Pi is not expressible as a ratio of integers.
> ∴ Pi is not a rational number.
> Pi is a number.
> ∴ There exists at least one nonrational number.

Each of the simple steps of reasoning which are linked together to form a complex argument is an argument in its own right. The complex argument above consists of two such steps. The first three statements make up the first, and the second three make up the second. The third statement is a component of both steps, functioning as the conclusion of the first and a premise of the second. With respect to the complex argument as a whole, however, it counts as a (nonbasic) premise.

SOLVED PROBLEMS

1.6 Rewrite the argument below in standard form.

> ①[You needn't worry about subzero temperatures in June even on the highest peaks.] ②[It never has gotten that cold in the summer months,] and (so) ③[it probably never will.]

Solution

'So' is a conclusion indicator, signaling that statement 3 follows from statement 2. But the ultimate conclusion is statement 1. Hence this is a complex argument with the following structure:

> It never has gotten below zero even on the highest peaks in the summer months.
> ∴ It probably never will.
> ∴ You needn't worry about subzero temperatures in June even on the highest peaks.

1.7 Rewrite the argument below in standard form:

> ①[Arthur said he will go to the party,] (which means that) ②[Judith will go too.] (So) ③[she won't be able to go to the movie with us.]

Solution

'Which means that' and 'so' are both conclusion indicators: the former signals a preliminary conclusion (statement 2) from which the ultimate conclusion (statement 3) is inferred. The argument has the following standard form:

> Arthur said he will go to the party.
> ∴ Judith will go to the party too.
> ∴ She won't be able to go to the movie with us.

1.4 ARGUMENT DIAGRAMS

Argument diagrams are a convenient way of representing inferential structure. To diagram an argument, circle the inference indicators and bracket and number each statement, as in Problems 1.2 to 1.7. If several premises function together in a single step of reasoning, write their numbers in a horizontal row, joined by plus signs, and underline this row of numbers. If a step of reasoning has only one premise, simply write its number. In either case, draw an arrow downward from the number(s) representing a premise (or premises) to the number representing the conclusion of the step. Repeat this procedure if the argument contains more than one step (a complex argument).

SOLVED PROBLEM

1.8   Diagram the argument below.

①[Today is either Tuesday or Wednesday.] But ②[it can't be Wednesday,] ⟨since⟩ ③[the doctor's office was open this morning,] and ④[that office is always closed on Wednesday.] ⟨Therefore,⟩ ⑤[today must be Tuesday.]

Solution

The premise indicator 'since' signals that statements 3 and 4 are premises supporting statement 2. The conclusion indicator 'therefore' signals that statement 5 is a conclusion from previously stated premises. Consideration of the context and meaning of each sentence reveals that the premises directly supporting 5 are 1 and 2. Thus the argument should be diagramed as follows:

$$\frac{3 \; + \; 4}{}$$
$$\downarrow$$
$$\frac{1 \; + \; 2}{}$$
$$\downarrow$$
$$5$$

The plus signs in the diagram mean "together with" or "in conjunction with," and the arrows mean "is intended as evidence for." Thus the meaning of the diagram of Problem 1.8 is: "3 together with 4 is intended as evidence for 2, which together with 1 is intended as evidence for 5."

An argument diagram displays the structure of the argument at a glance. Each arrow represents a single step of reasoning. In Problem 1.8 there are two steps, one from 3 and 4 to 2 and one from 1 and 2 to 5. Numbers toward which no arrows point represent basic premises. Numbers with arrows pointing both toward and away from them designate nonbasic premises. The number at the bottom of the diagram with one or more arrows pointing toward it but none pointing away represents the final conclusion.[5] The basic premises in Problem 1.8 are statements 1, 3, and 4; statement 2 is a nonbasic premise, and statement 5 is the final conclusion.

---

[5]Some authors allow diagrams that exhibit more than one final conclusion, but we will adopt the convention of splitting up such diagrams into as many separate diagrams as there are final conclusions (these may all have the same premises).

Argument diagrams are especially convenient when an argument has more than one step.

SOLVED PROBLEM

1.9 Diagram the following argument:

①[Watts is in Los Angeles] and ②[is ⟨therefore⟩ in the United States] and ⟨hence⟩ ③[is part of a fully industrialized nation.] ④[It is ⟨thus⟩ not a part of the third world,] ⟨since⟩ ⑤[the third world is made up exclusively of developing nations] and ⑥[developing nations are by definition not fully industrialized.]

Solution

The words 'therefore', 'hence', and 'thus' are conclusion indicators, signifying that the sentence following or containing them is a conclusion from previously stated premises. (2 and 3 are not complete sentences, since the subject term 'Watts' is missing. Yet it is clear that each expresses a statement; hence we bracket them accordingly.) 'Since' is a premise indicator, which shows that statements 5 and 6 are intended to support statement 4. The term 'thus' in statement 4 shows that 4 is also a conclusion from 3. Thus, 3, 5, and 6 function together as premises for 4. The argument can be diagramed as follows:

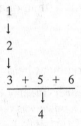

Because of the great variability of English grammar, there are no simple, rigorous rules for bracket placement. But there are some general principles. The overriding consideration is to bracket the argument in the way which best reveals its inferential structure. Thus, for example, if two phrases are joined by an inference indicator, they should be bracketed as separate units regardless of whether or not they are grammatically complete sentences, since the indicator signals that one expresses a premise and the other a conclusion. Problems 1.8 and 1.9 illustrate this principle.

It is also generally convenient to separate sentences joined by 'and', as we did with statements 3 and 4 in Problem 1.8 and statements 5 and 6 in Problem 1.9. This is especially important if only one of the two is a conclusion from previous premises (as will be the case with statements 2 and 3 in Problem 1.21, below), though it is not so crucial elsewhere. Later, however, we shall encounter contexts in which it is useful to treat sentences joined by 'and' as a single unit. 'And' usually indicates parallel function. Thus, for example, if one of two sentences joined by 'and' is a premise supporting a certain conclusion, the other is likely also to be a premise supporting that conclusion.

Some compound sentences, however, should never be bracketed off into their components, since breaking them up changes their meaning. Two common locutions which form compounds of this sort are 'either . . . or' and 'if . . . then'. (Sometimes the terms 'either' and 'then' are omitted.) Someone who asserts, for example, 'Either it will stop raining or the river will flood' is saying neither that it will stop raining nor that the river will flood. He or she is saying merely that one or the other will happen. To break this sentence into its components is to alter the thought. Similarly, saying 'If it doesn't stop raining, the river will flood' is not equivalent to saying that it will not stop raining and that the river will flood. The sentence means only that a flood will occur *if* it doesn't stop raining. This is a conditional statement that must be treated as a single unit.

Notice, by contrast, that if someone says '*Since* it won't stop raining, the river will flood', that person really is asserting both that it won't stop raining and that the river will flood. 'Since' is a premise indicator in this context, so the sentences it joins should be treated as separate units in argument analysis. Locutions like 'either . . . or' and 'if . . . then' are not inference indicators. Their function will be discussed in Chapters 3 and 4.

SOLVED PROBLEM

1.10 Diagram the argument below.

①[Either the UFOs are secret enemy weapons or they are spaceships from an alien world.] ②[If they are enemy weapons, then enemy technology is (contrary to current thinking) vastly superior to ours.] ③[If they are alien spacecraft, then they display a technology beyond anything we can even imagine.] In any case, (therefore) ④[their builders are more sophisticated technologically than we are.]

Solution

The conclusion indicator 'therefore' (together with the qualification 'in any case') signals that statement 4 is a conclusion supported by all the preceding statements. Note that these are bracketed without breaking them into their components. Thus the diagram is:

$$\frac{1 + 2 + 3}{\downarrow} \\ 4$$

In addition to 'either . . . or' and 'if . . . then', there are a variety of other locutions which join two or more sentences into compounds which should always be treated as single units in argument analysis. Some of the most common are:

Only if
Provided that
If and only if
Neither . . . nor
Unless
Until
When
Before

'Since' and 'because' also form unbreakable compounds when they are not used as premise indicators.

SOLVED PROBLEMS

1.11 Diagram the argument below.

①[I knew her even before she went to Nepal,] (so) ②[it was well before she returned that I first met her.] (Since) ③[you did not meet her until after she returned,] ④[I met her before you did.]

Solution

$$1 \\ \downarrow \\ \frac{2 + 3}{\downarrow} \\ 4$$

Notice that the compound sentences formed by 'before' and 'until' are treated as single units.

1.12   Diagram the argument below.

①[The check is void unless it is cashed within 30 days.] ②[The date on the check is September 2,] and ③[it is now October 8.] ⟨Therefore⟩ ④[the check is now void.] ⑤[You cannot cash a check which is void.] ⟨So⟩ ⑥[you cannot cash this one.]

Solution

$$\frac{1 \;+\; 2 \;+\; 3}{\downarrow}$$
$$\frac{4 \;+\; 5}{\downarrow}$$
$$6$$

Notice that premise 1, a compound sentence joined by 'unless', is treated as a single unit.

Often an argument is interspersed with material extraneous to the argument. Sometimes two or more arguments are intertwined in the same passage. In such cases we bracket and number all statements as usual, but only those numbers representing statements that are parts of a particular argument should appear in its diagram.

SOLVED PROBLEM

1.13   Diagram the argument below.

①[She could not have known that the money was missing from the safe,] ⟨since⟩②[she had no access to the safe itself.] ③[If she had known the money was missing, there is no reason to think that she wouldn't have reported it.] But ⟨since⟩①[she couldn't have known,] ④[there was nothing she could have done.] And ⑤[even if she could have done something, it was already too late to prevent the crime;] ⑥[the money was gone.] ⟨Therefore,⟩⑦[she bears no guilt in this incident.]

Solution

$$2$$
$$\downarrow$$
$$1$$
$$\downarrow$$
$$4$$
$$\downarrow$$
$$7$$

Notice that statement 1 occurs twice, the second time in a slightly abbreviated version. To prevent the confusion that might result if the same sentence had two numbers, we label it 1 in both its first and second occurrences. Statements 3, 5, and 6 make no direct contribution to the argument and thus are omitted from the diagram. However, 5 and 6 may be regarded as a separate argument inserted into the main line of reasoning, with 6 as the premise and 5 as the conclusion:

$$6$$
$$\downarrow$$
$$5$$

1.5   CONVERGENT ARGUMENTS

If an argument contains several steps of reasoning which all support the same (final or intermediate) conclusion, the argument is said to be *convergent*. Consider:

One should quit smoking. It is very unhealthy, and it is annoying to the bystanders.

Here the statements that smoking is unhealthy and that it is annoying function as independent reasons for the conclusion that one should quit smoking. We do not, for example, need to assume the first premise in order to understand the step from the second premise to the conclusion. Thus, we should not diagram this argument by linking the two premises and drawing a single arrow to the conclusion, as in the examples considered so far. Rather, each premise should have its own arrow pointing toward the conclusion. A similar situation may occur at any step in a complex argument. In general, therefore, a diagram may contain numbers with more than one arrow pointing toward them.

SOLVED PROBLEM

1.14 Diagram the argument below.

①[The Bensons must be home.] ②[Their front door is open.] ③[their car is in the driveway,] and ④[their television is on,] (since) ⑤[I can see its glow through the window.]

Solution

The argument is convergent. Statements 2, 3, and 4 function as independent reasons for the conclusion, statement 1. Each supports statement 1 separately, and must therefore be linked to it by a separate arrow.

Premises should be linked by plus signs, by contrast, when they do not function independently, i.e., when each requires completion by the others in order for the argument to make good sense.

SOLVED PROBLEM

1.15 Diagram the argument below.

①[Everyone at this party is a biochemist.] and ②[all biochemists are intelligent.] (Therefore,) (since) ③[Sally is at this party,] ④[Sally is intelligent.]

Solution

$$\frac{1 + 2 + 3}{\downarrow}$$
$$4$$

The argument is not convergent; each of its premises requires completion by the others. Taken by themselves, none of the premises would make good sense as support for statement 4.

Incidentally, note that the argument contains a premise indicator, 'since', immediately following a conclusion indicator, 'therefore'. This is a relatively common construction. It signals that the first statement following the premise indicator (in this case, 3) is a premise supporting the second (in this case, 4), and also that the second is supported by previously given premises.

Convergent arguments exhibit many different patterns. Sometimes separate lines of reasoning converge on intermediate conclusions, rather than on final conclusions. Sometimes they converge on both.

SOLVED PROBLEM

1.16   Diagram the argument below.

①[The Lions are likely to lose this final game,] ⟨for three reaons:⟩ ②[their star quarterback is sidelined with a knee injury,] ③[morale is low after two disappointing defeats,] and ④[this is a road game] and ⑤[they've done poorly on the road all season.] ⑥[If they lose this one, the coach will almost certainly be fired.] But ⟨that's not the only reason to think that⟩ ⑦[his job is in jeopardy.] ⟨For⟩ ⑧[he has been accused by some of the players of closing his eyes to drug abuse among the team,] and ⑨[no coach who lets his players use drugs can expect to retain his post.]

Solution

This argument exhibits a complex convergent structure:

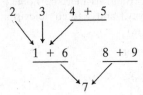

1.6   IMPLICIT STATEMENTS

It is often useful to regard certain arguments as incompletely expressed. Argument 1.1(*c*) and the argument of Problem 1.4, for instance, can be thought of as having unstated assumptions (see the footnotes concerning these arguments). There are also cases in which it is clear that the author wishes the audience to draw an unstated conclusion. For instance:

> One of us must do the dishes, and it's not going to be me.

Here the speaker is clearly suggesting that the hearer should do the dishes, since no other possibility is left open.

SOLVED PROBLEM

1.17   Complete and diagram the following incomplete argument:

①[It was certain that none of the President's top advisers had leaked the information,] and yet ②[it had indeed been leaked to the press.]

Solution

These two statements are premises which suggest the implicit conclusion:

③[Someone other than the President's top advisers leaked the information to the press.]

Thus the diagram is:

$$\frac{1 + 2}{\downarrow}$$
$$3$$

Implicit premises or conclusions should be "read into" an argument only if they are required to complete the arguer's thought. No statement should be added unless it clearly would be accepted by the arguer, since in analyzing an argument, it is the arguer's thought that we are trying to understand. The primary constraint governing interpolation of premises and conclusions is the *principle of charity*: in formulating implicit statements, give the arguer the benefit of the doubt; try to make the argument as strong as possible while remaining faithful to what you know of the arguer's thought. The point is to minimize misinterpretation, whether deliberate or accidental. (Occasionally we may have reason to restructure a bad argument in a way that corrects and hence departs from the arguer's thought. But in that case we are no longer considering the original argument; we are creating a new, though related, argument of our own.)

SOLVED PROBLEM

1.18 Complete and diagram the following incomplete argument:

①[Karla is an atheist,] (which just goes to show that) ②[you don't have to believe in God to be a good person.]

Solution

We first consider a solution which is incorrect. Suppose someone were to reply to this argument, "Well, that's a ridiculous thing to say; look, you're assuming that all atheists are good people." Now this alleged assumption is one way of completing the author's thought, but it is not a charitable one. This assumption is obviously false, and it is therefore unlikely to have been what the author had in mind. Moreover, the argument is not meant to apply to *all* atheists; there is no need to assume anything so sweeping to support the conclusion. What is in fact assumed is probably something more like:

③[Karla is a good person.]

This may well be true, and it yields a reasonably strong argument while remaining faithful to what we know of the author's thought. Thus a charitable interpretation of the argument is:

$$\frac{1 + 3}{\downarrow}$$
$$2$$

Sometimes, both the conclusion and one or more premises are implicit. In fact, an entire argument may be expressed by a single sentence.

SOLVED PROBLEMS

1.19 Complete and diagram the following incomplete argument.

①[If you were my friend, you wouldn't talk behind my back.]

Solution

This sentence suggests both an unstated premise and an unstated conclusion. The premise is:

②[You do talk behind my back.]

And the conclusion is:

③[You aren't my friend.]

Thus the diagram is:

$$\frac{1 + 2}{\downarrow}$$
$$3$$

1.20 Complete and diagram the following incomplete argument.

①[The liquid leaking from your engine is water.] ②[There are only three liquids in the engine: water, gasoline, and oil.] ③[The liquid that is leaking is not oil,] (because) ④[it is not viscous,] and ⑤[it is not gasoline,] (since) ⑥[it has no odor.]

Solution

The premise indicator 'because' signals that statement 4 is a premise supporting statement 3. But this step obviously depends on the additional assumption:

⑦[Oil is viscous.]

Likewise, the premise indicator 'since' shows that statement 6 supports statement 5, again with an additional assumption:

⑧[Gasoline has an odor.]

The conclusion of the argument is statement 1. Though no further inference indicators are present, it is clear that statements 2, 3, and 5 are intended to support statement 1. For the sake of completeness, we may also add the rather obvious assumption:

⑨[A liquid is leaking from your engine.]

The diagram is:

$$
\begin{array}{ccccc}
4 + 7 & & 6 + 8 \\
\downarrow & & \downarrow \\
9 + 2 + \quad 3 \quad + & & 5 \\
\downarrow \\
1
\end{array}
$$

Many arguments, of course, are complete as stated. The arguments of our initial examples and of Problems 1.8 and 1.10, for instance, have no implicit premises or conclusions. These are clear examples of completely stated arguments. In less clear cases, the decision to regard the argument as having an implicit premise may depend on the degree of rigor which the context demands. Consider, for instance, the argument of Problem 1.3. If we need to be very exacting—as is the case when we are formalizing arguments (see Chapters 3 and 6)—it may be appropriate to point out that the author makes the unstated assumption:

Borrowed money paid back in highly inflated dollars is less expensive in real terms than borrowed money paid back in less inflated dollars.

In ordinary informal contexts, however, this much rigor amounts to laboring the obvious and may not be worth the trouble.

We conclude this section with a complex argument that makes several substantial implicit assumptions.

SOLVED PROBLEM

1.21 This argument is from the great didactic poem *De rerum natura* (*On the Nature of the Universe*) by the Roman philosopher Lucretius. Diagram it and supply missing premises where necessary.

①[{The atoms that comprise spirit} are obviously far smaller than those of swift-flowing water or mist or smoke,] (since) ②[it far outstrips them in mobility] and ③[is moved by a far slighter impetus.] Indeed, ④[it is actually moved by images of smoke and mist.] So, for instance, ⑤[when we are sunk in sleep, we may see altars sending up clouds of steam and giving off smoke;] and (we cannot doubt that) ⑥[we are here dealing with images.] Now we see that ⑦[water flows out in all directions from a broken vessel and the moisture is dissipated, and mist and smoke vanish into thin air.] Be assured, (therefore,) that ⑧[spirit is similarly dispelled and vanishes far more speedily and is sooner dissolved into its component atoms once it has been let loose from the human frame.]

Solution

In order to grasp the general structure of the argument, we shall first diagram it without adding the implicit assumptions. The conclusion indicator 'therefore' signals that statement 8 is a conclusion from statement 7. The position of statement 8 at the end of the argument strongly suggests that it is the final conclusion. The premise indicator 'since' shows that statements 2 and 3 support statement 1. The rest of the structure must be inferred from more subtle clues.

Notice that the 'so' preceding 5 does not function here as a conclusion indicator. Rather, together with the phrase 'for instance', it signals that 5 is an example of the idea expressed in statement 4. Examples are often used to support the claims they exemplify, and so it is here: statement 5 is a premise supporting statement 4. Statement 4 bears a similar relation to statement 3; it is an example of the idea in statement 3 which is intended to support it. The word 'and' connecting 5 and 6 suggests parallel function, and indeed we can see that 6 works together with 5 in support of 4.

This leaves us with the question of how statement 1 is related to the final conclusion, statement 8. The connection may not be easy to see without some knowledge of Lucretius' theory of atoms. Lucretius believed that substances are fluid in proportion to the smallness and smoothness of their atoms. Keeping this in mind, and recalling that statement 7 supports statement 8, it becomes apparent that statements 1 and 7 work in conjunction to support statement 8. Thus, without adding any premises, we may diagram the argument as follows:

$$
\begin{array}{c}
\underline{5 \ + \ 6} \\
\downarrow \\
4 \\
\downarrow \\
\underline{2 \ + \ 3} \\
\downarrow \\
\underline{1 \ + \ 7} \\
\downarrow \\
8
\end{array}
$$

We now consider the matter of implicit premises. One of these has already been mentioned. We saw that in the step from statements 1 and 7 to statement 8, Lucretius was assuming something like this:

⑨[Substances are swift-flowing (mobile) in proportion to the smallness of their component atoms.]

(We omit mention of smoothness, since it plays no role in the argument.) It is evident that this same assumption is needed to complete the step from statements 2 and 3 to statement 1. The step from 4 to 3 depends on the assumption

⑩[Water, smoke, and mist are not moved by images of smoke or mist.]

in order to establish the comparison in 3 between the impetus necessary to move spirit and the impetus necessary to move water, smoke, or mist. And finally, the step from statements 5 and 6 to statement 4 obviously assumes that

⑪[Things seen in sleep move the spirit.]

Thus the complete argument is:

$$
\begin{array}{c}
\underline{5 \ + \ 6 \ + \ 11} \\
\downarrow \\
\underline{4 \ + \ 10} \\
\downarrow \\
\underline{2 \ + \ 3 \ + 9} \\
\downarrow \\
\underline{1 \ + \ 7 \ + \ 9} \\
\downarrow \\
8
\end{array}
$$

1.7 USE AND MENTION

In any subject matter which deals extensively with language, confusion can arise as to whether an expression is being *used* to say something or *mentioned* as the subject matter of what is being said. To prevent this confusion, when an expression is mentioned rather than used, it should be enclosed in quotation marks. As is customary in logic, we shall use single quotation marks for this purpose. The following sentences both correctly employ this convention and are both true:

> Socrates was a Greek philosopher.
> 'Socrates' is a name containing eight letters.

In the first, the name 'Socrates' is used to denote the person Socrates. In the second, it is merely mentioned, as the quotation marks indicate. By contrast, the sentences

> 'Socrates' was a Greek philosopher.
> Socrates is a name containing eight letters.

are both false. The first says of the name 'Socrates' that it was a Greek philosopher, and the second says of the person Socrates that he is a name containing eight letters.

SOLVED PROBLEM

1.22 Supply quotation marks in the following sentences in such a way as to make them true:

(a)   Bill's name is Bill.

(b)   The fact that $x + y = y + x$ is expressed by the equation $x + y = y + x$.

(c)   This sentence is part of Solved Problem 1.22.

(d)   The first letter of the alphabet is A.

(e)   A is a name of the first letter of the alphabet.

Solution

(a)   Bill's name is 'Bill'.

(b)   The fact that $x + y = y + x$ is expressed by the equation '$x + y = y + x$'.

(c)   This sentence is part of Solved Problem 1.22.

(d)   The first letter of the alphabet is 'A'.

(e)   "A" is a name of the first letter of the alphabet.

No quotation marks are needed in item (c). If we wrote, for example,

This sentence is a part of 'Solved Problem 1.22'.

we would be saying that the sentence in question is a part of the expression 'Solved Problem 1.22', an expression consisting of two words and a numerical label. But that is absurd. To understand items (d) and (e), note that what is printed below is the first letter of the alphabet:

A

To form a name of this letter we add quotation marks:

'A'

This is the name used in item (d) to mention the letter. Now to form a name of this name of the letter, we add a second set of quotation marks:

"A"

And this is what we use in item (e) to mention the name of the letter.

In logic and mathematics, letters themselves are sometimes used as names or variables standing for various objects. In such uses they may stand alone without quotation marks. In item (*b*), for example, the occurrences of the letters '*x*' and '*y*', without quotation marks, function as variables designating numbers.

Another point to notice about item (*b*) (and item (*d*)) is that the period at the end of the sentence is placed after the last quotation mark, not before, as standard punctuation rules usually dictate. In logical writing, punctuation that is not actually part of the expression being mentioned is placed outside the quotation marks. This helps avoid confusion, since the expression being mentioned is always precisely the expression contained within the quotation marks.

1.8 FORMAL VS. INFORMAL LOGIC

Logic may be studied from two points of view, the formal and the informal. *Formal logic* is the study of argument *forms*, abstract patterns common to many different arguments. An argument form is something more than just the structure exhibited by an argument diagram, for it encodes something about the internal composition of the premises and conclusion. A typical argument form is exhibited below:

> If *P*, then *Q*
> *P*
> ∴ *Q*

This is a form of a single step of reasoning with two premises and a conclusion. The letters '*P*' and '*Q*' are variables which stand for propositions (statements). These two variables may be replaced by any pair of declarative sentences to produce a specific argument. Since the number of pairs of declarative sentences is potentially infinite, the form thus represents infinitely many different arguments, all having the same structure. Studying the form itself, rather than the specific arguments it represents, allows one to make important generalizations which apply to all these arguments.

Informal logic is the study of particular arguments in natural language and the contexts in which they occur. Whereas formal logic emphasizes generality and theory, informal logic concentrates on practical argument analysis. The two approaches are not opposed, but rather complement one another. In this book, the approach of Chapters 1, 2, 7, and 8 is predominantly informal. Chapters 3, 4, 5, 6, 9, and 10 exemplify a predominantly formal point of view.

Supplementary Problems

I Some of the following are arguments; some are not. For those which are, circle all inference indicators, bracket and number statements, add implicit premises or conclusions where necessary, and diagram the argument.

   (1) You should do well, since you have talent and you are a hard worker.

   (2) She promised to marry him, and so that's just what she should do. So if she backs out, she's definitely in the wrong.

   (3) We need more morphine. We've got 32 casualties and only 12 doses of morphine left.

   (4) I can't help you if I don't know what's wrong—and I just don't know what's wrong.

   (5) If wishes were horses, then beggars would ride.

   (6) If there had been a speed trap back there, it would have shown up on this radar detector, but none did.

(7) The earth is approximately 93 million miles from the sun. The moon is about 250,000 miles from the earth. Therefore, the moon is about 250,000 miles closer to the sun than the earth is.

(8) She bolted from the room and then suddenly we heard a terrifying scream.

(9) I followed the recipe on the box, but the dessert tasted awful. Some of the ingredients must have been contaminated.

(10) Hitler rose to power because the Allies had crushed the German economy after World War I. Therefore if the Allies had helped to rebuild the German economy instead of crushing it, they would never have had to deal with Hitler.

(11) [The apostle Paul's] father was a Pharisee. . . . He [Paul] did not receive a classical education, for no Pharisee would have permitted such outright Hellenism in his son, and no man with Greek training would have written the bad Greek of the Epistles. (Will Durant, *The Story of Civilization*)

(12) The contestants will be judged in accordance with four criteria: beauty, poise, intelligence, and artistic creativity. The winner will receive $50,000 and a scholarship to attend the college of her choice.

(13) Capital punishment is not a deterrent to crime. In those states which have abolished the death penalty, the rate of incidence for serious crimes is lower than in those which have retained it. Besides, capital punishment is a barbaric practice, one which has no place in any society which calls itself "civilized."

(14) Even if he were mediocre, there are a lot of mediocre judges and people and lawyers. They are entitled to a little representation, aren't they, and a little chance? We can't have all Brandeises and Frankfurters and Cardozos and stuff like that there. (Senator Roman Hruska of Nebraska, defending President Richard Nixon's attempt to appoint G. Harrold Carswell to the Supreme Court in 1970)

(15) Neither the butler nor the maid did it. That leaves the chauffeur or the cook. But the chauffeur was at the airport when the murder took place. The cook is the only one without an alibi for his whereabouts. Moreover, the heiress was poisoned. It's logical to conclude that the cook did it.

(16) The series of integers (whole numbers) is infinite. If it weren't infinite, then there would be a last (or highest) integer. But by the laws of arithmetic, you can perform the operation of addition on any arbitrarily large number, call it n, to obtain $n + 1$. Since $n + 1$ always exceeds n, there is no last (or highest) integer. Hence the series of integers is infinite.

(17) The Richter scale measures the intensity of an earthquake in increments which correspond to powers of 10. A quake which registers 6.0 is 10 times more severe than one which measures 5.0; correspondingly, one which measures 7.0 releases 10 times more energy than a 6.0, or 100 times more than a 5.0. So a famous one such as the quake in San Francisco in 1906 (8.6) or Alaska in 1964 (8.3) is actually over a thousand times more devastating than a quake with a modest 5.0 reading on the scale.

(18) Can it be that there simply is no evil? If so, why do we fear and guard against something which is not there? If our fear is unfounded, it is itself an evil, because it stabs and wrings our hearts for nothing. In fact, the evil is all the greater if we are afraid when there is nothing to fear. Therefore, either there is evil and we fear it, or the fear itself is evil. (St. Augustine, *Confessions*)

(19) The square of any number n is evenly divisible by n. Hence the square of any even number is even, since by the principle just mentioned it must be divisible by an even number, and any number divisible by an even number is even.

(20) The count is 3 and 2 on the hitter. A beautiful day for baseball here in Beantown. Capacity crowd of over 33,000 people in attendance. There's the pitch, the hitter swings and misses, strike three. That's the tenth strikeout Roger Clemens has notched in this game. He has the hitters off stride and is pitching masterfully. He should be a candidate for the Cy Young award.

(21) Parents who were abused as children are themselves more often violent with their own children than parents who were not abused. This proves that being abused as a child leads to further abuse of the next generation. Therefore the only way to stop the cycle of child abuse is to provide treatment for abused children before they themselves become parents and perpetuate this sad and dangerous problem.

(22) Assume a perfectly square billiard table and suppose a billiard ball is shot from the middle of one side on a straight trajectory at an angle of 45 degrees to that side. Then the ball will hit the middle of an adjoining side at an angle of 45 degrees. Now the ball will always rebound at an angle equal to but in the opposite direction from the angle of its approach. Hence it will be reflected at an angle of 45 degrees and hit the middle of the side opposite from where it started. Thus by the same principle it will hit the middle of the next side at an angle of 45 degrees, and hence again it will return to the point from which it started.

II Supply quotation marks in the following sentences in such a way as to make them true.

(1) The capital form of x is X.

(2) The term man may designate either all human beings or only those who are adult and male.

(3) Love is a four-letter word.

(4) Rome is known by the name the Eternal City. The Vatican is in Rome. Therefore, the Vatican is in the Eternal City.

(5) Chapter 1 of this book concerns argument structure.

(6) In formal logic, the letters P and Q are often used to designate propositions.

(7) If we use the letter P to designate the statement It is snowing and Q to designate It is cold outside, then the argument It is snowing; therefore it is cold outside is symbolized as P; therefore Q.

Answers to Selected Supplementary Problems

I   (2) ①[She promised to marry him,] and ⓢⓞ ②[that's just what she should do.] ⓢⓞ ③[if she backs out, she's definitely in the wrong.]

$$
\begin{array}{c}
1 \\
\downarrow \\
2 \\
\downarrow \\
3
\end{array}
$$

(4) ①[I can't help you if I don't know what's wrong]—and ②[I just don't know what's wrong.] ③[I can't help you.]

$$
\begin{array}{c}
1 \; + \; 2 \\
\downarrow \\
3
\end{array}
$$

(5) Not an argument.

(10) ①[Hitler rose to power because the Allies had crushed the German economy after World War I.] Ⓣⓗⓔⓡⓔⓕⓞⓡⓔ, ②[if the Allies had helped to rebuild the German economy instead of crushing it, they would never have had to deal with Hitler.]

$$
\begin{array}{c}
1 \\
\downarrow \\
2
\end{array}
$$

(11) ①[The apostle Paul's father was a Pharisee.] ②[Paul did not receive a classical education,] ⓕⓞⓡ ③[no Pharisee would have permitted such outright Hellenism in his son,] and ④[no man with Greek training would have written the bad Greek of the Epistles.]

(16) ①[The series of integers (whole numbers) is infinite.] ②[If it weren't infinite, then there would be a last (or highest) integer.] But ③[by the laws of arithmetic, you can perform the operation of addition on any arbitrarily large number, call it n, to obtain $n + 1$.] (Since) ④[$n + 1$ always exceeds n,] ⑤[there is no last (or highest) integer.] (Hence) ①[the series of integers is infinite.]

$$3 + 4$$
$$\downarrow$$
$$2 + 5$$
$$\downarrow$$
$$1$$

(22) (Assume) ①[a billiard table is perfectly square] and (suppose) ②[a billiard ball is shot from the middle of one side on a straight trajectory at an angle of 45 degrees to that side.] (Then) ③[the ball will hit the middle of an adjoining side at an angle of 45 degrees.] Now ④[the ball will always rebound at an angle equal to but in the opposite direction from the angle of its approach.] (Hence) ⑤[it will be reflected at an angle of 45 degrees and hit the middle of the side opposite from where it started.] (Thus) by the same principle ⑥[it will hit the middle of the next side at an angle of 45 degrees,] and (hence) again ⑦[it will return to the point from which it started.]

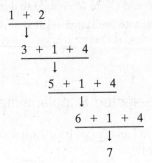

$$1 + 2$$
$$\downarrow$$
$$3 + 1 + 4$$
$$\downarrow$$
$$5 + 1 + 4$$
$$\downarrow$$
$$6 + 1 + 4$$
$$\downarrow$$
$$7$$

II   (1)   The capital form of 'x' is 'X'.

(3)   'Love' is a four-letter word.

(5)   Chapter 1 of this book concerns argument structure. (No quotation marks)

(7)   If we use the letter 'P' to designate the statement 'It is snowing' and 'Q' to designate 'It is cold outside', then the argument 'It is snowing; therefore it is cold outside' is symbolized as 'P; therefore Q'.

Chapter 2

Argument Evaluation

2.1 EVALUATIVE CRITERIA

Though an argument may have many objectives, its chief purpose is usually to demonstrate that a conclusion is true or at least likely to be true. Typically, then, arguments may be judged better or worse to the extent that they accomplish or fail to accomplish this purpose. In this chapter we examine four criteria for making such judgments: (1) whether all the premises are true; (2) whether the conclusion is at least probable, given the truth of the premises; (3) whether the premises are relevant to the conclusion; and (4) whether the conclusion is vulnerable to new evidence.

Not all of the four criteria are applicable to all arguments. If, for example, an argument is intended merely to show that a certain conclusion follows from a set of premises, whether or not these premises are true, then criterion 1 is inapplicable; and, depending on the case, criteria 3 and 4 may be inapplicable as well. Here, however, we shall be concerned with the more typical case in which it is the purpose of an argument to establish that its conclusion is indeed true or likely to be true.

2.2 TRUTH OF PREMISES

Criterion 1 is not by itself adequate for argument evaluation, but it provides a good start: no matter how good an argument is, it cannot establish the truth of its conclusion if any of its premises are false.

SOLVED PROBLEM

2.1 Evaluate the following argument with respect to criterion 1:

> Since all Americans today are isolationists, history will record that at the end of the twentieth century the United States failed as a defender of world democracy.

Solution

> The premise 'All Americans today are isolationists' is certainly false; hence the argument does not establish that the United States will fail as a defender of world democracy. This does not mean, of course, that the conclusion is false, but only that the argument is of no use in determining its truth or falsity. (One way to produce a better argument would be to make a careful study of the major forces currently shaping American foreign policy and to draw informed conclusions from that.)

Often the truth or falsity of one or more premises is unknown, so that the argument fails to establish its conclusion *so far as we know*. In such cases we lack sufficient information to apply criterion 1 reliably, and it may be necessary to suspend judgment until further information is acquired.

SOLVED PROBLEM

2.2 Evaluate the following argument with respect to criterion 1:

> There are many advanced extraterrestrial civilizations in our galaxy.
> Many of these civilizations generate electromagnetic signals powerful (and often) enough to be detected on earth.
> ∴ We have the ability to detect signals generated by extraterrestrial civilizations.

Solution

We do not yet know whether the premises of this argument are true. Hence we can do no better than to withhold judgment on it until we can reliably determine the truth or falsity of the premises. This argument should not convince anyone of the truth of its conclusion—at least not yet.

Criterion 1 requires only that the premises actually *be* true, but in practice an argument successfully communicates the truth of its conclusion only if those to whom it is addressed *know* that its premises are true. If an arguer knows that his or her premises are true but others do not, then to prove a conclusion *to them*, the arguer must provide further arguments to establish the premises.

SOLVED PROBLEM

2.3 A window has been broken. A little girl offers the following argument: "Billy broke the window. I saw him do it." In standard form:

> I saw Billy break the window.
> ∴ Billy broke the window.

Suppose we have reason to suspect that the child did not see this. Evaluate the argument with respect to criterion 1.

Solution

Even if the child is telling the truth, her argument fails to establish its conclusion to us, at least so long as we do not know that its premise is true. The best we can do for the present is to suspend judgment and seek further evidence.

Another limitation of criterion 1 is that the truth of the premises—or their being known to be true—is no guarantee that the conclusion be also true. It is a necessary condition for establishing the conclusion, but not a sufficient condition. In a good argument, the premises must also *support* the conclusion.

SOLVED PROBLEMS

2.4 Evaluate the following argument with respect to criterion 1:

> All acts of murder are acts of killing.
> ∴ Soldiers who kill in battle are murderers.

Solution

Since the premise is true, the argument satisfies criterion 1. It nevertheless fails to establish its conclusion, for the premise leaves open the possibility that some kinds of killing are not murder. Perhaps the killing done by soldiers in battle is of such a kind; the premise, at least, provides no good reason to think that it is not. Thus the premise, though true, does not adequately support the conclusion; the argument proves nothing.

2.5 Evaluate the following argument with respect to criterion 1:

> Snow is white.
> ∴ Whales are mammals.

Solution

Also in this case, the argument satisfies criterion 1: the premise is true. As a matter of fact the conclusion is true as well. Yet the argument does not itself establish the conclusion, for the premise does no job in supporting the conclusion.

These examples demonstrate the need for further criteria of argument evaluation, criteria to assess the degree to which a set of premises provides direct evidence for a conclusion. There are two main parameters one must take into account. One is probabilistic: the conclusion may be more or less probable relative to the premises. The other parameter is the relevance of the premises to the conclusion. These two parameters are respectively the concerns of our next two evaluative criteria.

2.3   VALIDITY AND INDUCTIVE PROBABILITY

Criterion 2 evaluates arguments with respect to the probability of the conclusion given the truth of the premises. In this respect, arguments may be classified into two categories: deductive and inductive. A *deductive* argument is an argument whose conclusion follows *necessarily* from its basic premises. More precisely, an argument is deductive if it is impossible for its conclusion to be false while its basic premises are all true. An *inductive* argument, by contrast, is one whose conclusion is not necessary relative to the premises: there is a certain probability that the conclusion is true if the premises are, but there is also a probability that it is false.[1]

The probability of a conclusion, given a set of premises, is called *inductive probability*. The inductive probability of a deductive argument is maximal, i.e., equal to 1 (probability is usually measured on a scale from 0 to 1). The inductive probability of an inductive argument is typically (perhaps always) less than 1.[2] Traditionally, the term 'deductive' is extended to include any argument which is intended or purports to be deductive in the sense defined above. It thus becomes necessary to distinguish between valid and invalid deductive arguments. *Valid* deductive arguments are those which are genuinely deductive in the sense defined above (i.e., their conclusions cannot be false so long as their basic premises are true). *Invalid* deductive arguments are arguments which purport to be deductive but in fact are not. (Some common kinds of "invalid deductive" arguments are discussed in Section 8.6.) Unless otherwise specified, however, we shall use the term 'deductive' in the narrower, nontraditional sense (i.e., as a synonym for 'valid' or 'valid deductive'). We adopt this usage because in practice there is frequently no answer to the question of whether or not the argument "purports" to be valid; hence, the traditional definition is in many cases simply inapplicable. Moreover, even where it can be applied it is generally beside the point; our chief concern in argument evaluation is with how well the premises actually support the conclusion (i.e., with the actual inductive probability and degree of relevance), not with how well someone claims they do.

SOLVED PROBLEM

2.6   Classify the following arguments as either deductive or inductive:

   (*a*)   No mortal can halt the passage of time.
            You are mortal.
      ∴   You cannot halt the passage of time.

---

[1]The distinction between inductive and deductive argument is drawn differently by different authors. Many define induction in ways that correspond roughly with what we, in Chapter 9, call Humean induction. Others draw the distinction on the basis of the purported or intended strength of the reasoning.

[2]This is a matter of controversy. According to some theories of inductive logic it is possible for the conclusion of an argument to be false while its premises are true and yet for the inductive probability of the argument to be 1. (See R. Carnap, *Logical Foundations of Probability*, 2d edn, Chicago, University of Chicago Press, 1962.)

(b) It is usually cloudy when it rains.

It is raining now.

∴ It is cloudy now.

(c) There are no reliably documented instances of human beings over 10 feet tall.

∴ There has never been a human being over 10 feet tall.

(d) Some pigs have wings.

All winged things sing.

∴ Some pigs sing.

(e) Everyone is either a Republican, a Democrat, or a fool.

The speaker of the House is not a Republican.

The speaker of the House is no fool.

∴ The speaker of the House is a Democrat.

(f) If there is a nuclear war, it will destroy civilization.

There will be a nuclear war.

∴ Civilization will be destroyed by a nuclear war.

(g) Chemically, potassium chloride is very similar to ordinary table salt (sodium chloride).

∴ Potassium chloride tastes like table salt.

Solution

(a)   Deductive

(b)   Inductive

(c)   Inductive

(d)   Deductive

(e)   Deductive

(f)   Deductive

(g)   Inductive

Problem 2.6 illustrates the fact that deductiveness and inductiveness are independent of the actual truth or falsity of the premises and conclusion; hence criterion 2 is independent of criterion 1 and is not by itself adequate for argument evaluation. Notice, for example, that each of the deductive arguments exhibits a different combination of truth and falsity. The premises and conclusion of Problem 2.6(a) are all true. All the statements in Problem 2.6(d), by contrast, are false. Problem 2.6(e) is a mix of truth and falsity; its first premise is surely false, but the truth and falsity of the others vary with time as House speakers come and go. None of the statements that make up Problem 2.6(f) is yet known to be true or to be false. Yet in items (e) and (f) alike the conclusion could not be false if the premises were true. Any combination of truth or falsity is possible in an inductive or a deductive argument, except that no deductive (valid) argument ever has true premises and a false conclusion, since by definition a deductive argument is one such that it is impossible for its conclusion to be false while its premises are true.

A deductive argument all of whose basic premises are true is said to be *sound*. A sound argument establishes with certainty that its conclusion is true. Argument 2.6(a), for example, is sound.

SOLVED PROBLEM

2.7 Evaluate the following argument with respect to criteria 1 and 2:

Everyone has one and only one biological father.

Full brothers have the same biological father.

No one is his own biological father.

∴ There is no one whose biological father is also his full brother.

Solution

The argument is sound. (Its assumptions are true and it is deductive.)

Notice that when we say it is impossible for the conclusion of a deductive argument to be false while the premises are true, the term "impossible" is to be understood in a very strong sense. It means not simply "impossible in practice," but *logically impossible*, i.e., impossible in its very conception.[3] The distinction is illustrated by the following problem.

SOLVED PROBLEM

2.8   Is the argument below deductive?

Tommy T. reads *The Wall Street Journal*.

∴ Tommy T. is over 3 months old.

Solution

Even though it is impossible in a practical sense for someone who is not older than 3 months to read *The Wall Street Journal*, it is still coherently conceivable; the idea itself embodies no contradiction. Thus it is logically possible (though not practically possible) for the conclusion to be false while the premise is true. In other words, the conclusion, though highly probable, is not absolutely necessary, given the premise. The argument is therefore not deductive (not valid).

On the other hand, the argument can be transformed into a deductive argument by the addition of a premise:

All readers of *The Wall Street Journal* are over 3 months old.

Tommy T. reads *The Wall Street Journal*.

∴ Tommy T. is over 3 months old.

Here it is not only practically impossible for the conclusion to be false while the premises are true; it is logically impossible. This new argument is therefore deductive.

As explained in Section 1.5, it is often useful to regard arguments like that of Problem 2.8 as incomplete and to supply the premise or premises needed to make them deductive.[4] In all such cases, however, one should ascertain that the author of the argument would have accepted (or wanted the audience to accept) the added premise as true. Supplying a premise not intended by the author unfairly distorts the argument. It is also useful to compare the argument of Problem 2.8 with the deductive arguments of Problem 2.6. In no context would any of these latter inferences require additional premises.

---

[3]Some authors define logical impossibility as violation of the laws of logic, but this presupposes some fixed conception of logical laws. Typically, these are taken to be the logical truths of formal predicate logic (see Chapter 6). But since we wish to discuss validity both in formal logical systems more extensive than predicate logic (see Chapter 11) and in informal logic, we require this broader and less precise notion.

[4]Some authors hold that all of what we are here calling "inductive arguments" are mere fragments which must be "completed" in this way before analysis, so that there are no genuine inductive arguments.

SOLVED PROBLEM

2.9   Add a premise to the following inductive argument so as to make it deductive:

> I haven't bought any food.
> You haven't bought any food.
> ∴ There won't be anything to eat tonight.

Solution

The argument is inductive, because it is conceivable that somebody else has bought food, or that there was already food in the house. It becomes deductive if we add the premise

> There won't be anything to eat tonight unless one of us bought some food.

This is legitimate, since it is clear that the arguer is making this implicit assumption.

It is not always obvious whether or not a particular argument is deductive. In Chapters 3 to 7 we discuss a variety of formal techniques for detecting deductiveness with clarity and precision. For the moment, let us simply observe that although deductive arguments provide the greatest certainty, in practice we must often settle for inductive reasoning. Unlike deductive arguments, whose inductive probability is always maximal, inductive arguments have a range of inductive probabilities and hence vary widely in reliability. When the inductive probability of an inductive argument is high, we say that the reasoning of the argument is *strong* or *strongly inductive*. When it is low, we say that the reasoning of the argument is *weak* or *weakly inductive*.

SOLVED PROBLEM

2.10   Of the two inductive arguments below, which has the higher inductive probability?

> (*a*)   Visitors to China almost never contract malaria there.
>   Jan is visiting China.
>   ∴   Jan will not contract malaria there.

> (*b*)   Slightly fewer than half the visitors to China contract minor digestive disorders
>   there.
>   Jan is visiting China.
>   ∴   Jan will not contract a minor digestive disorder there.

Solution

The phrase 'almost never' in the first premise of argument (*a*) lends a greater probability to its conclusion than the corresponding phrase 'slightly fewer than half' does to the conclusion of argument (*b*). Argument (*a*) therefore has the higher inductive probability.

This example illustrates once again that inductive probability, like deductive validity, is independent of the actual truth or falsity of the premises. In Problem 2.10 we can recognize differences in inductive probability without knowing whether or not the premises are true. (Jan is a person whose identity and activities are unknown to us, and we may have no information about the incidence of malaria or digestive problems in China.) If the premises of an inductive argument are not true, then of course the argument fails to establish its conclusion. But if they are all true, then (provided that there is no suppressed relevant evidence—see Section 2.5) the argument establishes that its conclusion has a degree of probability corresponding to the argument's inductive probability. (Of course, the arguer's audience must know the premises to be true in order to see this.)

SOLVED PROBLEM

2.11 Evaluate the reasoning of the following argument:

> I dream about monsters.
> My brother dreams about monsters.
> ∴ Everyone dreams about monsters.

Solution

This is a weak inductive argument; its inductive probability is lower than that of either of the arguments of Problem 2.8. It generalizes from a very small sample of two individuals to the largest possible population (everyone). In the absence of strong reason to think that the sample is typical (and this should not be presumed), the probability of the conclusion given the premises is quite low.

There is no sharp line between strong and weak inductive reasoning, since what counts as strong for one purpose may not be strong enough for another. If, for example, the conclusion that a certain valve will not malfunction over a 5-year period has a 0.9 probability, given certain premises, then we may regard this reasoning as strong. But if the valve is part of a nuclear power plant and the lives of thousands of people depend on its functioning correctly, then 0.9 may not be strong enough to satisfy us. Thus there is no simple answer to the question "How high an inductive probability must an argument have in order for the reasoning to be classified as strong?" Clearly, however, an argument is weak if its inductive probability is less than 0.5. For in that case the denial or negation of its conclusion is more probable, given the premises, than the conclusion itself is. (This is proved in Problem 10.27 of Chapter 10.)

It is, of course, sometimes expedient to act on a conclusion whose probability, given current evidence, is quite slim. Rescuers in mine disasters, for example, often rightly proceed as if the victims were still alive, even though the chances of this may be small. But that should not be confused with thinking that the conclusion (in this case that the victims are alive) is probable.

Because the information contained in the premises and conclusions of inductive arguments is rarely numerically quantifiable, it is not often possible to give a precise number as the inductive probability of a given argument. (Deductive arguments, of course, are an exception, since their inductive probabilities are always precisely 1.) Usually the most we can say for an inductive argument is that it is "fairly strong" or "fairly weak." Occasionally, however, when the premises and conclusion are themselves numerically precise, we can give meaningful numerical probabilities for inductive arguments. (We shall come back to this in Sections 9.2 and 9.3.)

SOLVED PROBLEM

2.12 Evaluate the inductive probability of the following argument:

> 90 percent of the cars in Ohio have had at least two owners.
> This car is from Ohio.
> ∴ This car has had more than one owner.

Solution

This is a strong inductive argument, since the premises provide solid reasons to think that the car in question is among those that changed ownership. More precisely, assuming there is no substantial suppressed evidence (e.g., that the car in question is a brand-new model), the probability of the conclusion given the premises is 0.9: there are 90/100 chances that the car has had more than one owner. (For a detailed discussion of this kind of computation, see Section 9.2.)

Thus far our examples have concerned only simple arguments, arguments consisting of a single step of reasoning. We now consider inductive probability for complex arguments, those with two or more steps (see Section 1.3). For this purpose, it is important to keep in mind that deductive validity and inductive probability are relations between the *basic* premises and the conclusion. Thus, for example, a deductive argument is one whose conclusion cannot be false while its *basic* premises are true. Nonbasic premises are not mentioned in this definition.

Arguments contain nonbasic premises (intermediate conclusions) primarily as a concession to the limitations of the human mind. We cannot grasp very intricate arguments in a single step; so we break them down into smaller steps, each of which is simple enough to be readily intelligible. However, for evaluative purposes we are primarily interested in the whole span of the argument—i.e., in the probability of the conclusion, given our starting points, the basic premises.

Nevertheless, each of the steps that make up a complex argument is itself an argument, and each has its own inductive probability. One might suspect, then, that there is a simple set of rules relating the inductive probabilities of the component steps to the inductive probability of the entire complex argument. (One obvious suggestion would be simply to calculate the inductive probability of the whole argument by multiplying the inductive probabilities of all its steps together.) But no such rule applies in all cases. The relation of the inductive probability of a complex argument to the inductive probabilities of its component steps is in general a very intricate affair. There are, however, a few helpful rules of thumb:

(1)   With regard to complex nonconvergent arguments, if one or more of the steps is weak, then usually the inductive probability of the argument as a whole is low.

(2)   If all the steps of a complex nonconvergent argument are strongly inductive or deductive, then (if there are not too many of them) the inductive probability of the whole is usually fairly high.

(3)   The inductive probability of a convergent argument (Section 1.4) is usually at least as high as the inductive probability of its strongest branch.

Yet, because the complex ways in which the information contained in some premises may conflict with or reinforce the information contained in others, each of these rules has exceptions. Rules 1 to 3 allow us to make quick judgments which are *usually* accurate. But the only way to *ensure* an accurate judgment of inductive probability in the cases mentioned in these rules is to examine directly the probability of the conclusion given the basic premises, ignoring the intermediate steps.

There is only one significant exceptionless rule relating the strength of reasoning of a complex argument to the strength of reasoning of its component steps:

(4)   If all the steps of a complex argument are deductive, then so is the argument as a whole.

It is not difficult to see why this is so. If each step is deductive, then the truth of the basic premises guarantees the truth of any intermediate conclusions drawn from them, and the truth of these intermediate conclusions guarantees the truth of intermediate conclusions drawn from them in turn, and so on, until we reach the final conclusion. Thus if the basic premises are true, the conclusion must be true, which is just to say that the complex argument as a whole is deductive.

SOLVED PROBLEMS

2.13   Diagram the following argument and evaluate it with respect to criterion 2.

① [All particles which cannot be decomposed by chemical means are either subatomic particles or atoms.] Now ② [the smallest particles of copper cannot be decomposed by chemical means,] yet ③ [they are not subatomic.] (Hence) ④ [the smallest particles of copper are atoms.] ⑤ [Anything whose smallest particles are atoms is an element.] (Thus). ⑥ [copper is an element.] And ⑦ [no elements are alloys.] (Therefore,) ⑧ copper is not an alloy.]

Solution

The argument is diagramed as follows:

$$
\begin{array}{c}
1 \quad + \quad 2 \quad + \quad 3 \\
\hline
\downarrow \quad \mathbf{D} \\
4 \quad + \quad 5 \\
\hline
\downarrow \quad \mathbf{D} \\
6 \quad + \quad 7 \\
\hline
\downarrow \quad \mathbf{D} \\
8
\end{array}
\qquad \boxed{\mathbf{D}}
$$

Each of the three steps is deductive. We indicate a deductive step on the diagram by placing a 'D' next to the arrow representing the step. Since each step is deductive, so is the argument as a whole (rule 4). We signify this by placing a 'D' in a box beside the diagram.

2.14   Diagram the argument below and evaluate it.

① [Random inspections of 50 coal mines in the United States revealed that 39 were in violation of federal safety regulations.] (Thus we may infer that) ② [a substantial percentage of coal mines in the United States are in violation of federal safety regulations.] (Since) ③ [all federal safety regulations are federal law,] (it follows that) ④ [a substantial percentage of coal mines in the United States are in violation of federal law.]

Solution

Here the diagram is:

$$
\begin{array}{c}
1 \\
\downarrow \quad \mathbf{I} \quad \text{(Strong)} \\
2 \quad + \quad 3 \qquad \boxed{\mathbf{I}\ \text{(Strong)}} \\
\downarrow \quad \mathbf{D} \\
4
\end{array}
$$

The 'I' next to the first arrow indicates that the step from statement 1 to statement 2 is inductive. The 'D' next to the second arrow indicates that the step from statements 2 and 3 to statement 4 is deductive. This makes the argument as a whole inductive, which we indicate by placing an 'I' in a box next to the diagram. The inductive probability of the first step and hence of the argument as a whole is fairly high; that is, the reasoning both of this step and of the argument as a whole is strong. The step from statement 1 to statement 2 is strong because, even though a sample of 50 may be rather small, statement 2 is a very cautious conclusion. It says only that a "substantial percentage" of mines are in violation, which is indeed quite likely, given statement 1. Had it said "most," the reasoning would be weaker; had it said "almost all," the reasoning would be even weaker. (For a more detailed discussion of the evaluation of this sort of inference, see Section 9.3.)

One can see clearly that the reasoning of the argument as a whole is strong by noting that the conclusion, statement 4, is quite likely, given the basic premises, statements 1 and 3. This accords with rule 2.

2.15   Diagram the argument below and evaluate it.

① [The MG Midget and the Austin-Healy Sprite are, from a mechanical point of view, identical in almost all aspects.] ② [Sprites have hydraulic clutches.] (Thus it seems safe to conclude that) ③ [Midgets do as well.] But ④ [hydraulic clutches are prone to malfunction due to leakage.] (Therefore) ⑤ [both Sprites and Midgets are poorly designed cars.]

Solution

The diagram is:

$$
\begin{array}{c}
\underline{1 \quad + \quad 2} \\
\downarrow \qquad \text{I (Strong)} \\
\underline{2 \quad + \quad 3 \quad + \quad 4} \qquad \boxed{\textbf{I (Weak)}} \\
\downarrow \qquad \text{I (Weak)} \\
5
\end{array}
$$

Statement 3 is reasonably probable, though not certain, given 1 and 2, so that the first step is reasonably strong. But 5 is not very likely, given 3 and 4. Statement 5 says that each car as a whole is poorly designed, whereas statements 3 and 4 tell us at most that one part (the clutch) is poorly designed. Actually, they don't even tell us that much, since the fact that hydraulic clutches in general are prone to leakage does not guarantee that the particular clutches found in these two cars are poorly designed. Thus the second step is very weak. For the same reason it is clear that the probability of statement 5, given the basic premises of statements 1, 2, and 4, is low, so that the reasoning of the argument as a whole is quite weak. This accords with rule 1.

2.16   Diagram the following argument and evaluate it.

① [Mrs Compson is old and frail,] and ② [it is unlikely that anyone in her physical condition could have delivered the blows that killed Mr. Smith.] Moreover, ③ [two reasonably reliable witnesses who saw the murderer say that she was not Mrs. Compson.] And finally, ④ [Mrs. Compson had no motive to kill Mr. Smith,] and ⑤ [she would hardly have killed him without a motive.] ⟨Thus⟩ ⑥ [she is innocent of Mr. Smith's murder.]

Solution

$$
\begin{array}{c}
\underline{1 + 2} \qquad\qquad 3 \qquad\qquad \underline{4 + 5} \\
\diagdown \; \text{I (Strong)} \; \downarrow \; \text{I (Strong)} \diagup \text{I (Strong)} \qquad \boxed{\textbf{I (Very strong)}} \\
\rightarrow 6 \leftarrow
\end{array}
$$

This argument is convergent. Each step is strongly inductive; and when taken together, the steps reinforce one another. The inductive probability of the whole argument is therefore (in accord with rule 3) greater than the inductive probability of any of its component steps; its reasoning is quite strong.

In convergent arguments, unlike nonconvergent ones, a single weak step generally does not lessen the strength of the whole. For example, if we added the weak step

Mrs. Compson denies being the murderer.
∴ She is innocent of Mr. Smith's murder.

as an additional branch to the argument above, the overall inductive probability of this argument would remain about the same. This is because *in a convergent argument no single branch is crucial to the derivation of the conclusion.* In nonconvergent arguments, by contrast, each step is crucial, so that (as in Problem 2.15) one weak step usually drastically weakens the argument as a whole. This is the rationale behind our first rule of thumb. There are exceptions, however, as the following problem illustrates.

SOLVED PROBLEM

2.17   Diagram the argument below and evaluate its reasoning.

① [All your friends are misfits.] (Thus,) (since) ② [Jeff is a misfit,] ③ [he must be one of your misfit friends.] But ④ [misfits can't be good friends.] And (so) ⑤ [Jeff is not a good friend to you.]

Solution

$$
\begin{array}{ccc}
1 & + & 2 \\
\hline
& \downarrow \quad \textbf{I} & \text{(Weak)} \\
3 & + & 4 \qquad \boxed{\textbf{D}} \\
\hline
& \downarrow \quad \textbf{D} \\
& 5 &
\end{array}
$$

The step from statements 1 and 2 to statement 3 may appear to be deductive, but it is not. It is perfectly possible for all of someone's friends to be misfits and for Jeff to be a misfit and yet for Jeff not to be that person's friend. In fact, statement 3 is not very probable at all, given statements 1 and 2. The step from statements 3 and 4 to statement 5, however, is clearly deductive. And, surprisingly, so is the argument as a whole; for if the basic premises 2 and 4 are true, then 5 must be true as well. That is, if Jeff is a misfit and misfits can't be good friends, then (regardless of whether or not all your friends are misfits—a now superfluous premise) Jeff is not a good friend to you. (He may not be a friend at all.)

Despite its overall deductiveness, this argument is unacceptably flawed by its faulty initial step and hence would be objectionable as a means of establishing its conclusion.

2.4   RELEVANCE

Not every argument with true premises and high inductive probability is a good argument, even if all of its component steps have high inductive probabilities as well. A conclusion may be probable or certain, given a set of premises, even though the premises are irrelevant to the conclusion. But any argument which lacks relevance (regardless of its inductive probability) is useless for demonstrating the truth of its conclusion. For this reason, it is said to commit a *fallacy of relevance*.

Relevance is the concern of our third evaluative criterion for arguments. Like inductive probability, it is a matter of degree. Among the examples of simple arguments given thus far in this chapter, the premises are highly relevant to the conclusion in Problems 2.2, 2.3, 2.6 (all seven arguments), 2.7, 2.10 (both arguments), and 2.12–2.16.

SOLVED PROBLEM

2.18   Evaluate the argument below with respect to criteria 2, inductive probability, and 3, degree of relevance.

I abhor the idea of an infinitely powerful creator.
∴ God does not exist.

Solution

One's likes or dislikes have nothing to do with the actual existence of God; hence the premise is hardly relevant. It is difficult to assign any clear inductive probability to an argument like this, but certainly we should judge it not to be high.

Relevance and inductive probability do not always vary together. Some arguments exhibit high inductive probability with low relevance or low inductive probability with high relevance. Perhaps the simplest cases of high inductive probability with low relevance occur among arguments whose conclusions are logically necessary. A *logically necessary* statement is a statement whose very conception or meaning requires its truth; its falsehood, in other words, is logically impossible. Here are some examples:

> Either something exists, or nothing at all exists.
> $2 + 2 = 4$.
> No smoker is a nonsmoker.
> If it is raining, then it is raining.
> Everything is identical with itself.

(One important class of logically necessary statements, tautologies, will be studied in Chapter 3.)

Logically necessary statements have the peculiar property that if one occurs as the conclusion of an argument, then the argument is automatically deductive, regardless of the nature of the premises. This follows from the definition of deduction. A deductive argument is one whose conclusion cannot (i.e., logically cannot) be false while its premises are true. But logically necessary statements cannot be false under any conditions. Hence, trivially, if we take a logically necessary statement as a conclusion, then for any set of true premises we supply, the conclusion cannot be false.

SOLVED PROBLEM

2.19 Evaluate the inductive probability and degree of relevance of the following argument:

> Some sheep are black.
> Some sheep are white.
> ∴ If something is a cat, then it is a cat.

Solution

This preposterously artifical argument has a logically necessary conclusion and is therefore deductive, though its premises are wholly irrelevant to its conclusion.

Such an argument is, of course, useless as a means of demonstrating the truth of its conclusion, since the premises, being irrelevant to the conclusion, provide no reason to believe it. But since the conclusion is logically necessary, no further reason is needed to believe it; its truth is obvious as soon as it is understood.

Intuitively, lack of relevance is signaled by a kind of oddity or discontinuity which we feel in the inference from premises to conclusion. Where the premises are highly relevant, by contrast, the inference is usually natural and obvious.

SOLVED PROBLEM

2.20 Evaluate the inductive probability and degree of relevance of the argument below.

> All of Fred's friends go to Freeport High.
> All of Frieda's friends go to Furman High.
> Nobody goes to both Freeport and Furman.
> ∴ Fred and Frieda have no friends in common.

Solution

The argument is deductive, and so its inductive probability is 1. Its premises are highly relevant to its conclusion.

Having a logically necessary conclusion is not the only way an argument can be deductive and yet lack relevance. This may also occur if an argument has inconsistent premises. A set of statements is *inconsistent* if it is logically impossible for all of them to be true simultaneously. Each of the following sets of statements, for example, is inconsistent:

(*a*)   All butterflies are insects.
        Some butterflies are not insects.

(*b*)   Jim is taller than Bob.
        Bob is taller than Sally.
        Sally is taller than Jim.

(*c*)   This pole is either positively or negatively charged.
        It is not positively charged.
        It is not negatively charged.

(*d*)   Today is both Wednesday and not Wednesday.
        Today I play golf.

Case (*d*) is slightly different from the other cases. In all the others, the statements are inconsistent because they are in logical conflict. There is no conflict between the two statements of case (*d*). Rather, the first of these two contradicts itself. Hence this pair is inconsistent simply because the first statement can't be true under any circumstances (and hence they can't both be true).

Any argument with inconsistent premises is deductive, regardless of what the conclusion says. Again this follows from the definition of deduction. An argument is deductive if it is impossible for its premises all to be true while its conclusion is false. Thus since it is impossible (under any conditions) for inconsistent premises all to be true, it is clearly also impossible for these premises to be true while some conclusion is false. Hence, *any conclusion follows deductively from inconsistent premises.*

SOLVED PROBLEM

2.21   Evaluate the inductive probability and degree of relevance of the argument below.

> This book has more than 900 pages.
> This book has fewer than 800 pages.
> ∴ This is a very profound book.

Solution

Since it is logically impossible for the book to have more than 900 and fewer than 800 pages, it is clearly impossible for both premises to be true while the conclusion is false. Therefore the argument is deductive.[5] The premises, however, are wholly irrelevant to the conclusion. (The first is also false, if 'this book' designates the book you are now reading.)

Note that although any argument with inconsistent premises is deductive, no such argument is sound, since inconsistent premises cannot all be true. Hence *no conclusion can ever be proved by deducing it from inconsistent premises.*

Just as the premises of some deductive arguments are irrelevant to their conclusions, so too the premises of some strongly inductive arguments exhibit little relevance. This occurs primarily when the

---

[5]Here we would like to add, "... and hence its inductive probability is 1." But unfortunately matters are not so simple. Under some interpretations of probability, the inductive probability of an argument with inconsistent premises is undefined (see Section 10.3). Hence under these interpretations, arguments with inconsistent premises are exceptions to the rule that the inductive probability of a deductive argument is 1. (They are the only exceptions.) This, however, is essentially a matter of convention and convenience; nothing substantial turns on it. It is simply easier to state the laws of probability if this particular exception is made.

conclusion is a very weak statement. A statement is weak if it is logically probable—i.e., probable even in the absence of evidence. (See Section 9.1.) As a result, it will also be probable given most sets of irrelevant or weakly relevant premises.

SOLVED PROBLEM

2.22   Evaluate the inductive probability and degree of relevance of the argument below.

> You haven't proved that there are exactly 2,000,001 angels dancing on the head of this pin.
> ∴ There are are not exactly 2,000,001 angels dancing on the head of this pin.

Solution

The fact that one *hasn't* proved a proposition P is slightly relevant, but not strongly relevant, to the truth of P. One may not be competent to produce such a proof, or perhaps the effort has never been made. Yet it is highly probable in a logical sense that there are not exactly 2,000,001 angels dancing on the head of the pin. As a result, this conclusion is highly probable given the premise. That is, the argument is inductively strong. It is nevertheless a bad argument, because the premise is not strongly relevant to the conclusion.

A good argument, then, requires not only true premises (criterion 1) and high inductive probability (criterion 2), but also a high degree of relevance (criterion 3). Many treatments of logic tend to slight relevance as a factor in argument evaluation because it is difficult to characterize precisely. Some logicians have argued that relevance is a purely subjective notion and therefore not a proper subject matter for logic. Yet clearly any account of argument evaluation which ignores relevance is incomplete.

In recent years a formal discipline called *relevance logic* has emerged. Relevance logic is the study of the relation of *entailment*. A set of premises is said to *entail* a conclusion if the premises deductively imply the conclusion and in addition are relevant to it. In relevance logic, therefore, deductiveness and relevance are studied in combination as a single relation. Here, however, we shall follow the more standard approach of *classical logic*, in which inductive probability and relevance are considered as separate factors in argument evaluation. We shall come back to issues of relevance in Chapter 8.

2.5   THE REQUIREMENT OF TOTAL EVIDENCE

One crucial respect in which inductive arguments differ from deductive arguments is in their vulnerability to new evidence. A deductive argument remains deductive if new premises are added (regardless of the nature of the premises). An inductive argument, by contrast, can be either strengthened or weakened by the addition of new premises. Consequently, the probability of a conclusion inferred inductively from true premises may be radically altered by the acquisition of new evidence, while the certainty of a conclusion inferred deductively from true premises remains unassailable.

SOLVED PROBLEMS

2.23   Evaluate the inductive probability and degree of relevance of the argument below.

> Very few Russians speak English well.
> Sergei is Russian.
> ∴ Sergei does not speak English well.

Solution

This argument is strongly inductive (i.e., has a high inductive probability) and its premises are quite relevant to its conclusion.

2.24 Evaluate the inductive probability and degree of relevance of the argument below.

> Very few Russians speak English well.
> Sergei is Russian.
> Sergei is an exchange student at an American university.
> Exchange students at American universities almost always speak English well.
> ∴ Sergei does not speak English well.

Solution

This argument is obtained from the previous one by adding two more premises. Once again the premises are quite relevant, but now the inductive probability of the argument is quite low. The premises of this new argument are in conflict. The first two support the conclusion; the second two are evidence against it. As a result, the inductive probability of this argument is considerably less than that of the original argument in Problem 2.23. Indeed, it seems more reasonable, given this evidence, to draw the opposite conclusion—namely, that Sergei does speak English well.

Because of its conflicting premises, the argument of Problem 2.24 would not be an effective tool for demonstrating the truth of its conclusion to an audience. Hence we would not expect to encounter such an argument in practice. Rather, we should think of the addition of premises exhibited in Problem 2.24 as representing our acquisition of new evidence about Sergei. As the evidence available to us increases, the probability of the proposition that Sergei does not speak English well, relative to the available evidence, may fluctuate considerably.

Problem 2.24 shows how this probability may diminish. The next example shows how it may increase.

SOLVED PROBLEM

2.25 Evaluate the inductive probability and degree of relevance of the argument below.

> Very few Russians speak English well.
> Sergei is Russian.
> Sergei is an exchange student at an American university.
> Exchange students at American universities almost always speak English well.
> Sergei is a deaf-mute.
> ∴ Sergei does not speak English well.

Solution

The argument is now deductive and hence its inductive probability is 1. The premises are all relevant to the conclusion, though the first four are now superfluous in establishing it, since the conclusion follows deductively from the final premise, 'Sergei is a deaf-mute', alone.

Once a deductive argument has been achieved, no further additions can alter the inductive probability; it remains fixed at 1. If all the premises are true (and remain true), then the conclusion is certainly true;

no further evidence can decrease its certainty. Indeed, the argument remains deductive even if we add the premise 'Sergei speaks perfect English', for that premise is inconsistent with the premise 'Sergei is a deaf-mute', and as we saw in Section 2.4, inconsistent premises deductively imply any conclusion. (Of course, in this case the premises cannot both be true.)

In summary, inductive arguments, unlike deductive arguments, can be converted into arguments with higher or lower inductive probability by the addition of certain premises. Thus in inductive reasoning the choice of premises is crucial. Using one portion of the available evidence as premises may make a conclusion seem extremely probable, while using another portion may make it seem extremely improbable. By selectively manipulating the evidence in this way, we may be able to make a conclusion appear as probable or as improbable as we like, even though all the assumptions we make may be true.

Now, this selective manipulation of the evidence is of course illegitimate. It is precisely this illegitimacy that defines the concern of our fourth criterion of argument evaluation, which is called the *requirement of total evidence* or the *total evidence condition*. It stipulates that if an argument is inductive, its premises must contain all known evidence that is relevant to the conclusion. Inductive arguments which fail to meet this requirement, particularly if the evidence omitted tells strongly against the conclusion, are said to commit the *fallacy of suppressed evidence*.

SOLVED PROBLEM

2.26 Evaluate the following argument with respect to criteria 1, 2, 3, and 4:

> Most cats do well in apartments.
> They are very affectionate, and love being petted.
> ∴ This cat will make a good pet.

Solution

This argument fares well with respect to the first three criteria: the premises are true and relevant to the conclusion, and the inductive probability is certainly high. (Notice that this is a convergent argument, each branch providing independent good evidence for the conclusion.) However, if the arguer is withholding that the cat in question has lived most of its life in a cat shelter, where it became aggressive and dirty, then the argument is irremediably flawed by a fallacy of suppressed evidence.

Fallacies of suppressed evidence may be committed either intentionally or unintentionally. If the author of the argument intentionally omits known relevant information, the fallacy is a deliberate deception. The omission of relevant evidence which the author knows may, however, be a simple blunder; the author may simply have forgotten to consider some of the available relevant facts. It may also happen that an author has included among the premises all the relevant information he or she knows, but that others know relevant information of which the author is unaware. Here again, the argument commits a fallacy of suppressed evidence, but again the fallacy is unintentional. The author has done his or her best with the available information.

It should be noted that even if an inductive argument meets the requirement of total evidence, it may still lead us from true premises to a false conclusion. *Inductive arguments provide no guarantees*. It is possible, though (we hope) unlikely, that the entire body of known relevant evidence is misleading.

Suppressed evidence should not be confused with implicit premises (Section 1.6). Implicit premises are assumptions that the author of an argument intends the audience to take for granted. Suppressed evidence, by contrast, is information that the author has deliberately concealed or unintentionally omitted. Implicit assumptions are part of the author's argument. Suppressed evidence is not.

SOLVED PROBLEM

2.27 Compare the inductive probability and degree of relevance of the two arguments below:

(a) Students with a solid background in logic typically do well in Law School.
   ∴ Louise will do well in Law School.

(b) Only one third of the students do well in Law School.
   ∴ Louise will not do well in Law School.

Solution

As it stands, (a) is a very weak inductive argument, with a premise of dubious relevance. However, it seems safe to suppose that the arguer is relying on an implicit assumption about Louise:

Louise has a solid background in logic.

On this understanding, the inductive probability of argument (a) is actually quite high. By contrast, argument (b) appears to have a fairly strong inductive probability. If, however, it is a known fact that students with a solid background in logic typically do well in Law School, and if it is also a fact that Louise has a solid background in logic, then the evaluation of (b) would change substantially. In that case (b) would be a bad argument by criterion 4: it would commit the fallacy of suppressed evidence.

Some logicians argue that the requirement of total evidence is too stringent, that no argument can incorporate all the known evidence to a given conclusion. With respect to arguments involving very complex issues, this may well be true. For such arguments, the requirement of total evidence is a theoretical ideal, seldom (or perhaps never) attained in practice. Consequently, inductive arguments on very complex issues will generally suffer to some degree from the fallacy of suppressed evidence. The best arguments will be those which minimize suppressed evidence and suppress no evidence that drastically affects the probability of the conclusion. In many simple cases, however, the requirement of total evidence is quite stringently met.

SOLVED PROBLEM

2.28 The question arises (say, in a medical emergency) of whether a boy has type AB blood. Suppose nothing specific is known about the boy's blood type (he has never had a blood test). Evaluate the following argument:

Type AB blood is relatively rare.
   ∴ The boy does not have type AB blood.

Solution

This is a reasonably good inductive argument with respect to all four criteria. The premise is true and relevant, the inductive probability is fairly high, and all the known relevant evidence is contained in the premise.

Of course, in a real medical emergency we would like more data than this argument offers us. Typically, we would not be satisfied with a mere statistical inference; instead we would make more evidence available by performing a blood test. We could then use the results as premises to construct a much more reliable argument. In general, if little evidence is available, the rational thing to do is to obtain more evidence, if possible, before drawing any conclusions. But this is not always possible. We may not have the equipment or expertise to perform a blood test, and yet human needs may force us to draw a conclusion anyway. In such a situation, the argument above would represent the best we could do, and it would be logical to regard the conclusion as probably true.

Supplementary Problems

I   Apply criterion 2 to the following arguments; i.e., for each, determine whether its reasoning is deductive or inductive and, if it is inductive, whether it is strong or weak (i.e., whether its inductive probability is high or low).

(1)   All Arcturians are good kissers.
        Some Arcturians have several mouths.
    ∴   Some things with several mouths are good kissers.

(2)   All smokers contract emphysema.
        All those who contract emphysema suffer painful deaths.
    ∴   All smokers suffer painful deaths.

(3)   Nobody told me that the meeting was today.
    ∴   I had no way of knowing that it was today.

(4)   All Americans like hamburgers.
        Joe is American.
    ∴   Joe likes hamburgers.

(5)   Almost all Americans like hamburgers.
        Joe is American.
    ∴   Joe likes hamburgers.

(6)   Most Americans like hamburgers.
        Joe is American.
    ∴   Joe likes hamburgers.

(7)   Many Americans like hamburgers.
        Joe is American.
    ∴   Joe likes hamburgers.

(8)   Some Americans like hamburgers.
        Joe is American.
    ∴   Joe likes hamburgers.

(9)   No Americans like hamburgers.
        Joe is American.
    ∴   Joe likes hamburgers.

(10) Most people have two legs.
        Most people have two arms.
    ∴   Some people have two arms and two legs.

(11) If either Nation A or Nation B launches a nuclear attack, it will lead to massive destruction.
        Neither side wants this destruction.
    ∴   Neither side will launch a nuclear attack.

(12) Numbers are not physical objects.
        Numbers exist.
    ∴   It is not true that all existing things are physical objects.

(13) All arguments are either inductive or deductive.
        What you are now reading is an argument.
        This argument is not inductive.
    ∴   This argument is deductive.

(14) We don't know that machines can think like humans.
    ∴   We know that machines can't think like humans.

(15)   After sniffing pepper, I always sneeze.
      Yesterday I sniffed pepper.
∴ Yesterday I sneezed.

(16)   We saw an eagle in the park.
      There are only two species of eagles in the park, bald and golden.
      The golden eagle is commonly seen in the park.
      The bald eagle is rarely seen in the park.
∴ The eagle we saw in the park was a golden eagle.

(17)   Conservatives are always strong proponents of law and order.
      Adams is a strong proponent of law and order.
∴ Adams is conservative.

(18)   You're not convinced that I'm innocent.
∴ You must think that I'm guilty.

(19)   The fortune-teller said that Anne would be murdered, but not by a man or boy.
∴ Anne will be murdered by a woman or girl.

(20)   Aspirin does not cure arthritis.
      Drug X is just aspirin.
∴ Drug X does not cure arthritis.

(21)   All life forms observed thus far on earth are DNA-based.
∴ All earthly life forms are DNA-based.

(22)   All life forms observed thus far on earth are DNA-based.
∴ All life forms in the universe are DNA-based.

(23)   Joe was the only person near the victim at the time of the murder.
∴ Joe is the murderer.

(24)   All human creations will eventually perish.
      Whatever perishes is ultimately meaningless.
∴ All human creations are ultimately meaningless.

(25)   There are more people in the world than hairs on any one person's head.
∴ At least two people's heads have an equal number of hairs.

(26)   There are more people in the world than hairs on any one person's head.
      No one is bald.
∴ At least two people's heads have an equal number of hairs.

(27)   God made the universe.
      God is perfectly good.
      Whatever is made by a perfectly good being is perfectly good.
      Whatever is perfectly good contains no evil.
∴ The universe contains no evil.

(28)   If there were more than one null set (set with no members), then there would be more than one set
      with exactly the same members.
      No two sets have exactly the same members.
∴ There is at most one null set.

(29)   Jody has a high fever, purple splotches on her tongue, and severe headaches, but no other
      symptoms.
      Jeff has the same set of symptoms, and no others.
∴ Jody and Jeff have the same disease.

(30)   We priced bicycle helmets at a number of retail outlets.
      We found none for under $25 that passed the safety tests.
∴ If you buy a new bicycle helmet that passes the safety tests from a retail outlet, you'll have to pay
      at least $25.

II   For each of the following arguments, circle inference indicators, bracket and number each statement, and diagram the argument. Place a 'D' or an 'I' next to each arrow of the diagram to indicate whether the corresponding step of reasoning is deductive or inductive. If it is inductive, indicate its strength. If the argument is complex, place a 'D' or an 'I' in a box next to the diagram to indicate whether the argument as a whole is deductive or inductive. Again if it is inductive, indicate its strength.

(1)   There is no greatest prime number. But of all the prime numbers we shall have ever thought of, there certainly is a greatest. Hence there are primes greater than any we shall have ever thought of. (Bertrand Russell, "On the Nature of Acquaintance")

(2)   Since you said you would meet me at the drive-in and you weren't there, you're a liar. So I can't believe anything you say. So I can't possibly feel comfortable with you.

(3)   Argument 2 is unsound, for two reasons: (i) I was at the drive-in, but you must have missed me, and so one of your premises is false, and (ii) your reasoning is invalid.

(4)   The square of any integer n is evenly divisible by n. Hence the square of any even number is even, since by the principle just mentioned it must be divisible by that even number, and any number divisible by an even number is even.

(5)   When he was 40, De Chirico abandoned his *pittura metafisica*; he turned back to traditional modes, but his work lost depth. Here is certain proof that there is no "back to where you came from" for the creative mind whose unconscious has been involved in the fundamental dilemma of modern existence. (Aniela Jaffe, "Symbolism in the Visual Arts")

(6)   Since habitual overeating contributes to several debilitating diseases, it can contribute to the destruction of your health. But your health is the most important thing you have. So you should not habitually overeat.

(7)   The forecast calls for rain, the sky looks very threatening, and the barometer is falling rapidly, all of which are phenomena strongly correlated with rain. Therefore it's going to rain. But if it rains, we'll have to cancel the picnic. So it looks as if the picnic will be canceled.

(8)   There is no way to tell whether awareness continues after death, so we can only conclude that it does not. But we are nothing more than awareness, since without awareness we experience nothing, not even blackness. Thus we do not survive death. Any moral system based on the certainty of reward or punishment in the hereafter is therefore fundamentally mistaken.

(9)   All citizens of voting age have the right to vote unless they are mentally disabled or have been convicted of a crime. Jim is a citizen of voting age, and yet he has said he did not have the right to vote. He's not mentally disabled. So either what he said is false, or he's been convicted of a crime. But he also told me he'd never been arrested, and it's impossible to be convicted of a crime if you've never been arrested. Thus at least one of the things he said is false.

(10)   Just as without heat there would be no cold, without darkness there would be no light, and without pain there would be no pleasure, so too without death there would be no life. Thus it is clear that our individual deaths are absolutely necessary for the life of the universe as a whole. Death should therefore be a happy end toward which we go voluntarily, rather than an odious horror which we selfishly and futilely fend off with our last desperate ounce of energy.

III   Provide a rough estimate of the inductive probability and degree of relevance of each of the following arguments.

(1)   Roses are red.
Violets are blue.
∴   The next rose I see will not have exactly 47 petals.

(2)   All my friends say that snorting a little nutmeg now and then is good for you.
∴   Snorting a little nutmeg now and then is good for you.

(3)   All my friends say that snorting a little nutmeg now and then is good for you.
My friends are never wrong.
∴   Snorting a little nutmeg now and then is good for you.

(4)   Mr. Plotz owns a summer home in New Hampshire.
       He also owns his family residence in Washington, D.C.
  ∴   He owns at least two homes.

(5)   I looked cross-eyed at a toad once, and that very same day I broke my toe.
  ∴   Looking cross-eyed at a toad is bad luck.

(6)   Sue is an intellectual.
       Sara is not an intellectual.
  ∴   No one both is and is not an intellectual.

(7)   Albert wears ridiculous-looking clothes.
       Albert is always bumping into things.
  ∴   Albert is stupid.

(8)   I both love you and do not love you.
  ∴   I love you.

(9) $2 + 2 = 4$
       $4 = 2^2$
  ∴ $2 + 2 = 2^2$

(10) $2 + 3 = 5$
       $3 + 7 = 10$
  ∴ $2 + 2 = 2^2$

IV Evaluate each of the following arguments with respect to the four criteria discussed in this chapter by answering the following questions:

(1)   Are the premises known to be true?
(2)   How high is the argument's inductive probability?
(3)   How relevant are the premises to the conclusion?
(4)   If the argument is inductive, is any evidence suppressed?

On the basis of your answers to questions 1 to 4, assess the degree to which the argument accomplishes the goal of demonstrating the truth of its conclusion.

(1)   Very few presidents of the United States have been Hollywood actors.
       Ronald Reagan was a president of the United States.
  ∴   Ronald Reagan was not a Hollywood actor.

(2)   If Topeka is in the United States, then it is either in the continental United States, in Alaska, or in Hawaii.
       Topeka is not in Alaska.
       Topeka is not in Hawaii.
       Topeka is in the United States.
  ∴   Topeka is in the continental United States.

(3)   Human beings are vastly superior, both intellectually and culturally, to modern apes.
  ∴   Human beings and modern apes did not evolve from common ancestors.

(4)   The enemies have possessed nuclear weapons for many years.
       They have never used nuclear weapons in battle.
  ∴   The enemies will use their nuclear weapons in battle soon.

(5)   Of all the known planets, only one, Earth, is inhabited.
       At least nine planets are known.
  ∴   The proportion of inhabited planets in the universe at large is not high.

(6)   For any integer n, the number of positive integers smaller than n is finite.
       For any integer n, the number of positive integers larger than n is infinite.
       Any infinite quantity is larger than any finite quantity.
  ∴   For any integer n, there are more positive integers larger than n than there are positive integers smaller than n.

(7) Without exception, all matter thus far observed has mass.
     There is matter in galaxies beyond the reach of our observation.
∴   The matter in these unobserved galaxies has mass.

(8) No promise is ever broken.
     John F. Kennedy promised that the United States would put a man on the moon by 1970.
∴   The United States did put a man on the moon by 1970.

(9) No human being has ever reached the age of 100 years.
∴   No one now alive will reach the age of 100 years.

(10) Many people find the idea of ghosts and poltergeists fascinating.
∴   Ghosts and poltergeists exist.

Answers to Selected Supplementary Problems

I (5) Inductive (strong).

(10) Deductive. Since most means "more than half," the two groups mentioned in the premises cannot fail to overlap if the premises are true. Therefore, given the premises, there must be some people (i.e., at least one person) with two arms and two legs.

(15) Inductive. The premises do not say that the sneezing occurs immediately; hence it may be that the sniffing occurred shortly before midnight yesterday though a sneeze did not result until shortly after midnight. The strength of this inference is difficult to estimate, since the argument itself provides no guidelines to how long a delay we might imagine. (Could one have sniffed pepper yesterday and not sneeze until a week hence? This is at least logically possible.) But if for practical purposes we discount such wild possibilities, then the reasoning is fairly strong.

(20) The word 'just' means "nothing more than"; hence the argument is deductive.

(25) Inductive. Suppose, for example, that there were only two people left in the world, one totally bald and one with just a single hair. Then the premise would be true and the conclusion false, so that the argument is not deductive. The argument is, however, fairly strong, since under most conceivable circumstances which make the premise true the conclusion would be true as well.

(30) Inductive (moderately strong).

II (5) ① [When he was 40, De Chirico abandoned his *pittura metafisica*; he turned back to traditional modes, but his work lost depth.] ⟨Here is certain proof that⟩ ② [there is no "back to where you came from" for the creative mind whose unconscious has been involved in the fundamental dilemma of modern existence.]

$$
\begin{array}{c}
1 \\
\downarrow \quad \mathbf{I}\ (\text{Weak}) \\
2
\end{array}
$$

Comment: The example of a single artist hardly suffices to establish so general a thesis as statement 2 with any substantial degree of probability. Premise 1 could be broken up into two or three separate statements or kept as a single unit as we have done here; this is a matter of indifference.

(10) Just as ① [without heat there would be no cold,] ② [without darkness there would be no light,] and ③ [without pain there would be no pleasure,] ⟨so⟩ too ④ [without death there would be no life.] ⟨Thus it is clear that⟩ ⑤ [our individual deaths are absolutely necessary for the life of the universe as a whole.] ⑥ [Death should ⟨therefore⟩ be a happy end toward which we go voluntarily, rather

than an odious horror which we selfishly and futilely fend off with our last desperate ounce of energy.]

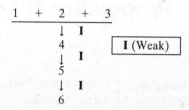

Comment: All three steps in this argument suffer from vagueness, so that it is difficult to evaluate any of them accurately. The first is not strong, because it establishes no clear and significant parallel between the pairs of opposites mentioned in the premises (statements 1, 2, and 3) and the pair mentioned in the conclusion (statement 4). (That conclusion may well be true, but it is not extremely likely, given just the information contained in the premises.) The second step is also not as strong as it may at first appear. It may be true in general that without death there would be no life, but this does not by itself imply that all living things must die or that we in particular must die. Likewise, the third inference is hardly airtight. It seems to require the additional assumption that we should happily and voluntarily accept what is necessary for the life of the universe as a whole, but the truth of this assumption is far from obvious. With the addition of this assumption, the final inference would be deductive, though based on at least one dubious premise. Taken just as it stands it is not deductive, and its inductive strength is not clearly determinable.

III   (5)   Low inductive probability, low relevance.

(10)   Since the conclusion is logically necessary, the argument is deductive and has maximal inductive probability. But (in contrast to the previous problem) the premises are not directly relevant to the conclusion.

IV   (5)   The premises are true and relevant to the conclusion. The argument is at best moderately inductive, since the nine known planets constitute a very small (and nonrandom) sample upon which to base the extensive generalization of the conclusion. We know of no suppressed contrary evidence. Thus the argument provides some evidence for its conclusion, though this evidence is far from conclusive.

(10)   The premise is true, but it lacks relevance, and the inductive probability of the argument is low. This is a very bad argument.

Chapter 3

Propositional Logic

3.1 ARGUMENT FORMS

This chapter begins our treatment of formal logic. Formal logic is the study of *argument forms*, abstract patterns of reasoning shared by many different arguments. The study of argument forms facilitates broad and illuminating generalizations about validity and related topics. We shall initially focus on the notion of deductive validity, leaving inductive arguments to a later treatment (Chapters 8 to 10). Specifically, our concern in this chapter will be with the idea that a valid deductive argument is one whose conclusion cannot be false while the premises are all true (see Section 2.3). By studying argument forms, we shall be able to give this idea a very precise and rigorous characterization.

We begin with three arguments which all have the same form:

(1) Today is either Monday or Tuesday.

    Today is not Monday.

∴ Today is Tuesday.

(2) Either Rembrandt painted the *Mona Lisa* or Michelangelo did.

    Rembrandt didn't do it.

∴ Michelangelo did.

(3) Either he's at least 18 or he's a juvenile.

    He's not at least 18.

∴ He's a juvenile.

It is easy to see that these three arguments are all deductively valid. Their common form is known by logicians as *disjunctive syllogism*, and can be represented as follows:

Either P or Q.

It is not the case that P.

∴ Q.

The letters 'P' and 'Q' function here as placeholders for declarative[1] sentences. We shall call such letters *sentence letters*. Each argument which has this form is obtainable from the form by replacing the sentence letters with sentences, each occurrence of the same letter being replaced by the same sentence. Thus, for example, argument 1 is obtainable from the form by replacing 'P' with the sentence 'Today is Monday' and 'Q' with the sentence 'Today is Tuesday'. The result,

Either today is Monday or today is Tuesday.

It is not the case that today is Monday.

∴ Today is Tuesday.

is for our present purposes a mere grammatical variant of argument 1. We can safely ignore such grammatical variations here, though in some more sophisticated logical and philosophical contexts they must be reckoned with. An argument obtainable in this way from an argument form is called an *instance* of that form.

---

[1]From now on we shall omit the qualification 'declarative', since we shall only be concerned with sentences that can be used to express premises and conclusions of arguments (and by definition these can only be expressed by declarative sentences; see Section 1.1).

SOLVED PROBLEM

3.1　Identify the argument form common to the following three arguments:

(*a*)　If today is Monday, then I have to go to the dentist.
Today is Monday.
∴ I have to go to the dentist.

(*b*)　If you have good grades, you are eligible for a scholarship.
You have good grades.
∴ You are eligible for a scholarship.

(*c*)　I passed the test, if you did.
You passed the test.
∴ I passed the test.

Solution

The three arguments have the following form (known by logicians as *modus ponens*, or "assertive mode"):

If P, then Q.
P.
∴ Q.

The word 'then' is missing in the first premise of argument (*b*), but this is clearly immaterial: the meaning would not change by rewriting the premise as

If you have good grades, then you are eligible for a scholarship.

Likewise, the first premise of argument (*c*) is just a grammatical variant of the following:

If you passed the test, then I passed the test.

Notice that one can detect more than one form in a particular argument, depending on how much detail one puts into representing it. For example, the following argument is clearly an instance of disjunctive syllogism, like arguments 1 to 3:

(4)　Either we can go with your car, or we can't go at all.
We can't go with your car.
∴ We can't go at all.

We can see that 4 has the form of a disjunctive syllogism by replacing 'P' with 'We can go with your car' and 'Q' with 'We can't go at all'. However, one can also give the following, more detailed representation of argument 4:

Either P, or it is not the case that Q.
It is not the case that P.
∴ It is not the case that Q.

It is clear that argument 4 is an instance of this form, too, since it can be obtained by replacing 'P' with 'We can go with your car' and 'Q' with 'We can go at all'. Arguments 1 to 3, by contrast, are not substitution instances of this argument form. As will become clear, recognizing the appropriate argument forms is a crucial step towards applying formal logic to everyday reasoning.

SOLVED PROBLEM

3.2 Identify an argument form common to the following two arguments:

(a) If Murphy's law is valid, then anything can go wrong.
But not everything can go wrong.
∴ Murphy's law is not valid.

(b) If you passed the test and Jane also passed, then so did Olaf.
Olaf did not pass.
∴ It's false that both you and Jane passed the test.

Solution

One form common to both argument is the following (known by logicians as *modus tollens*, or "denying mode"):

If P, then Q.
It is not the case that Q.
∴ It is not the case that P.

Argument (b) can also be analyzed as having the following form:

If P and Q, then R.
It is not the case that R.
∴ It is not the case that P and Q.

This form, however, is not common to argument (a), since in (a) there are no sentences to replace 'P' and 'Q' as required. It should also be noted that the two arguments have the following form in common:

P.
Q.
∴ R.

However, this form is common to all nonconvergent arguments with two premises; it exhibits no logically interesting feature, so it may be ignored for all purposes.

This problem shows that one can obtain instances of an argument form by replacing its sentence letters with sentences of arbitrary complexity. If every instance of an argument form is valid, then the argument form itself is said to be *valid*; otherwise the argument form is said to be *invalid*. (Thus, one invalid instance is enough to make the argument form invalid.) Disjunctive syllogism, for example, is a valid argument form: *every* argument of this form is such that if its premises were true its conclusion would have to be true. (Of course, not all instances of disjunctive syllogism are sound; some—e.g., argument 2 above—have one or more false premises.) The argument forms *modus ponens* and *modus tollens* in Problems 3.1 and 3.2 are likewise valid. By contrast, the following form (known as *affirming the consequent*) is invalid:

If P, then Q.
Q.
∴ P.

Though some instances of this form are valid arguments, others are not. Here is an instance which is valid—and indeed sound:

(5) If April precedes May, then April precedes May and May follows April.
April precedes May and May follows April.
∴ April precedes May.

The conclusion of this argument follows necessarily from its premises, both of which are true. But here is an argument of the same form which is invalid:

(6) If you are dancing on the moon, then you are alive.
     You are alive.
∴   You are dancing on the moon.

The premises are true, but the conclusion is false; hence the argument is invalid.

Since any form that has even one invalid instance is invalid, the invalidity of argument 6 proves the invalidity of affirming the consequent. Though affirming the consequent also has valid instances (such as argument 5), these are not valid *as a result of* being instances of affirming the consequent. Indeed, the reason for the validity of 5 is that its conclusion follows validly from the second premise alone; the first premise is superfluous and could be omitted from the argument without loss.

3.2 LOGICAL OPERATORS

The domain of argument forms studied by logicians is continuously expanding. In this chapter we shall be concerned only with a modest selection, namely those forms consisting of sentence letters combined with one or more of the following five expressions: 'it is not the case that', 'and', 'either . . . or', 'if . . . then', and 'if and only if'. These expressions are called *logical operators* or *connectives*. This modest beginning is work enough, however; for very many different forms are constructible from these simple expressions, and some of them are among the most widely used patterns of reasoning.

The operator 'it is not the case that' prefixes a sentence to form a new sentence, which is called the *negation* of the first. Thus the sentence 'It is not the case that he is a smoker' is the negation of the sentence 'He is a smoker'. There are many grammatical variations of negation in English. For example, the sentences 'He is a nonsmoker', 'He is not a smoker', and 'He is no smoker' are all ways of expressing the negation of 'He is a smoker'. The particles 'un-', 'ir-', 'in-', 'im-', and 'a-', used as prefixes to words, may also express negation, though they may express other senses of opposition as well. Thus 'She was unmoved' is another way of saying 'It is not the case that she was moved', and 'It is impossible' says the same thing as 'It is not the case that it is possible'. But 'It is immoral' does not mean "It is not the case that it is moral." 'Immoral' means "wrong," and 'moral' means "right," but these two classifications are not exhaustive, for some actions (e.g., scratching your nose) are amoral—i.e., neither right nor wrong, but morally neutral. These actions are not moral, but they are not immoral either; so 'not moral' does not mean the same thing as 'immoral'. True negation allows no third or neutral category. Thus care must be used in treating particles like those just mentioned as expressions of negation.

The other four operators each join two statements into a compound statement. We shall call them *binary operators*.

A compound consisting of two sentences joined by 'and' (or 'both . . . and') is called a *conjunction*, and its two component sentences are called *conjuncts*. Conjunction may also be expressed in English by such words as 'but', 'yet', 'although', 'nevertheless', 'whereas', and 'moreover', which share with 'and' the characteristic of affirming both the statements they join—though they differ in expressing various shades of attitude toward the statements thus asserted.

A compound statement consisting of two statements joined by 'either . . . or' is called a *disjunction* (hence the name 'disjunctive syllogism' for the argument form discussed above). The two statements are called *disjuncts*. Thus the first premise of argument 1, 'Today is either Monday or Tuesday'—which for logical purposes is the same as 'Either today is Monday or today is Tuesday'—is a disjunction whose disjuncts are 'Today is Monday' and 'Today is Tuesday'. The term 'either' is often omitted. The same proposition can thus be expressed even more simply as 'Today is Monday or Tuesday'.

Statements formed by 'if . . . then' are called *conditionals*. The statement following 'if' is called the *antecedent*; the other statement is the *consequent*. In the sentence 'If you touch me, then I'll scream', 'You touch me' is the antecedent and 'I'll scream' is the consequent. The word 'then' may be omitted.

Conditionals can also be expressed in reverse order; e.g., 'I'll scream if you touch me'. This is a mere grammatical variant of the previous sentence, and its antecedent and consequent are the same.

Lastly, statements formed by 'if and only if' are called *biconditionals*. There is no special name for their components, except for *left-hand side* and *right-hand side* of the biconditional. A biconditional may be regarded as a conjunction of two conditionals. To see this, consider the sentence:

It is a triangle if and only if it is a three-sided polygon.

This is obviously just a variant of

It is a triangle if it is a three-sided polygon, and it is a triangle only if it is a three-sided polygon.

which in turn is a variant of

If it is a three-sided polygon, then it is a triangle; and it is a triangle only if it is a three-sided polygon.

But what does 'It is a triangle only if it is a three-sided polygon' mean? From a logical point of view, 'only if' is in fact just another way of expressing a conditional. Statements of the form '*P* only if *Q*' mean just "If *P*, then *Q*." (In an 'only if' statement, it is therefore the consequent, not the antecedent, that follows 'if'.) Thus the statement 'There's fire only if there's oxygen' means "If there's fire, then there's oxygen"; and the statement 'It is a triangle only if it is a three-sided polygon' means "If it is a triangle, then it is a three-sided polygon." Therefore our biconditional is just another way of saying:

If it is a three-sided polygon, then it is a triangle; and if it is a triangle, then it is a three-sided polygon.

Since the order of conjuncts clearly makes no difference to meaning, this may be rewritten as:

If it is a triangle, then it is a three-sided polygon; and if it is a three-sided polygon, then it is a triangle.

This example illustrates a general rule: Statements of the form '*P* if and only if *Q*' are equivalent to (i.e., true under the same conditions as) statements of the form 'if *P*, then *Q*; and if *Q*; then *P*'. That is why they are called biconditionals.

SOLVED PROBLEM

3.3 Classify the following compound sentences in terms of the logical operators used to construct them; in each case, name the component sentence(s) to which the operator is applied.

(*a*)   Today is Monday only if yesterday was Sunday.

(*b*)   Today is Monday if yesterday was Sunday.

(*c*)   If today is Monday, yesterday was Sunday.

(*d*)   She lied, if today is Monday.

(*e*)   It's her birthday if and only if it's Monday.

(*f*)   It is either Monday or Monday.

(*g*)   John and Mary are competent.

(*h*)   He is incompetent.

(*i*)   He is not very competent.

(*j*)   He is competent but too old.

(*k*)   He is either competent or corrupt.

(*l*)   If he is competent, she is a liar.

(*m*)   Either he or she is very corrupt.

(*n*)   Only if she is a liar is he competent.

(*o*)   Although he is very competent, she avoids him.

Solution

(a) Conditional; antecedent: 'Today is Monday'; consequent: 'Yesterday was Sunday'.

(b) Conditional; antecedent: 'Yesterday was Sunday'; consequent: 'Today is Monday'.

(c) Conditional; antecedent: 'Today is Monday'; consequent: 'Yesterday was Sunday'.

(d) Conditional; antecedent: 'Today is Monday'; consequent: 'She lied'.

(e) Biconditional; left-hand side: 'It's her birthday'; right-hand side: 'It's Monday'.

(f) Disjunction; first disjunct: 'It is Monday'; second disjunct: 'It is Monday'.

(g) Conjunction; first conjunct: 'John is competent'; second disjunct: 'Mary is competent'.

(h) Negation; negated sentence: 'He is competent'.

(i) Negation; negated sentence: 'He is very competent'.

(j) Conjunction; first conjunct: 'He is competent'; second conjunct: 'He is too old'.

(k) Disjunction; first disjunct: 'He is competent'; second disjunct: 'He is corrupt'.

(l) Conditional; antecedent: 'He is competent'; consequent: 'She is a liar'.

(m) Disjunction; first disjunct: 'He is very corrupt'; second disjunct: 'She is very corrupt'.

(n) Conditional; antecedent: 'He is competent'; consequent: 'She is a liar'.

(o) Conjunction; first conjunct: 'He is very competent'; second conjunct: 'She avoids him'.

To facilitate recognition and comparison of argument forms, in formal logic each logical operator is represented by a special symbol:[2]

Logical Operator	Symbol
It is not the case that	$\sim$
And	&
Either ... or	$\vee$
If ... then	$\rightarrow$
If and only if	$\leftrightarrow$

Thus, for example, the argument form disjunctive syllogism may be expressed as:

$$P \vee Q$$
$$\sim P$$
$$\therefore Q$$

It is also customary to write argument forms horizontally, with the premises separated by commas:

$$P \vee Q, \sim P \vdash Q$$

In this horizontal format, we use the symbol '$\vdash$' instead of the triple dots. This symbol is called a *turnstile* or an *assertion sign*.

---

[2]Some authors use different symbols. Here are some of the most common alternatives:

Logical Operator	Alternative Symbol(s)
It is not the case that	$-$ or $\neg$
And	$\cdot$ or $\wedge$
Either ... or	none
If ... then	$\supset$
If and only if	$\equiv$

SOLVED PROBLEM

3.4 Express the following argument forms in symbolic notation, using the horizontal format:

(*a*) If P, then Q.
       It is not the case that P.
    $\therefore$ It is not the case that Q.

(*b*) P and Q.
          P.
    $\therefore$ It is not the case that Q.

(*c*) P if and only if Q.
       It is not the case that Q.
    $\therefore$ It is not the case that P.

(*d*) P.
    $\therefore$ Either P or Q.

(*e*) If P then Q.
       If Q then R.
    $\therefore$ If P then R.

Solution

(*a*) $P \rightarrow Q, \sim P \vdash \sim Q$.

(*b*) $P \& Q, P \vdash \sim Q$.

(*c*) $P \leftrightarrow Q, \sim Q \vdash \sim P$.

(*d*) $P \vdash P \vee Q$.

(*e*) $P \rightarrow Q, Q \rightarrow R \vdash P \rightarrow R$.

3.3 FORMALIZATION

The language consisting of the symbolic notation introduced in the previous section is called the *language of propositional logic*. We shall now examine the syntax of this language first by showing how the forms of various English sentences may be expressed as symbolic formulas (i.e., how these sentences may be *formalized*), and then by stating explicit grammatical rules (*formation rules*) for the language itself.

The process of formalization converts an English sentence or argument into a sentence form or argument form, a structure composed of sentence letters and logical operators. The sentence letters have no meaning in themselves; but in the context of a particular problem, they may be interpreted as expressing definite propositions or statements. This interpretation, however, is inessential to the form. In a different problem, the same sentence letters may stand for different statements. Whenever we talk about the meaning of a sentence letter, we are speaking of its meaning *under the particular interpretation specified by the problem at hand*.

The formalization of simple English sentences is quite easy. If we interpret the sentence letter 'M', for example, as 'Today is Monday', then the sentence 'Today is not Monday' will be formalized simply as '$\sim M$'. But where English sentences contain several logical operators, formalization requires care. Suppose, for example, that we wish to formalize the sentence 'Today is not both Monday and Tuesday'. We cannot simply write '$\sim M \& T$'. The operator '$\sim$', like the negative sign in algebra, applies to the smallest possible part of the formula. In the algebraic formula '$-1 + 3$', for example, the '$-$' sign applies just to '1', so that the whole formula denotes the number 2. Similarly, in '$\sim M \& T$', the '$\sim$' sign applies just to 'M', so that '$\sim M \,\&\, T$' means "Today is not Monday, and today is Tuesday," which is not what we wanted to say. We can, however, extend the part of the formula to which the operator applies in each case by adding brackets. In the algebraic case, this yields the formula '$-(1 + 3)$', which denotes the negative number -4. In the logical case, it yields '$\sim(M \,\&\, T)$', which means "It is not the case that today is (both) Monday and Tuesday," which is precisely what we wanted to say.

To take a different example, suppose we wish to formalize the sentence 'Either today is Monday, or today is Tuesday and election day'. This is a disjunction whose second disjunct is the conjunction 'Today is Tuesday and election day'. It is formalized as '$M \lor (T \,\&\, E)$'. If we leave out the brackets and simply write '$M \lor T \,\&\, E$', the meaning is not clear. For this could be read as a conjunction whose first conjunct is the disjunction 'Today is Monday or Tuesday', so that it would express the sentence 'Today is Monday or Tuesday, and it's election day', which is quite different from our original statement.

SOLVED PROBLEM

3.5 Interpreting the sentence letter 'R' as 'It is raining' and the letter 'S' as 'It is snowing', express the form of each of the following English sentences in the language of propositional logic:

(a) It is raining.

(b) It is not raining.

(c) It is either raining or snowing.

(d) It is both raining and snowing.

(e) It is raining, but it is not snowing.

(f) It is not both raining and snowing.

(g) If it is not raining, then it is snowing.

(h) It is not the case that if it is raining then it is snowing.

(i) It is not the case that if it is snowing then it is raining.

(j) It is raining if and only if it is not snowing.

(k) It is neither raining nor snowing.

(l) If it is both snowing and raining, then it is snowing.

(m) If it's not raining, then it's not both snowing and raining.

(n) Either it's raining, or it's both snowing and raining.

(o) Either it's both raining and snowing or it's snowing but not raining.

Solution

(a) R

(b) $\sim R$

(c) $R \lor S$

(d) $R \,\&\, S$

(e) $R \,\&\, \sim S$

(f) $\sim(R \,\&\, S)$

(g) $\sim R \to S$

(h) $\sim(R \to S)$

(i) $\sim(S \to R)$

(j) $R \leftrightarrow \sim S$

(k) $\sim(R \vee S)$ or, alternatively, $\sim R \& \sim S$; these two ways of formalizing the statement are
        equivalent and equally correct (compare Problem 3.8 below).

(l) $(S \& R) \rightarrow S$

(m) $\sim R \rightarrow \sim(S \& R)$

(n) $R \vee (S \& R)$

(o) $(R \& S) \vee (S \& \sim R)$

Observe that these formulas are all constructed from the following three sets of symbols:

Sentence letters: Any capital letter counts as a sentence letter; occasionally we may add numerical
subscripts to produce additional sentence letters. Thus, for example, 'S_1', 'S_2', 'S_3', etc., are all sentence
letters distinct from 'S'.

Logical operators: $\sim$, $\&$, $\vee$, $\rightarrow$, $\leftrightarrow$

Brackets: (,)

These three sets of symbols constitute the *vocabulary* of the language of propositional logic. The
vocabulary of a formal language is usually divided into *logical* and *nonlogical* symbols. The logical
symbols of our formal language are the logical operators and brackets; the nonlogical symbols are the
sentence letters. We have already pointed out that nonlogical symbols have different interpretations in
different contexts; the sentence letter 'P', for example, may stand for 'Today is Tuesday' in one problem
and 'The princess dines' in another. By contrast, the function or interpretation of logical symbols
always remains fixed.

A *formula* of the language of propositional logic is any sequence of elements of the vocabulary.
Thus the answers to Problem 3.5 are all formulas, but so are such nonsense sequences as '((&(P'. To
distinguish these nonsense sequences from meaningful formulas, we introduce the concept of a
grammatical or *well-formed* formula—wff, for short. This concept is defined by the following rules,
called *formation rules*, which constitute the grammar of the formal language. The rules use Greek
letters (which do not belong to the vocabulary of the formal language) to denote arbitrary formulas.

(1)   Any sentence letter is a wff.

(2)   If ϕ is a wff, then so is $\sim\phi$.

(3)   If ϕ and ψ are wffs, then so are $(\phi \& \psi)$, $(\phi \vee \psi)$, $(\phi \rightarrow \psi)$, and $(\phi \leftrightarrow \psi)$.

Anything not asserted to be a wff by these three rules is not a wff.

Complex wffs are built up from simple ones by repeated application of the formation rules. Thus,
for example, by rule 1 we see the 'P' and 'Q' are both wffs. It follows from this by rule 3 that '$(P \& Q)$'
is a wff. Hence, by rule 2, '$\sim(P \& Q)$' is a wff. Or again, by rule 1, 'P' is a wff, whence it follows by 2
that '$\sim P$' is a wff, and again by 2 that '$\sim \sim P$' is a wff. (We can go on adding as many negation signs
as we like; indeed, '$\sim \sim \sim \sim \sim \sim \sim P$' is a wff!)

Notice that rule 3 stipulates that each time we introduce one of the binary operators we also
introduce a corresponding pair of brackets. Thus, whereas '$(P \& \sim Q)$' is a wff, for example, '$P \& \sim Q$'
is not. However, pairs of brackets which enclose everything else in the formula are not actually
necessary to make the meaning of the formula clear. Thus we will adopt the unofficial convention that
such outer brackets may be omitted, even though officially the resulting formula is not a wff. This
convention was used implicitly in the solutions to Problems 3.4 and 3.5. If we had been following the
formation rules to the letter, 3.5(c), for example, would have been '$(R \vee S)$', rather than '$R \vee S$'.
Omission of outer brackets is the only deviation from the formation rules that we will allow.

SOLVED PROBLEM

3.6 Use the formation rules to determine which of the following formulas are wffs and which are not. Explain your answer.

(a) $\sim\sim\sim R$

(b) $(\sim R)$

(c) PQ

(d) $P \rightarrow Q$

(e) $(P \rightarrow Q)$

(f) $\sim(P \rightarrow Q)$

(g) $((P \,\&\, Q) \rightarrow R)$

(h) $(P \,\&\, Q) \rightarrow R$

(i) $\sim(\sim P \,\&\, \sim Q)$

(j) $((P \,\&\, Q) \vee (R \,\&\, S))$

(k) $((P) \rightarrow (Q))$

(l) $(P \vee Q \vee R)$

(m) $(\sim P \leftrightarrow (Q \,\&\, R))$

(n) $\sim(P \leftrightarrow (Q \,\&\, R))$

(o) $\sim\sim(P \,\&\, P)$

Solution

(a)   'R' is a wff, by rule 1; so '$\sim\sim\sim R$' is a wff, by three applications of rule 2.

(b)   Not a wff. Brackets are introduced only with binary operators (rule 3).

(c)   Not a wff. Two or more sentence letters can produce a wff only in combination with a binary operator (rule 3).

(d)   Not officially a wff; outer brackets are missing. But we shall use such formulas "unofficially."

(e)   'P' and 'Q' are wffs, by rule 1; so '$(P \rightarrow Q)$' is a wff, by rule 3. This is the "official" version of formula (d).

(f)   This is a wff, by application of rule 2 to formula (e).

(g)   'P', 'Q', and 'R' are all wffs, by rule 1. Thus '$(P \,\&\, Q)$' is a wff, by rule 3, and so '$((P \,\&\, Q) \rightarrow R)$' is a wff, by a second application of rule 3.

(h)   Not officially a wff. This is the result of dropping the outer brackets from formula (g).

(i)   'P' and 'Q' are wffs, by rule 1; hence '$\sim P$' and '$\sim Q$' are wffs, by rule 2. But then '$(\sim P \,\&\, \sim Q)$' is a wff, by 3, and so '$\sim(\sim P \,\&\, \sim Q)$' is a wff, by 2.

(j)   'P', 'Q', 'R', and 'S' are all wffs, by 1. Hence '$(P \,\&\, Q)$' and '$(R \,\&\, S)$' are wffs, by 3, and so '$((P \,\&\, Q) \vee (R \,\&\, S))$' is a wff, again by 3.

(k)   Not a wff. No rule allows us to surround sentence letters with brackets.

(l)   Not a wff. Rule 3 allows us to combine only two sentence letters at a time.

(m)   'P', 'Q', and 'R' are wffs, by 1. So '$\sim P$' is a wff, by 2, and '$(Q \,\&\, R)$' is a wff, by 3. Hence '$(\sim P \leftrightarrow (Q \,\&\, R))$' is a wff, by 3.

(n)   As in part (m), 'P' and '$(Q \,\&\, R)$' are wffs. Therefore '$(P \leftrightarrow (Q \,\&\, R))$' is a wff, by 3. So '$\sim(P \leftrightarrow (Q \,\&\, R))$' is a wff, by 2.

(o)   'P' is a wff; hence '$(P \,\&\, P)$' is a wff, by 3, and so '$\sim\sim(P \,\&\, P)$' is a wff, by two applications of rule 2.

Sentence letters are called *atomic* wffs; all other wffs are said to be *molecular* or *compound*. A *subwff* is a part of a wff which is itself a wff. Thus 'P' is a subwff of '$\sim(P \& Q)$' and '$\sim R$' is a subwff of '$\sim \sim R$'. Each wff is regarded as a subwff of itself.

A particular occurrence of an operator in a wff, together with the part of the wff to which that occurrence of the operator applies, is called the *scope* of that occurrence of the operator. Equivalently, we may say that the scope of an occurrence of an operator in a wff is the smallest subwff that contains that occurrence. Thus in the formula '$(\sim P \& (Q \rightarrow \sim R))$', the scope of the first occurrence of '$\sim$' is '$\sim P$', the scope of the second occurrence of '$\sim$' is '$\sim R$', the scope of '$\rightarrow$' is '$(Q \rightarrow R)$', and the scope of '$\&$' is the whole formula. And in the formula '$\sim(P \& (Q \vee R))$', the scope of '$\vee$' is '$(Q \vee R)$', the scope of '$\&$' is '$(P \& (Q \vee R))$', and the scope of '$\sim$' is the whole formula.

Each wff has exactly one operator whose scope is the entire wff. This is called the *main operator* of that wff. A wff whose main operator is '$\&$' (regardless of how many other operators it contains) is called a *conjunction*; a wff whose main operator is '$\sim$' is a *negation*; and so on.

Having rigorously defined our formal language, we can now use it to display the forms of English arguments. Ultimately, this will lead us to a method for proving deductive validity.

SOLVED PROBLEM

3.7 Formalize the following arguments in a horizontal format, using the sentence letters indicated. Use premise and conclusion indicators to distinguish premises from conclusions (see Section 1.2). Omit outer brackets, according to our "unofficial" convention, stated above.

(a) If God exists, then life is meaningful. God exists. Therefore life is meaningful. (L, G)

(b) God does not exist. For if God exists, life is meaningful. But life is not meaningful. (G, L)

(c) If the plane had not crashed, we would have had radio contact with it. We had no radio contact with it. Therefore, it crashed. (C, R)

(d) Since today is not Thursday, it must be Friday. For it is either Thursday or Friday. (T, F)

(e) If today is Thursday, then tomorrow is Friday. If tomorrow is Friday, then the day after tomorrow is Saturday. Consequently, if today is Thursday, then the day after tomorrow is Saturday. (T, F, S)

(f) It is on a weekend if and only if it is on either a Saturday or Sunday. Therefore, it is on a weekend, since it is on a Sunday. (W, S_1, S_2)

(g) It is on a weekend if it is on either a Saturday or a Sunday. But it is not on a weekend. So it is not on a Saturday and it is not on a Sunday. (W, S_1, S_2)

(h) It is on a weekend only if it is on either a Saturday or a Sunday. It is not on a Saturday. It is not on a Sunday. Therefore, it is not on a weekend. (W, S_1, S_2)

(i) The grant proposal is in the mail. If the referees receive it by Friday, they'll give it due consideration. Therefore, they'll give it due consideration, since if the proposal is in the mail, they will receive it by Friday. (M, F, C)

(j) She's either not at home or not answering the phone. But if she's not at home, then she's been kidnapped. And if she's not answering the phone, she is in some other danger. Therefore, either she has been kidnapped or she's in some other danger. (H, A, K, D)

Solution

(a) $G \rightarrow L, G \vdash L$

> (b) $G \to L, \sim L \vdash \sim G$
>
> (c) $\sim C \to R, \sim R \vdash C$
>
> (d) $T \lor F, \sim T \vdash F$
>
> (e) $T \to F, F \to S \vdash T \to S$
>
> (f) $W \leftrightarrow (S_1 \lor S_2), S_2 \vdash W$
>
> (g) $(S_1 \lor S_2) \to W, \sim W \vdash \sim S_1 \,\&\, \sim S_2$
>
> (h) $W \to (S_1 \lor S_2), \sim S_1, \sim S_2 \vdash \sim W$
>
> (i) $M, F \to C, M \to F \vdash C$
>
> (j) $\sim H \lor \sim A, \sim H \to K, \sim A \to D \vdash K \lor D$

3.4 SEMANTICS OF THE LOGICAL OPERATORS

At this point we give a rigorous formulation of the intended interpretation (semantics) of the five logical operators. The semantics of an expression is its contribution to the truth or falsity of sentences in which it occurs (more precisely, to the truth or falsity of the statements expressed by those sentences[3]). The truth or falsity of a statement is also called the *truth value* of the statement. So, in particular, the semantics of a logical operator is given by a rule for determining the truth value of any compound statement involving that operator on the basis of the truth values of the components.

Note that in describing these semantic rules we shall assume the *principle of bivalence*, to the effect that *true* and *false* are the only truth values and that in every possible situation each statement has one and only one of them. Philosophers have argued that certain kinds of statements (e.g., vague or spurious statements, statements about the future, statements about infinite processes, or paradoxical statements such as 'I am lying') may have truth values other than true and false, or no truth value at all, and hence are not bivalent. Some have also argued that certain statements may have both truth values. These views cannot be examined here. But it is important to note that *classical logic*, the kind of logic presented in this book, applies only to bivalent statements. This is a limitation; on the other hand, nonclassical logics are invariably more complicated than classical logic and are best understood in contrast to classical logic, so it is convenient to master classical logic thoroughly before attempting the study of nonclassical logics.[4]

Let us then assume the principle of bivalence. The semantic rule for negation is simple. The negation of a statement ϕ is true if ϕ is false and false if ϕ is true. (This applies regardless of whether ϕ is atomic or compound.) Using the abbreviations 'T' for "true" and 'F' for "false," we may summarize this rule as follows:

ϕ	$\sim\phi$
T	F
F	T

This summary is called a *truth table*. Under 'ϕ', two possibilities are listed: either ϕ is true or ϕ is false. The entries under '$\sim\phi$' indicate the truth value of $\sim\phi$ in each case. In this way, each horizontal line of truth values represents a class of possible situations. The first line represents situations in which ϕ is true. In these situations $\sim\phi$ is false. The second represents situations in which ϕ is false. In these, $\sim\phi$ is true. Given bivalence, these are the only possibilities; hence the table completely describes the truth value of $\sim\phi$ in every possible situation.

---

[3]Recall that the same sentence can be used to express different statements in different circumstances—even statements that disagree in their being true or false (see Chapter 1, footnote 2). However, where there is no danger of confusion, we shall suppress the distinction and speak freely of the truth value of a sentence (or of a wff).

[4]Some important families of nonclassical logics are surveyed in D. Gabbay and F. Guenthner (eds.), *Handbook of Philosophical Logic, Volume 3: Non-Classical Logics*, Dordrecht, Reidel, 1986.

The truth table for conjunction is equally straightforward. A conjunction is true if both of its conjuncts are true, and false otherwise. Thus the table is:

ϕ	ψ	$\phi \,\&\, \psi$
T	T	T
T	F	F
F	T	F
F	F	F

Since conjunction operates on two statements, there are four kinds of possible situations to consider: those in which ϕ and ψ are both true, those in which ϕ is true and ψ is false, those in which ϕ is false and ψ is true, and those in which both ϕ and ψ are false. These are represented, respectively, by the four horizontal lines of truth values. The column under '$\phi \,\&\, \psi$' lists the truth value of $\phi \,\&\, \psi$ in each.

A disjunction is true if at least one of its disjuncts is true, and false otherwise:

ϕ	ψ	$\phi \lor \psi$
T	T	T
T	F	T
F	T	T
F	F	F

There is also a sense of disjunction in which 'either P or Q' means 'either P or Q, *but not both*'. This is the *exclusive* sense of disjunction, as opposed to the *inclusive* sense characterized by the true table above. The operator '$\lor$' symbolizes the inclusive sense. If we use 'xor' for the exclusive sense of 'or', we may write its truth table as follows:

ϕ	ψ	$\phi \text{ xor } \psi$
T	T	F
T	F	T
F	T	T
F	F	F

In logic, the inclusive sense of 'or' is standard. But in English, the exclusive sense is quite common, especially when 'or' is preceded by 'either'. However, there is no need to add 'xor' to our stock of logical operators, since statements of the form 'P xor Q' may be formalized as '$(P \lor Q) \,\&\, \sim(P \,\&\, Q)$' (see Problem 3.11, below). Thus, for example, if the boss says, 'You may take either Thursday or Friday off', what's meant is undoubtedly the exclusive disjunction 'You may take either Thursday or Friday off, *but not both*'. The best formalization is therefore '$(T \lor F) \,\&\, \sim(T \,\&\, F)$'. By contrast, the statement 'She must be intelligent or rich' seems not to exclude the possibility that the person referred to is *both* intelligent and rich, and hence is best formalized as the inclusive disjunction '$I \lor R$'. We shall treat English disjunctions as inclusive unless there is some compelling reason not to. (Of course, we could as well use 'xor' instead of '$\lor$' and express every statement of the form '$P \lor Q$' as '$(P \text{ xor } Q)$ xor $(P \,\&\, Q)$'. In this sense, our preference for '$\lor$' is dictated merely by standard practice.)

Of all the logical operators, '$\rightarrow$' is least like its English translations in meaning. The meanings of conditionals in English are still a matter of dispute. It is widely held that there are several different kinds of conditionals, which assert different relationships between antecedent and consequent. The conditional expressed by our symbol '$\rightarrow$' is called the *material conditional*. What is asserted by '$P \rightarrow Q$' is precisely this: It is not the case that P and not Q. Thus if someone says, "If Paul goes, then Quinton will go too," using 'If . . . then' in the material sense, then that person is saying that it is not the case that

Paul will go and Quinton won't. This statement has the form '~ $(P \& \sim Q)$', and since it has the same meaning as '$P \rightarrow Q$', it is true under precisely the same circumstances. We can therefore obtain the truth table for '$P \rightarrow Q$' by finding the truth table for '$\sim(P \& \sim Q)$'. We can do this by using the tables for '~' and '&'.

To construct a truth table for a complex wff, we find the truth values for its smallest subwffs and then use the truth tables for the logical operators to calculate values for increasingly larger subwffs, until we obtain the values for the whole wff. The smallest subwffs of '$\sim(P \& \sim Q)$' are 'P' and 'Q', and so we copy the columns for the sentence letters 'P' and 'Q' under the occurrences of these sentence letters in the formula:

P	Q	$\sim(P \& \sim Q)$	
T	T	T	T
T	F	T	F
F	T	F	T
F	F	F	F

The next-smallest subwff of the formula is '$\sim Q$'. The truth values for '$\sim Q$' are just the opposite of the truth values for 'Q'; we write them under the '~' in '$\sim Q$':

P	Q	$\sim(P \& \sim Q)$	
T	T	T	F T
T	F	T	T F
F	T	F	F T
F	F	F	T F

Actually, we could have saved a bit of work and clutter by skipping the first step in the case of '$\sim Q$'. That is, we could have written the reverse of the 'Q' column directly under the '~' in '$\sim Q$' without first writing the column for 'Q'. In succeeding problems, we sometimes skip this first step for negated sentence letters.

Now the formula '$P \& \sim Q$', being a conjunction, is true only if both conjuncts are true. The only case in which this occurs is the second line of the table. Hence this formula is true on the second line and false on the others, a fact which we record by writing these truth values under '&':

P	Q	$\sim(P$	$\&$	$\sim$	$Q)$
T	T	T	F	F	T
T	F	T	T	T	F
F	T	F	F	F	T
F	F	F	F	T	F

Finally, since '$\sim(P \& \sim Q)$' is the negation of '$P \& \sim Q$', its truth values in each of the four kinds of situations are just the reverse of those listed under '&'. We write them under the initial negation sign and circle them to indicate that they are the truth values for the whole formula:

P	Q	$\sim$	$(P$	$\&$	$\sim$	$Q)$
T	T	(T)	T	F	F	T
T	F	F	T	T	T	F
F	T	T	F	F	F	T
F	F	(T)	F	F	T	F

This, then, is the truth table for the material conditional:

φ	ψ	φ → ψ
T	T	T
T	F	F
F	T	T
F	F	T

The material conditional is false if its antecedent is true and its consequent false; otherwise it is true. It is therefore true whenever its antecedent is false. It is also true whenever its consequent is true. As a result, material conditionals often have a paradoxical flavor. The sentence 'If you're dead, then you're alive', for example, is currently true, taking 'you' to refer to you and reading 'if ... then' as the material conditional. Since you are in fact alive (otherwise you wouldn't be reading this), the antecedent is false and the consequent true, which makes the whole conditional true. Similarly, the sentences 'If you are alive, then you are reading this book) (antecedent and consequent both true) and 'If you are dead, then you can run at the speed of light' (antecedent and consequent both false) are, oddly, true when 'if ... then' is understood in the material sense. Such oddities reveal the disparity between the material conditional and conditionals as ordinarily understood in English. Yet the material conditional is by far the simplest kind of conditional, and it is the only one whose meaning can be represented on a truth table. Moreover, experience has shown that it is adequate for most logical and mathematical purposes, so that it has been accepted as the standard conditional of logic and mathematics. The material conditional is the only kind of conditional considered in this chapter, though at the end of Section 10.7 we briefly consider two others.

As for the biconditional, we have seen that a statement of the form '$P \leftrightarrow Q$' means the same thing as '$(P \to Q) \& (Q \to P)$', where '→' is the material conditional. Accordingly, '↔' is called the *material biconditional*. (The meanings of other kinds of biconditionals are similarly related to the corresponding conditionals, but we shall not discuss other kinds of biconditionals here.) The truth table for '$P \leftrightarrow Q$' may thus be obtained by constructing a truth table for '$(P \to Q) \& (Q \to P)$' from the tables for the material conditional and conjunction. We begin by copying the columns from beneath the letters 'P' and 'Q' at left below the occurrences of these letters in the formula:

P	Q	(P → Q)	&	(Q → P)
T	T	T T		T T
T	F	T F		F T
F	T	F T		T F
F	F	F F		F F

Now the truth values for '$P \to Q$' and '$Q \to P$' may be computed from the truth table for the material conditional. A material conditional is false when its antecedent is true and its consequent false; otherwise, it is true. Hence '$P \to Q$' is false at the second line and '$Q \to P$' is false at the third, and both are true at all the other lines. We record this by writing the appropriate truth values under the respective occurrences of '→':

P	Q	(P	→	Q)	&	(Q	→	P)
T	T	T	T	T		T	T	T
T	F	T	F	F		F	T	T
F	T	F	T	T		T	F	F
F	F	F	T	F		F	T	F

To complete the table, we calculate the truth values for the whole conjunction, using the values for '$P \rightarrow Q$' and '$Q \rightarrow P$'. These are written under '&' and circled:

P	Q	$(P$	$\rightarrow$	$Q)$	&	$(Q$	$\rightarrow$	$P)$
T	T	T	T	T	(T)	T	T	T
T	F	T	F	F	F	F	T	T
F	T	F	T	T	F	T	F	F
F	F	F	T	F	(T)	F	T	F

The circled values represent the truth table for the biconditional. The biconditional, then, is true if its two components have the same truth value and false if their truth values differ:

ϕ	ψ	$\phi \leftrightarrow \psi$
T	T	T
T	F	F
F	T	F
F	F	T

Note that this way of defining the semantic rules for conditionals and biconditionals suggests that these operators are not, strictly speaking, necessary: every statement of the form '$P \rightarrow Q$', and consequently every statement of the form '$P \leftrightarrow Q$', can be expressed in terms of '$\sim$', '&', and '$\vee$'. Hence, just as it was not necessary to add a sixth logical operator 'xor' to express the exclusive sense of 'or', likewise one could do without the two operators '$\rightarrow$' and '$\leftrightarrow$' and keep only '$\sim$', '&', and '$\vee$'. The only reasons why we have all five operators in our formal language are perspicuity and notational convenience, but in principle three operators would be enough. Indeed, it turns out that only one of '&' and '$\vee$' would be needed besides '$\sim$', since the other could always be expressed indirectly: '$P \vee Q$' means the same thing as 'It is not the case that not P and not Q', which has the form '$\sim(\sim P \& \sim Q)$', and '$P \& Q$' means the same thing as 'It is not the case that not P or not Q', which has the form '$\sim(\sim P \vee \sim Q)$'.

There are other combinations of logical operators that would serve our purpose of formalizing arguments in propositional logic, but we need not consider the alternatives here. The important thing is that our operators, though not equally necessary, jointly provide a sufficient apparatus for that purpose. And this is indeed the case. It can be shown that as long as we have negation together with either conjunction, disjunction, or material conditional, we can express *every* other logical operator, at least every logical operator whose meaning can be represented by a truth table. This fact (whose proof is beyond the scope of this book) is called the *functional completeness* of our set of logical operators.

SOLVED PROBLEM

3.8 Show that there is no need for a logical operator corresponding to the English 'neither ... nor', since every statement of the form

neither P nor Q

can be formalized using the operators already available.

Solution

A sentence of the form 'neither P nor Q' is true just in case both P and Q are false. This amounts to saying that it is true just in case '$\sim P$' and '$\sim Q$' are both true, i.e., just in case the conjunction '$(\sim P \& \sim Q)$' is true. Hence there is no need for a special symbol in order to express this operator: negation and conjunction are enough. (Another way of expressing 'neither P nor Q' is '$\sim(P \lor Q)$', which uses only negation and disjunction; see Problem 3.5(k).)

It is remarkable that if we did add the 'neither … nor' operator, then our five operators would all become redundant. For we could express every statement of the form '$\sim P$' as 'neither P nor P' and every statement of the form '$P \& Q$' as 'neither $\sim P$ nor $\sim Q$', and we have seen that this gives us sufficient resources to express the other operators as well.

3.5 TRUTH TABLES FOR WFFs

Our presentation of the semantic rules for '$\rightarrow$' and '$\leftrightarrow$' in Section 3.4 exemplified the construction of truth tables for complex wffs. Here we discuss this procedure more systematically.

The number of lines in a truth table is determined by the number of sentence letters in the formula or formulas to be considered. If there is only one sentence letter, there are only two possibilities: the sentence it stands for is either true or false. Hence the table will have two lines. If there are two sentence letters, there are four possible combinations of truth and falsity and the table has four lines. In general, if the number of sentence letters is n, the number of lines is 2^n. Thus if a formula contains three different sentence letters, its truth table has $2^3 = 8$ lines, and so on.

To set up a truth table for a formula, write the formula at the upper right side of the table and list the sentence letters it contains in alphabetical order to the left. Now where n is the number of sentence letters, write beneath the rightmost of them a column of 2^n alternating T's and F's, beginning with T. Then, under the next letter to the left (if any remains), write another column of 2^n T's and F's, again beginning with T, but alternating every two lines. Repeat this procedure, moving again to the left and doubling the alternation interval each time, until each sentence letter has a column of T's and F's beneath it. If, for example, the formula contains three sentence letters, P, Q, and R, the left side of the table looks like this:

P	Q	R
T	T	T
T	T	F
T	F	T
T	F	F
F	T	T
F	T	F
F	F	T
F	F	F

Finally, using the truth tables for the logical operators, calculate the truth values of the formula by determining the values for it smallest subwffs first and then obtaining the values for larger and larger subwffs, until the values for the whole formula are discovered. *The column for any wff or subwff is always written under its main operator.* Circle the column under the main operator of the entire wff to show that the entries in it are the truth values for the whole formula.

SOLVED PROBLEMS

3.9 Construct a truth table for the formula

$\sim\sim P$

Solution

P	$\sim$	$\sim$	P
T	(T)	F	T
F	(F)	T	F

The table has two lines, since there is only one sentence letter. We copy the column for the sentence letter 'P' beneath the occurrence of 'P' in the formula. The negation sign to its immediate left reverses the values in this column, and the negation sign to *its* left (which is the main operator of the whole formula) reverses them again, so that '$\sim \sim P$' has the same truth value as 'P' in every possible situation.

3.10   Construct a truth table for the formula

$$\sim P \vee Q$$

Solution

P	Q	$\sim P$	$\vee$	Q
T	T	F	T	T
T	F	F	F	F
F	T	T	T	T
F	F	T	T	F

We begin by copying the 'Q' column under 'Q' and the reverse of the 'P' column under the negation sign in '$\sim P$' (skipping the unnecessary step of copying the 'P' column under 'P'). We then determine the truth values for '$\sim P \vee Q$' from those for '$\sim P$' and 'Q', using the disjunction table. Since a disjunction is false if and only if both disjuncts are false, and since '$\sim P$' and 'Q' are both false only at line 2, '$\sim P \vee Q$' is false at line 2 and true at all the other lines. We circle the column containing the values for '$\sim P \vee Q$'. Note that these values are the same as for '$P \rightarrow Q$'. This shows that for logical purposes '$\sim P \vee Q$' and '$P \rightarrow Q$' are synonymous, just as '$\sim (P \& \sim Q)$' and '$(P \rightarrow Q)$' are.

3.11   Construct a truth table for the formula

$$(P \vee Q) \& \sim (P \& Q)$$

Solution

P	Q	$(P$	$\vee$	$Q)$	$\&$	$\sim$	$(P$	$\&$	$Q)$
T	T	T	T	T	F	F	T	T	T
T	F	T	T	F	T	T	T	F	F
F	T	F	T	T	T	T	F	F	T
F	F	F	F	F	F	T	F	F	F

This formula is the formalization of the exclusive disjunction of P with Q (see Section 3.4). In fact, the circled column is the correct truth table for exclusive disjunction. The truth values for the formula are entered in the following order. First, copy the columns for the sentence letters 'P' and 'Q' under the occurrences of these letters in the formula. Next, using the truth tables for disjunction and conjunction, determine the values for the formulas '$P \vee Q$' and '$P \& Q$' and write these under '$\vee$' and the second occurrence of '$\&$', respectively. The truth values for '$\sim (P \& Q)$' are just the reverse of those for '$P \& Q$'; write them under '$\sim$'. Finally, the truth values for the entire formula are determined from the truth values for '$P \vee Q$' and '$\sim (P \& Q)$', using the conjunction table. These are written under the main operator '$\&$' and circled.

SOLVED PROBLEM

3.12   Construct a truth table for the formula

$$P \vee \sim P$$

Solution

P	P	$\vee$	~P
T	T	(T)	F
F	F	(T)	T

The truth table shows that the formula '$P \vee {\sim}P$' is true in every possible situation.

Formulas which, like '$P \vee {\sim}P$', are true at every line of the truth table are called *tautologies*, as are the statements they represent. Their truth tables show that tautologies are true in all possible circumstances. Thus tautologousness is a kind of logical necessity, the kind generated by the semantics of the operators of propositional logic.

SOLVED PROBLEM

3.13 Construct a truth table for the formula

$$P \mathbin{\&} {\sim}P$$

Solution

P	P	&	~P
T	T	(F)	F
F	F	(F)	T

The table shows that '$P \mathbin{\&} {\sim}P$' is false in every possible situation.

Any formula whose truth table contains only F's under its main operator is called a *contradiction* and is said to be *truth-functionally inconsistent*, as are all specific statements of the same form. Truth-functional inconsistency is one kind of inconsistency, the kind generated by the operators of propositional logic. Since it is the only kind that concerns us in this chapter, we will frequently refer to it here simply as "inconsistency"; but it should be kept in mind that not all inconsistency is truth-functional. The statement 'George is not identical to himself', for example, is inconsistent, but its inconsistency is due to the semantics of the expression 'is identical to' (see Section 6.7) as well as to the semantics of 'not', and hence it is not purely truth-functional.

Formulas which are true at some lines of their truth tables and false at others are said to be *truth-functionally contingent*, as are the statements they represent. A truth-functionally contingent statement is one that could be either true or false, *so far as the operators of propositional logic are concerned*. Not all truth-functionally contingent statements, however, are genuinely contingent, i.e., capable of being either true or false, depending on the facts. The statement 'Jim is a bachelor and Jim (the same Jim) is married', for example, has the propositional form '$B \mathbin{\&} M$', and hence is truth-functionally contingent, as this truth table reveals:

B	M	B	&	M
T	T	T	(T)	T
T	F	T	F	F
F	T	F	F	T
F	F	F	(F)	F

But this statement is inconsistent, not contingent; what it asserts is impossible in a logical or conceptual sense. Its inconsistency, however, is non-truth-functional, being a consequence of the semantics of the expressions 'is a bachelor' and 'is married', in addition to the logical operator 'and'.

To summarize: If a wff is tautologous, then any statement of the same form is logically necessary and must always be true. If a wff is truth-functionally inconsistent, then any statement of the form is inconsistent and must always be false. But if a wff is truth-functionally contingent, then it is contingent *only so far as the operators represented in the wff are concerned*. Some specific statements of the form will be genuinely contingent, while others will be non-truth-functionally necessary or inconsistent as a result of factors not represented in the wff.

SOLVED PROBLEMS

3.14 Construct a truth table to determine whether the following wff is tautologous, inconsistent, or truth-functionally contingent:

$$(\sim P \lor \sim Q) \leftrightarrow (P \& Q)$$

Solution

P	Q	($\sim P$	$\lor$	$\sim Q$)	$\leftrightarrow$	(P	$\&$	Q)
T	T	F	F	F	F	T	T	T
T	F	F	T	T	F	T	F	F
F	T	T	T	F	F	F	F	T
F	F	T	T	T	F	F	F	F

Because the column under the main operator '$\leftrightarrow$' consists entirely of F's, the wff is inconsistent.

3.15 Construct a truth table to determine whether the following wff is tautologous, inconsistent, or truth-functionally contingent:

$$P \rightarrow (Q \lor \sim R)$$

Solution

P	Q	R	P	$\rightarrow$	(Q	$\lor$	$\sim R$)
T	T	T	T	T	T	T	F
T	T	F	T	T	T	T	T
T	F	T	T	F	F	F	F
T	F	F	T	T	F	T	T
F	T	T	F	T	T	T	F
F	T	F	F	T	T	T	T
F	F	T	F	T	F	F	F
F	F	F	F	T	F	T	T

This wff is truth-functionally contingent, because both T's and F's occur in the column under the main operator '$\rightarrow$'.

3.16 Construct a truth table to determine whether the following wff is tautologous, inconsistent, or truth-functionally contingent:

$$((P \& Q) \& (R \& S)) \rightarrow P$$

Solution

P	Q	R	S	((P	&	Q)	&	(R	&	S))	→	P
T	T	T	T	T	T	T	T	T	T	T	T	T
T	T	T	F	T	T	T	F	T	F	F	T	T
T	T	F	T	T	T	T	F	F	F	T	T	T
T	T	F	F	T	T	T	F	F	F	F	T	T
T	F	T	T	T	F	F	F	T	T	T	T	T
T	F	T	F	T	F	F	F	T	F	F	T	T
T	F	F	T	T	F	F	F	F	F	T	T	T
T	F	F	F	T	F	F	F	F	F	F	T	T
F	T	T	T	F	F	T	F	T	T	T	T	F
F	T	T	F	F	F	T	F	T	F	F	T	F
F	T	F	T	F	F	T	F	F	F	T	T	F
F	T	F	F	F	F	T	F	F	F	F	T	F
F	F	T	T	F	F	F	F	T	T	T	T	F
F	F	T	F	F	F	F	F	T	F	F	T	F
F	F	F	T	F	F	F	F	F	F	T	T	F
F	F	F	F	F	F	F	F	F	F	F	T	F

The wff is tautologous, because only T's occur in the column under the main connective '→'.

3.6   TRUTH TABLES FOR ARGUMENT FORMS

We are now finally in a position to provide a rigorous account of deductive validity based on the semantics of the logical operators. An argument form is valid if and only if all its instances are valid. An instance of a form is valid if it is impossible for its conclusion to be false while its premises are true, i.e., if there is no possible situation in which its conclusion is false while its premises are true. Now a truth table is in effect an exhaustive list of possible situations. Hence if we put not just a single wff on a truth table but an entire argument form, we can use the table to determine whether or not that form is valid.

If the form turns out to be valid, then (since by definition a valid form is one all instances of which are valid; see Section 3.1) any instance of it must be valid as well. Hence we can use truth tables to establish the validity not only of argument forms, but also of specific arguments. Consider, for example, the disjunctive syllogism

Either the princess or the queen attends the ceremony.
The princess does not attend.
∴ The queen attends.

We may formalize this as

$P \vee Q, {\sim}P \vdash Q.$

We can then construct a truth table for this form as follows:

P	Q	P	∨	Q,	~P	⊢ Q
T	T	T	T	T	F	T
T	F	T	T	F	F	F
F	T	F	T	T	T	T
F	F	F	F	F	T	F

This table is computed in the same way as tables for single wffs, but it displays three separate wffs, instead of just one. We may think of these wffs as expressing an abstract argument form, a structure with many instances, or we may give them a specific interpretation, as when we stipulate that 'P' means "the princess attends the ceremony" and 'Q' means "the queen attends the ceremony." Under this specific interpretation, this sequence of wffs expresses only the argument above, which is just one instance of the abstract form.

Let us consider the wffs under this specific interpretation first. The four lines of the truth table then stand for four distinct possibilities: either the princess and the queen both attend, the princess attends but the queen doesn't, the queen attends but the princess doesn't, or neither attends. Under the presupposition of bivalence, *these are the only possible situations*. In only one of these, the third, are the premises both true; but in this situation the conclusion is true as well. Thus there is no possible situation in which the premises are true and the conclusion false; the table shows that the argument is valid.

Indeed, it shows that every argument of this form is valid. Since any such argument is composed of sentences P and Q which (again by the principle of bivalence) are either true or false, and since the table shows that no matter what combination of truth and falsity they exhibit, there is no possible situation in which the premises are both true and the conclusion is false, so that no instance of the form can be invalid. Hence the validity of the form itself is apparent.

SOLVED PROBLEM

3.17 Construct a truth table for the following form and use it to verify that the form is valid:

$$P \rightarrow Q, Q \rightarrow R \vdash P \rightarrow R$$

Solution

P	Q	R	P	$\rightarrow$	$Q,$	Q	$\rightarrow$	R	$\vdash$	P	$\rightarrow$	R
T	T	T	T	T	T	T	T	T		T	T	T
T	T	F	T	T	T	T	F	F		T	F	F
T	F	T	T	F	F	F	T	T		T	T	T
T	F	F	T	F	F	F	T	F		T	F	F
F	T	T	F	T	T	T	T	T		F	T	T
F	T	F	F	T	T	T	F	F		F	T	F
F	F	T	F	T	F	F	T	T		F	T	T
F	F	F	F	T	F	F	T	F		F	T	F

There are four possible situations in which both premises are true, corresponding to the first, fifth, seventh, and eighth lines of the table. In all these situations the conclusion is also true. Therefore the form is valid.

If a form is *invalid*, its truth table shows its validity by exhibiting one or more lines in which all the premises get the value T while the conclusion gets the value F. Such lines are called *counterexamples*. The existence of even one counterexample is sufficient to establish that the argument is invalid.

SOLVED PROBLEM

3.18 Construct a truth table for the following form and use it to show that the form is invalid:

$$P \rightarrow Q, Q \vdash P$$

Solution

P	Q		P	→	Q,	Q	⊢	P
T	T		T	T	T	T		T
T	F		T	F	F	F		T
F	T		F	T	T	T		F	×
F	F		F	T	F	F		F

The table shows that there are two kinds of possible situations in which the premises are both true, those represented by the first and third lines. At the first line, the conclusion is also true; but at the third line it is false. The third line is therefore a counterexample (we write an '×' on its right to indicate this). It is in effect a prescription for constructing instances of the form with true premises and a false conclusion. Any instance constructed from sentences P and Q, where P is false and Q is true, will fill the bill. For example, if we interpret 'P' as "whales are fish" and 'Q' as "whales inhabit bodies of water," we obtain:

> If whales are fish, they inhabit bodies of water.
> Whales inhabit bodies of water.
> ∴ Whales are fish.

This is an instance of the form with true premises and a false conclusion (whales are mammals, not fish). Since a form is invalid if it has even one such instance, this form is clearly invalid. It is in fact the form known as *affirming the consequent* which was discussed at the end of Section 3.1.

In summary: To determine whether an argument form of propositional logic is valid, put the entire form on a truth table, making as many lines as determined by the number of distinct sentence letters occurring in the relevant formulas. *If the table displays no counterexample, then the form is valid (and hence so is any instance of it). If the table displays one or more counterexamples, then the form is invalid.* Since invalid forms may have valid as well as invalid instances, the truth table test does not establish the invalidity of specific arguments. If we formalize an argument and then show that the resulting form is invalid, we are not thereby entitled to infer that the argument is invalid. But if a truth table shows a form to be invalid, then it shows that none of its instances is valid solely in virtue of having that form. Any valid instances must derive their validity at least in part from some feature of the argument which has been lost in the process of formalization. Argument 5 of Section 3.1, for example, is valid despite being an instance of affirming the consequent, which is invalid by the truth table test; when it is formalized as affirming the consequent (i.e., as $P \rightarrow Q$, $Q \vdash P$), the information that the conclusion follows from the second premise is lost.

In Section 3.1 we have seen that a particular argument may in fact be an instance of several forms, some of which are valid and some of which are not. But if it is an instance of *any* valid form, then it is valid. For example, the argument

> If she loves me, then she doesn't hate me.
> It's not true that she doesn't hate me.
> ∴ She doesn't love me.

is an instance of each of the following forms, only the first two of which are valid:

> $L \rightarrow \sim H, \sim \sim H \vdash \sim L$
> $L \rightarrow D, \sim D \vdash \sim L$
> $L \rightarrow D, N \vdash \sim L$
> $L \rightarrow D, N \vdash S$
> $I, N \vdash S$

In the third case, for example, we have formalized 'She loves me' as 'L', 'She doesn't hate me' as 'D', and 'It's not true that she doesn't hate me' as 'N'. And even this list is not complete. The reader can probably discover several more forms of which this argument is an instance. In formalizing an argument, we generally select the form which displays the most logical structure (in this case the first form), since if the argument is valid in virtue of any of its forms, it will be valid in virtue of that one. However, if a form with less structure is valid (as is the second form on this list), then it too is an adequate formalization for demonstrating the validity of the argument.

SOLVED PROBLEMS

3.19 Construct a truth table for the following form, and use the table to determine whether the form is valid:

$$P \rightarrow Q, P \rightarrow {\sim}Q \vdash {\sim}P$$

Solution

P	Q	P	→	Q,	P	→	~Q⊢~P		
T	T	T	T	T	T	T	F	F	F
T	F	T	F	F	T	T	T	T	F
F	T	F	T	T	F	T	T	F	T
F	F	F	T	F	F	T	T	T	T

The only possible situations in which the premises are both true are those represented by the third and fourth lines of the table. But in these situations the conclusion is also true; hence the form is valid.

3.20 Construct a truth table for the following form, and use the table to determine whether the form is valid:

$$P \rightarrow Q \vdash {\sim}(Q \rightarrow P)$$

Solution

P	Q	P	→	Q⊢~	(Q	→	P)	
T	T	T	T	T	F	T	T	T	×
T	F	T	F	F	F	F	T	T	
F	T	F	T	T	T	T	F	F	
F	F	F	T	F	F	F	T	F	×

The table displays two kinds of counterexamples. The first is when '*P*' and '*Q*' are both true (first line of the table); the second when they are both false (last line of the table). Hence the form is invalid.

3.21 Construct a truth table for the following form, and use the table to determine whether the form is valid:

$$P \vee Q, Q \vee R \vdash P \vee R$$

Solution

P	Q	R	P	∨	Q,	Q	∨	R⊢P	∨	R	
T	T	T	T	T	T	T	T	T	T	T	T
T	T	F	T	T	T	T	T	F	T	T	F
T	F	T	T	T	F	F	T	T	T	T	T
T	F	F	T	T	F	F	F	F	T	T	F
F	T	T	F	T	T	T	T	T	F	T	T
F	T	F	F	T	T	T	T	F	F	F	F	×
F	F	T	F	F	F	F	T	T	F	T	T
F	F	F	F	F	F	F	F	F	F	F	F

The form is invalid, because in situations in which '*P*' and '*R*' are false and '*Q*' is true (line 6 of the table), the premises are both true while the conclusion is false.

3.22 Construct a truth table for the following form, and use the table to determine whether the form is valid:

$$P, \sim P \vdash Q$$

Solution

P	Q	P,	$\sim P$	$\vdash Q$
T	T	T	F	T
T	F	T	F	F
F	T	F	T	T
F	F	F	T	F

Since the premises are mutually inconsistent, there is no possible situation in which both are true. Hence there are no counterexamples; the form is valid. However, notice that every argument of this form is unacceptably flawed as a means of proving its conclusion: in the terminology of Chapter 2, criterion 1 (truth of premises) is violated, so the argument cannot be sound. In addition, such an argument may well violate criterion 3 (relevance). Compare Problem 2.21.

3.23 Construct a truth table for the following form, and use the table to determine whether the form is valid:

$$R \vdash P \leftrightarrow (P \vee (P \& Q))$$

Solution

P	Q	R	R⊦P	↔	(P	∨	(P	&	Q))
T	T	T	T T	T	T	T	T	T	T
T	T	F	F T	T	T	T	T	T	T
T	F	T	T T	T	T	T	T	F	F
T	F	F	F T	T	T	T	T	F	F
F	T	T	T F	T	F	F	F	F	T
F	T	F	F F	T	F	F	F	F	T
F	F	T	T F	T	F	F	F	F	F
F	F	F	F F	T	F	F	F	F	F

The conclusion of this argument is a tautology, so there is no situation in which the premise is true *and* the conclusion false. The argument is therefore valid. As in the previous example, however, arguments of this form lack relevance, since their validity is totally independent of the relationship between premise and conclusion (compare Problem 2.19).

3.7 REFUTATION TREES

Truth tables provide a rigorous and complete test for the validity or invalidity of propositional logic argument forms, as well as a test for tautologousness, truth-functional contingency, and inconsistency of wffs. Indeed, they constitute an *algorithm*, the sort of precisely specifiable test which can be performed by a computer and which always yields an answer after a finite number of finite operations. When there is an algorithm for determining whether or not the argument forms expressible in a formal system are valid, that system is said to be *decidable*. Thus the truth tables ensure the decidability of predicate logic. But they are cumbersome and inefficient, especially for problems involving more than two or three sentence letters. Refutation trees, the topic of this section, provide a more efficient algorithm for performing the same tasks.

Given a list of wffs, a refutation tree is an exhaustive search for ways in which all the wffs on the list can be true. To test an argument form for validity using a refutation tree, we construct a list consisting of its premises and the *negation* of its conclusion. The search is carried out by breaking down the wffs on the list into sentence letters or their negations. If we find any assignment of truth and falsity to sentence letters which makes all the wffs on the list true, then under that assignment the premises of the form are true while its conclusion is false. Thus we have refuted the argument form; it is invalid. If the search turns up no assignment of truth and falsity to sentence letters which makes all the wffs on the list true, then our attempted refutation has failed; the form is valid. To illustrate, we examine some simple examples.

SOLVED PROBLEMS

3.24 Construct a refutation tree to show that the form '$P \& Q \vdash \sim \sim P$' is valid.

Solution

We begin by forming the list consisting of the premise and the negation of the conclusion:

$$P \& Q$$
$$\sim \sim \sim P$$

Now the premise is true if and only if 'P' and 'Q' are both true. Hence we can without distortion replace '$P \& Q$' by these two sentence letters. We show this by writing 'P' and 'Q' at the bottom of the list and checking off the formula '$P \& Q$' to indicate that we are finished with it. A checked formula is in effect eliminated from the list:

$$\sqrt{\quad} \quad P \& Q$$
$$\qquad \sim \sim \sim P$$
$$\qquad P$$
$$\qquad Q$$

Moreover '$\sim \sim \sim P$' is true if and only if the simpler formula '$\sim P$' is true; hence we can check '$\sim \sim \sim P$' and replace it by '$\sim P$':

$$\sqrt{\quad} \quad P \& Q$$
$$\sqrt{\quad} \quad \sim \sim \sim P$$
$$\qquad P$$
$$\qquad Q$$
$$\qquad \sim P$$

We have now broken down the original list of formulas into a list of sentence letters or negations of sentence letters, all of which must be true if all the members of our original list are to be true. But among these sentence letters and their negations are both 'P' and '$\sim P$', which cannot both be true. Hence it is impossible for everything on the finished list to be true. We express this by writing 'X' at the bottom of the list.

$$\sqrt{\quad} \quad P \& Q$$
$$\sqrt{\quad} \quad \sim \sim \sim P$$
$$\qquad P$$
$$\qquad Q$$
$$\qquad \sim P$$
$$\qquad X$$

The refutation tree is now complete. Our search for a refutation of the original argument form has failed; hence this form is valid.

3.25 Construct a refutation tree to show that the form '$P \lor Q, \sim P \vdash Q$' is valid.

Solution

As before, we begin with a list consisting of the premises followed by the negation of the conclusion:

$$P \lor Q$$
$$\sim P$$
$$\sim Q$$

Since both '$\sim P$' and '$\sim Q$' are negations of sentence letters, they cannot be analyzed further; but '$P \lor Q$' is true if and only if either 'P' or 'Q' is true. To represent the fact that '$P \lor Q$' can be true in either of these two ways, we check '$P \lor Q$' and "branch" the tree, like this:

The tree now contains two paths, each starting with the checked formula '$P \lor Q$'. The first branches to the left and ends with 'P'; the second branches to the right and ends with 'Q'. The three formulas on the initial list can be true if and only if all the formulas on one or both of these paths can be true. But the first path contains both 'P' and '$\sim P$', and the second contains both 'Q' and '$\sim Q$'. Hence not all the formulas on either path can be true. As in the previous problem, we indicate this by ending each path with an X:

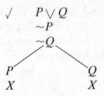

This is the finished tree. Since the attempted refutation fails along both paths, the original argument is valid.

3.26 Construct a refutation tree to determine whether the following form is valid:

$$P \lor Q, P \vdash \sim Q$$

Solution

Again, we form a list consisting of the premises and the negation of the conclusion:

$$P \lor Q$$
$$P$$
$$\sim \sim Q$$

'$\sim \sim Q$' is equivalent to the simpler formula 'Q', so we check it and write 'Q' at the bottom of the list. Then, as in the previous problem, we check '$P \lor Q$' and show its truth possibilities by branching the tree:

The tree is now finished. All formulas on the tree have been broken down into sentence letters or their negations. Moreover, both paths represent possible truth value assignments, since neither contains both a sentence letter and its negation. Since each path contains both '*P*' and '*Q*', each represents an assignment on which both these letters are true. Moreover, the tree indicates that under this truth value assignment all three formulas on the original list are true, i.e., that the premises of the form '$P \lor Q, P \vdash \sim Q$' are true while its conclusion is false. Hence this form is invalid.

The reader may wish to confirm the findings of these three problems by constructing truth tables for their respective forms. We now consider more systematically the procedure by which these problems were solved.

A refutation tree is an analysis in which a list of statements is broken down into sentence letters or their negations, which represent ways in which the members of the original list may be true. Since the ways in which a statement may be true depend on the logical operators it contains, formulas containing different logical operators are broken down differently. All wffs containing logical operators fall into one of the following ten categories:

Negation	Negated negation
Conjunction	Negated conjunction
Disjunction	Negated disjunction
Conditional	Negated conditional
Biconditional	Negated biconditional

Corresponding to each category is a rule for extending refutation trees. Problems 3.24 to 3.26 illustrated four of these rules. In order to state them, we need first to define the concept of an *open path*. An open path is any path of a tree which has not been ended with an '*X*'. Paths which have been ended with an '*X*' are said to be *closed*. The four rules may now be stated as follows:

Negation (~): If an open path contains both a formula and its negation, place an '*X*' at the bottom of the path.

The idea here is that any path which contains both a formula and its negation is not a path all of whose formulas can be true, which is what we are searching for in constructing a refutation tree. Hence we can close this path as a failed attempt at refutation. The negation rule was used in Problems 3.24 and 3.25.

Negated Negation (~ ~): If an open path contains an unchecked wff of the form $\sim \sim \phi$, check it and write ϕ at the bottom of every open path that contains this newly checked wff.

This rule was used in Problems 3.24 and 3.26.

Conjunction (&): If an open path contains an unchecked wff of the form $\phi \& \psi$, check it and write ϕ and ψ at the bottom of every open path that contains this newly checked wff.

This rule was used in Problem 3.24.

Disjunction ($\lor$): If an open path contains an unchecked wff of the form $\phi \lor \psi$, check it and split the bottom of each open path containing this newly checked wff into two branches, at the end of the first of which write ϕ and at the end of the second of which write ψ.

This rule was used in Problems 3.25 and 3.26.

A path is *finished* if it is closed or if the only unchecked wffs it contains are sentence letters or their negations, so that no more rules apply to its formulas. A tree is finished if all its paths are finished. If all the paths of a finished tree are closed (as in Problems 3.24 and 3.25), then the original formulas from which the tree is constructed cannot be true simultaneously. Thus, if the list is constructed from an argument form by negating its conclusion, that form is valid. On the other hand, if one or more of the paths of a finished tree are open (as in Problem 3.26), then the original formulas from which the tree is constructed *can* be true simultaneously. If the list is constructed from an argument form by negating its conclusion, this means that the form is invalid.

Indeed, the finished tree displays more than just the validity or invalidity of the argument form. *Each open path of the finished tree is a prescription for constructing counterexamples.* The only unchecked formulas on a finished open path are sentence letters or their negations. Any situation in which the unnegated sentence letters on the path are true and the negated ones are false is a counterexample. For example, the finished tree of Problem 3.26 shows two open paths, each containing both 'P' and 'Q'; thus, any situation in which 'P' and 'Q' are both true is a counterexample to the form 'P ∨ Q, P ⊢ ~Q'.

It is useful to annotate trees by numbering the lines they contain and indicating which rules and lines have been used to add formulas to the tree. We number lines in a column to their left and indicate the lines from which they are derived and the rules used to derive them to their immediate right. Rules are designated by the signs for the connectives they employ. Thus, for example, the annotated version of the tree of Problem 3.25 is as follows:

$$
\begin{array}{llll}
1 & \checkmark & P \vee Q & \\
2 & & \sim P & \\
3 & & \sim Q & \\
 & & \diagup \quad \diagdown & \\
4 & P \quad 1\vee & \quad Q \quad 1\vee \\
5 & X \quad 2,4\sim & \quad X \quad 3,4\sim
\end{array}
$$

We now state and illustrate the remaining six rules for generating refutation trees. Together with the four rules already given, they enable us to construct a tree for any finite set of wffs of propositional logic.

Conditional (→): If an open path contains an unchecked wff of the form φ → ψ, check it and split the bottom of each open path containing this newly checked wff into two branches, at the end of the first of which write ~φ and at the end of the second of which write ψ.

This rule is based on the fact that φ → ψ is true if and only if either φ is false or ψ is true (see the truth table for the material conditional).

SOLVED PROBLEM

3.27 Construct a refutation tree to determine whether the following form is valid:

$$P \rightarrow Q, Q \rightarrow R, P \vdash R$$

Solution

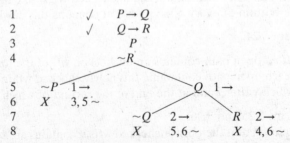

We begin by writing the premises and then the negation of the conclusion (lines 1 to 4). The conditional rule is then applied to line 1 to obtain line 5. The left branch closes at 6, by the negation rule, but the right branch remains open, and so the conditional rule is applied to 2 to obtain line 7. The negation rule then closes the two remaining paths. Since the finished tree is closed, the attempted refutation has failed and the form is valid.

<u>Biconditional (↔):</u>   If an open path contains an unchecked wff of the form φ ↔ ψ, check it and split the bottom of each open path containing this newly checked wff into two branches, at the end of the first of which write both φ and ψ, and at the end of the second of which write both ~φ and ~ψ.

This rule is an expression of the fact that φ ↔ ψ is true if and only if φ and ψ are either both false or both true.

SOLVED PROBLEM

3.28   Construct a refutation tree to determine whether the following form is valid:

$P \leftrightarrow Q, \sim P \vdash \sim Q$

Solution

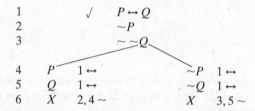

     Notice that it is not necessary to apply the negated negation rule in line 3 in this problem. The tree closes even without this move, since lines 3 and 5 are one the negation of the other. The form is valid.

<u>Negated Conjunction (~ &):</u>   If an open path contains an unchecked wff of the form ~(φ & ψ), check it and split the bottom of each open path containing this newly checked wff into two branches, at the end of the first of which write ~φ and at the end of the second of which write ~ψ.

This rule depends upon the fact that ~(φ & ψ) is true if and only if either φ or ψ is false.

SOLVED PROBLEM

3.29   Construct a refutation tree to determine whether the following form is valid:

$\sim(P \& Q) \vdash \sim P \& \sim Q$

Solution

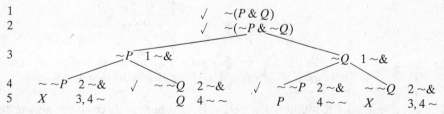

    Again, we begin with the premise and the negation of the conclusion. We analyze these by two steps of negated conjunction (lines 3 and 4). Two of the four paths then close, but two remain open, even after the applications of negated negation at line 5. Since no more rules apply, the tree is finished. And since there are two open paths, the form is invalid. The open paths indicate that situations in which 'P' is false and 'Q' true or 'Q' false and 'P' true are counterexamples.

<u>Negated Disjunction ($\sim \vee$)</u>: If an open path contains an unchecked wff of the form $\sim(\phi \vee \psi)$, check it and write both $\sim\phi$ and $\sim\psi$ at the bottom of every open path that contains this newly checked wff.

This rule is an expression of the fact that $\sim(\phi \vee \psi)$ is true if and only if both ϕ and ψ are false.

SOLVED PROBLEM

3.30 Construct a refutation tree to determine whether the following form is valid:

$$P \rightarrow Q \vdash P \vee Q$$

Solution

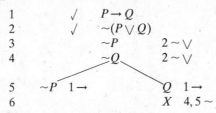

1	√	$P \rightarrow Q$	
2	√	$\sim(P \vee Q)$	
3		$\sim P$	$2 \sim \vee$
4		$\sim Q$	$2 \sim \vee$
5	$\sim P$ $1 \rightarrow$		Q $1 \rightarrow$
6			X $4,5\sim$

The negated disjunction rule is applied to line 2 to yield lines 3 and 4. The open path in the finished tree indicates that the form is invalid, and that any situation in which 'P' and 'Q' are both false is a counterexample.

<u>Negated Conditional ($\sim \rightarrow$)</u>: If an open path contains an unchecked wff of the form $\sim(\phi \rightarrow \psi)$, check it and write both ϕ and $\sim\psi$ at the bottom of every open path that contains this newly checked wff.

A negated conditional $\sim(\phi \rightarrow \psi)$ is true if and only if the conditional $\phi \rightarrow \psi$ is false, hence if and only if ϕ is true and ψ is false; that is the justification for this rule.

SOLVED PROBLEM

3.31 Construct a refutation tree to determine whether the following form is valid:

$$\sim P \rightarrow \sim Q \vdash P \rightarrow Q$$

Solution

1	√	$\sim P \rightarrow \sim Q$	
2	√	$\sim(P \rightarrow Q)$	
3		P	$2 \sim \rightarrow$
4		$\sim Q$	$2 \sim \rightarrow$
5	√ $\sim \sim P$ $1 \rightarrow$		$\sim Q$ $1 \rightarrow$
6	P $5 \sim \sim$		

The tree has two open paths, each indicating that the premise is true and the conclusion false when 'P' is true and 'Q' false. Hence the form is invalid.

<u>Negated Biconditional ($\sim \leftrightarrow$)</u>: If an open path contains an unchecked wff of the form $\sim(\phi \leftrightarrow \psi)$, check it and split the bottom of each open path containing this newly checked wff into two branches, at the end of the first of which write both ϕ and $\sim\psi$, and at the end of the second of which write both $\sim\phi$ and ψ.

This rule is an expression of the fact that a negated biconditional $\sim(\phi \leftrightarrow \psi)$ is true if and only if the biconditional is false, hence if and only if ϕ is true and ψ false or ϕ is false and ψ true.

SOLVED PROBLEM

3.32 Construct a refutation tree to determine whether the following form is valid:

$$P, P \rightarrow Q \vdash P \leftrightarrow Q$$

Solution

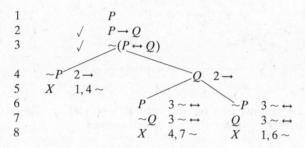

1	P
2	$\checkmark$ $P \rightarrow Q$
3	$\checkmark$ $\sim(P \leftrightarrow Q)$

4    $\sim P$ 2→ Q 2→
5   X 1,4 $\sim$
6   P 3 $\sim \leftrightarrow$ $\sim P$ 3 $\sim \leftrightarrow$
7                            $\sim Q$ 3 $\sim \leftrightarrow$ Q 3 $\sim \leftrightarrow$
8   X 4,7 $\sim$ X 1,6 $\sim$

Since the finished tree is closed, the form is valid.

Refutation trees are useful for purposes other than testing the validity of argument forms. A list of wffs is *truth-functionally consistent* if the tree beginning with those formulas (and including no other formulas except for those obtained by applying the rules) contains at least one finished open path. Thus, a consistent list of wffs is one all members of which can be true simultaneously, since a finished open path represents a way to make all the formulas on the list true. If a finished tree contains no open paths, the list of formulas from which it is constructed is inconsistent.

The list may consist of just one formula. If the finished tree for a single formula contains no open paths, then the formula is truth-functionally inconsistent. If it contains more than one open path, then the formula is either tautologous or truth-functionally contingent.

Indeed, refutation trees can also be used to test specifically for tautologousness. *A wff is tautologous if and only if its negation is truth-functionally inconsistent.* Therefore for any wff ϕ, ϕ is tautologous if and only if all the paths on the finished tree for $\sim\phi$ are closed (i.e., if and only if $\sim\phi$ is truth-functionally inconsistent).

SOLVED PROBLEMS

3.33 Construct a refutation tree to determine whether the following wff is tautologous:

$$(P \rightarrow Q) \vee (P \,\&\, \sim Q)$$

Solution

1	$\checkmark$	$\sim((P \rightarrow Q) \vee (P \,\&\, \sim Q))$	
2	$\checkmark$	$\sim(P \rightarrow Q)$	1 $\sim\vee$
3	$\checkmark$	$\sim(P \,\&\, \sim Q)$	1 $\sim\vee$
4		P	2 $\sim\rightarrow$
5		$\sim Q$	2 $\sim\rightarrow$

6    $\sim P$ 3 $\sim \&$ $\sim \sim Q$ 3 $\sim \&$
7   X 4,6 $\sim$ X 5,6 $\sim$

We construct the tree for the negation of the wff in question. Since all paths close, our attempt to find a way to make its negation true has failed; it is therefore tautologous.

3.34 Construct a refutation tree to determine whether the following wff is tautologous:

$$\sim(Q \rightarrow (P \& \sim P))$$

Solution

```
1        √      ~ ~(Q → (P & ~P))
2        √      Q → (P & ~P)        1 ~ ~

3     ~Q  2→              P & ~P   2→
4                            P      3 &
5                           ~P      3 &
6                            X      4, 5 ~
```

The left-branching path remains open, showing that '$\sim \sim(Q \rightarrow (P \& \sim P))$' is true if '$Q$' is false. Hence '$\sim(Q \rightarrow (P \& \sim P))$' is not tautologous.

The reader should keep the following points in mind when constructing refutation trees:

(1) *The rules for constructing trees apply only to whole formulas, not to mere subformulas.* Thus, for example, the use of negated negation in the following tree is impermissible:

```
1        √      P & ~ ~Q
2        √      P & Q        1 ~ ~ (incorrect)
3               P            2 &
4               Q            2 &
```

Although removing double negations from subformulas does not produce wrong answers, it is unnecessary and makes trees more difficult to read. More serious problems may result from trying to apply some of the rules for binary operators to subformulas, since we have not defined a consistent procedure for doing so.

(2) *The order in which rules are applied makes no difference to the final answer, but it is usually most efficient to apply nonbranching rules first.* After branching rules have been applied, further steps may require formulas to be written at the bottoms of several paths, which may necessitate more writing than if nonbranching rules had been applied first. Consider, for example, what happens if we apply the branching rule '$\rightarrow$' first in Problem 3.30. Then the tree is:

```
1        √      P → Q
2        √      ~(P ∨ Q)

3     ~P  1→              Q     1→
4     ~P  2 ~∨           ~P     2 ~∨
5     ~Q  2 ~∨           ~Q     2 ~∨
6                         X      3, 5 ~
```

Here the formulas '$\sim P$' and '$\sim Q$' must each be written twice, whereas if we apply the nonbranching rule negated disjunction first, as in Problem 3.30, we need write them only once.

(3) *The open paths of a finished tree for an argument form display all the counterexamples to that form.* This is true even if not all the sentence letters of the form occur among the unchecked formulas of some open paths. Consider, for example, the invalid form '$P \rightarrow Q \vdash P$'. Its tree is:

```
1        √      P → Q
2               ~P

3     ~P  1→      Q  1→
```

Both paths are open. The right-branching path indicates that situations in which '*P*' is false while '*Q*' is true are counterexamples; on the other hand, the left-branching path indicates simply that situations in which '*P*' is false are counterexamples. The letter '*Q*' does not occur among the unchecked formulas of this path. This shows that the falsity of '*P*' is by itself sufficient for a counterexample, i.e., that any situation in which '*P*' is false is a counterexample, regardless of the truth value of '*Q*'. Thus the tree indicates that there are two kinds of counterexamples to the form: situations in which '*P*' is false and '*Q*' true, and situations in which both are false.

Table 3-1   Refutation Tree Rules

Negation ($\sim$):   If an open path contains both a formula and its negation, place an '*X*' at the bottom of the path.

Negated Negation ($\sim\sim$):   If an open path contains an unchecked wff of the form $\sim\sim\phi$, check it and write ϕ at the bottom of every open path that contains this newly checked wff.

Conjunction (&):   If an open path contains an unchecked wff of the form $\phi\ \&\ \psi$, check it and write ϕ and ψ at the bottom of every open path that contains this newly checked wff.

Negated Conjunction ($\sim$&):   If an open path contains an unchecked wff of the form $\sim(\phi\ \&\ \psi)$, check it and split the bottom of each open path containing this newly checked wff into two branches, at the end of the first of which write $\sim\phi$ and at the end of the second of which write $\sim\psi$.

Disjunction ($\vee$):   If an open path contains an unchecked wff of the form $\phi\vee\psi$, check it and split the bottom of each open path containing this newly checked wff into two branches, at the end of the first of which write ϕ and at the end of the second of which write ψ.

Negated Disjunction ($\sim\vee$):   If an open path contains an unchecked wff of the form $\sim(\phi\vee\psi)$, check it and write both $\sim\phi$ and $\sim\psi$ at the bottom of every open path that contains this newly checked wff.

Conditional ($\rightarrow$):   If an open path contains an unchecked wff of the form $\phi\rightarrow\psi$, check it and split the bottom of each open path containing this newly checked wff into two branches, at the end of the first of which write $\sim\phi$ and at the end of the second of which write ψ.

Negated Conditional ($\sim\rightarrow$):   If an open path contains an unchecked wff of the form $\sim(\phi\rightarrow\psi)$, check it and write both ϕ and $\sim\psi$ at the bottom of every open path that contains this newly checked wff.

Biconditional ($\leftrightarrow$):   If an open path contains an unchecked wff of the form $\phi\leftrightarrow\psi$, check it and split the bottom of each open path containing this newly checked wff into two branches, at the end of the first of which write both ϕ and ψ, and at the end of the second of which write both $\sim\phi$ and $\sim\psi$.

Negated Biconditional ($\sim\leftrightarrow$):   If an open path contains an unchecked wff of the form $\sim(\phi\leftrightarrow\psi)$, check it and split the bottom of each open path containing this newly checked wff into two branches, at the end of the first of which write both ϕ and $\sim\psi$, and at the end of the second of which write both $\sim\phi$ and ψ.

Supplementary Problems

I Formalize the following statements, using the interpretation indicated below:

Sentence Letter	Interpretation
P	Pam is going.
Q	Quincy is going.
R	Richard is going.
S	Sara is going.

(1) Pam is not going.

(2) Pam is going, but Quincy is not.

(3) If Pam is going, then so is Quincy.

(4) Pam is going if Quincy is.

(5) Pam is going only if Quincy is.

(6) Pam is going if and only if Quincy is.

(7) Neither Pam nor Quincy is going.

(8) Pam and Quincy are not both going.

(9) Either Pam is not going or Quincy is not going.

(10) Pam is not going if Quincy is.

(11) Either Pam is going, or Richard and Quincy are going.

(12) If Pam is going, then both Richard and Quincy are going.

(13) Pam is staying, but Richard and Quincy are going.

(14) If Richard is going, then if Pam is staying, Quincy is going.

(15) If neither Richard nor Quincy is going, then Pam is going.

(16) Richard is going only if Pam and Quincy are staying.

(17) Richard and Quincy are going, although Pam and Sara are staying.

(18) If either Richard or Quincy is going, then Pam is going and Sara is staying.

(19) Richard and Quincy are going if and only if either Pam or Sara is going.

(20) If Sara is going, then either Richard or Pam is going, and if Sara is not going, then both Pam and Quincy are going.

II Determine which of the following formulas are wffs and which are not. Explain your answer.

(1) $\sim(\sim P)$

(2) $P \sim Q$

(3) $(P \rightarrow P)$

(4) $P \rightarrow P$

(5) $\sim \sim \sim (\sim P \,\&\, Q)$

(6) $((P \rightarrow Q))$

(7) $\sim (P \,\&\, Q) \,\&\, \sim R$

(8) $(P \leftrightarrow (P \leftrightarrow (P \leftrightarrow P)))$

(9) $(P \rightarrow (Q \rightarrow (R \,\&\, S)))$

(10) $(P \rightarrow (Q \lor R \lor S))$

III Determine whether the following formulas are tautologous, truth-functionally contingent, or inconsistent, using either truth tables or refutation trees.

(1) $P \rightarrow P$

(2) $P \rightarrow \sim P$

(3) $\sim(P \rightarrow P)$

(4) $P \rightarrow Q$

(5) $(P \vee Q) \rightarrow P$

(6) $(P \& Q) \rightarrow P$

(7) $P \leftrightarrow \sim(P \vee Q)$

(8) $\sim((P \& Q) \leftrightarrow (P \vee Q))$

(9) $(P \& Q) \& \sim(P \vee R)$

(10) $(P \rightarrow (Q \& R)) \rightarrow (P \rightarrow R)$

IV Test the following forms for validity, using either truth tables or refutation trees.

(1) $\sim P \vdash P \rightarrow \sim P$

(2) $P \vee Q \vdash P \& Q$

(3) $P \rightarrow \sim Q \vdash \sim(P \& Q)$

(4) $P \vdash (P \rightarrow (Q \& P)) \rightarrow (P \& Q)$

(5) $P \vee Q, \sim P, \sim Q \vdash R$

(6) $(Q \& R) \rightarrow P, \sim Q, \sim R \vdash \sim P$

(7) $\sim(P \vee Q), R \leftrightarrow P \vdash \sim R$

(8) $\sim(P \& Q), R \leftrightarrow P \vdash \sim R$

(9) $P \leftrightarrow Q, Q \leftrightarrow R \vdash P \leftrightarrow R$

(10) $P \rightarrow (R \vee S), (R \& S) \rightarrow Q \vdash P \rightarrow Q$

V Formalize the following arguments, using the interpretation given below, and test their forms for validity using either truth tables or refutation trees.

Sentence Letter	Interpretation
C	Argument form F has a counterexample.
I	The premises of argument form F are inconsistent.
O	The finished tree for argument form F contains an open path.
V	Argument form F is valid.

(1) If argument form F is valid, then its finished tree contains no open paths. Hence if its finished tree contains an open path, it is invalid.

(2) If the premises of argument form F are inconsistent, then F is valid. Therefore, if the premises of form F are not inconsistent, then form F is invalid.

(3) If argument form F has a counterexample, then its premises are not inconsistent. For it is not the case both that it has a counterexample and that its premises are consistent.

(4) Either argument form F has a counterexample or it is valid, but not both. Hence form F is valid if and only if it has no counterexample.

(5) The premises of argument form F are inconsistent. If the premises of F are inconsistent, then F is valid. F is valid if and only if its finished tree contains no open path. If its finished tree contains no open path, then it has no counterexample. Form F has a counterexample. Therefore, the premises of form F are not inconsistent.

Answers to Selected Supplementary Problems

I   (5) $P \rightarrow Q$.

     (10) $Q \rightarrow \sim P$.

     (15) $\sim(R \vee Q) \rightarrow P$, or equivalently $(\sim R \,\&\, \sim Q) \rightarrow P$.

     (20) $(S \rightarrow (R \vee P)) \,\&\, (\sim S \rightarrow (P \,\&\, Q))$.

II   (5)   'P' and 'Q' are wffs by rule 1 in Section 3.3; so '$\sim P$' is a wff by rule 2. Hence '$(\sim P \,\&\, Q)$' is a wff by rule 3, whence by three applications of rule 2 it follows that '$\sim \sim \sim (\sim P \,\&\, Q)$' is a wff.

     (10)   Not a wff. No rule allows us to disjoin three statements at once.

III   (2)   Truth-functionally contingent

     (4)   Truth-functionally contingent

     (6)   Tautologous

     (8)   Truth-functionally contingent

     (10)   Tautologous

IV   (2)   Invalid

     (4)   Valid

     (6)   Invalid

     (8)   Invalid

     (10)   Invalid

V   (2)   Invalid form (indeed, the argument itself is invalid)

     (4)   Valid form

Chapter 4

The Propositional Calculus

4.1 THE NOTION OF INFERENCE

Chapter 3 dealt with propositional logic from a semantic point of view, using techniques for testing the deductive validity of argument forms based on the intended interpretation of the logical operators. In this chapter we present a different method for establishing deductive validity, which makes no explicit reference to the notion of a truth value. It is based on the idea that if an argument is deductively valid, one should be able to *infer* or *derive* the conclusion from the premises, that is, to show how the conclusion actually *follows from* the premises. Typically, this is how one proceeds when one offers an argument to support a certain statement. One does not simply list the premises and conclusion, and then issue a challenge to describe a possible situation in which the conclusion is false while the premises are true. Rather, one tries to show how the conclusion can actually be reached through a finite number of successive steps of reasoning, each of which is fully explicit and indisputable. Obviously this inferential process is not captured by the method of truth tables, nor by the technique of refutation trees.

The key to proving validity by means of step-by-step deductions lies in mastering the principles which can be appealed to in making the successive steps. These principles are called *rules of inference*. In this chapter we shall formulate them with reference to the language of propositional logic, though it will be clear that the same approach can be extended to cover other patterns of reasoning (see Chapters 7 and 11). Overall we shall formulate ten such rules: two—an introduction and an elimination rule—for each of the five logical operators. The elimination rule for an operator is used to reason from premises in which that operator is the main operator. The introduction rule for an operator is used to derive conclusions in which it is the main operator. These rules form a system called the *propositional calculus*. (The term means simply "system for performing calculations with propositions"; it indicates no close relationship with the differential or integral calculus of advanced mathematics.) A *deduction* in the propositional calculus is thus a series of formulas in the language of propositional logic, each of which is either used as a premise or obtained by applying a rule of introduction or a rule of elimination. A deduction is also called a *derivation* or *proof*; we shall use these terms synonymously.

Other versions of the propositional calculus employ different rules from the ones used here. However, all commonly used sets of rules are equivalent, in the sense that they establish the validity of exactly the same argument forms, namely all and only the valid argument forms expressible in the language of propositional logic. This means that if a form can be shown to be valid by the semantic methods of Chapter 3, then it is also provable by the rules of the calculus. We can also express this fact by saying that the calculus is *complete*. In addition, the calculus is *valid*, i.e., it does not allow one to generate an argument which is not valid. These properties of the calculus are of course crucial if we want to say that the propositional calculus provides an alternative but equivalent account of the notion of deductive validity. Their demonstrations, however, are beyond the scope of this book.[1]

4.2 NONHYPOTHETICAL INFERENCE RULES

In this section we introduce eight of the ten basic inference rules. The other two (the conditional and negation introduction rules) are different in character and will be discussed in Section 4.3.

---

[1] The validity and completeness of a propositional calculus very similar to the one presented here are demonstrated in E. J. Lemmon, *Beginning Logic*, Indianapolis, Hackett, 1978, pp. 75–91.

To illustrate the concept of a derivation, consider the following argument form:

$$P, Q \to R, P \to Q \vdash R$$

This form is valid (see Problem 3.7(i)); hence the conclusion should be derivable from the premises. The derivation is as follows:

$$
\begin{array}{lll}
1 & P & A \\
2 & Q \to R & A \\
3 & P \to Q & A \\
\therefore\ 4 & Q & 1,3 \to E \\
\therefore\ 5 & R & 2,4 \to E \\
\end{array}
$$

We list the three assumptions on the first three lines of the derivation, number each line, and add the label 'A' to indicate that each is an assumption (a premise of the argument). Then we deduce the conclusion 'R' by two steps of reasoning. The first step is from lines 1 and 3 to line 4; the second is from 2 and 4 to 5. The numbers at the right denote the lines from which 4 and 5 are derived. The two steps have the same form; each is an instance of the first of our ten inference rules, conditional elimination, which we designate by '$\to E$':

> Conditional Elimination ($\to E$): From a conditional and its antecedent we may infer its consequent.

This rule is also known as *modus ponens* (see Problem 3.1). Its validity is obvious; and since a complex argument consisting entirely of valid steps must itself be valid (see Section 2.3), this complex derivation proves the validity of our original argument form.

Note that the rule says nothing about the complexity of the conditional to which it is applied: the antecedent and the consequent may be atomic (sentence letters) or they may themselves be compound wffs.

SOLVED PROBLEM

4.1 Prove:

$$\sim P \to (Q \to R), \sim P, Q \vdash R$$

Solution

$$
\begin{array}{lll}
1 & \sim P \to (Q \to R) & A \\
2 & \sim P & A \\
3 & Q & A \\
4 & Q \to R & 1, 2 \to E \\
5 & R & 3, 4 \to E \\
\end{array}
$$

(We omit the triple dots in front of conclusions here and in all subsequent derivation problems, since the citation of line numbers at right is sufficient to indicate that they are conclusions.) The first premise is a conditional whose antecedent is negated and whose consequent is itself a conditional. Line 2 contains its antecedent. Thus the derivation of its consequent at 4 is clearly an instance of conditional elimination, as is the step from lines 3 and 4 to line 5.

Notice also that we drop the outer brackets from the expression '$(Q \to R)$' at step 4, where we detach it by conditional elimination from the assumption given in line 1. This is a result of the convention for dropping outer brackets (see Section 3.3), as is the fact that we write line 1 as '$\sim P \to (Q \to R)$' rather than '$(\sim P \to (Q \to R))$'. We shall employ this convention throughout this chapter.

Our second inference rule is the negation elimination rule, which we designate by the notation '~E'. This rule allows us to reason from premises which are negations of negations:

Negation Elimination (~E):   From a wff of the form ~~ϕ, we may infer ϕ.

(We use the Greek letter 'ϕ' to indicate that this rule is perfectly general, applying to all wffs whether atomic or compound; see Section 3.3.) The validity of negation elimination follows immediately by the truth table for '~' (see Problem 3.9). Invariably the negation of a false statement is true, and the negation of a true statement is false. Thus if we begin with a true doubly negated sentence (e.g., 'It is not the case that Richard Nixon was not president') and remove one of the negations ('Richard Nixon was not president') we get a falsehood. If we remove both, however ('Richard Nixon was president'), the result is again true. Thus any inference from a doubly negated statement to the result of removing both negations is valid. The following problem employs ~E:

SOLVED PROBLEM

4.2   Prove:

$$\sim P \rightarrow \sim\sim Q, \sim\sim\sim P \vdash Q$$

Solution

1	$\sim P \rightarrow \sim\sim Q$	A
2	$\sim\sim\sim P$	A
3	$\sim P$	2 ~E
4	$\sim\sim Q$	1, 3 →E
5	Q	4 ~E

The assumption in step 1 is a conditional with a negated antecedent and a doubly negated consequent. We derive its antecedent at line 3 by applying ~E to line 2 ('~~~P' is the double negation of '~P'). This allows us to deduce its consequent at line 4 by →E. Then another step of ~E yields our conclusion.

It is important to note that ~E would not permit us to reason from line 1 to '~$P \rightarrow Q$', though that would in fact be a valid argument (check this by computing the truth table). Line 1 is a conditional, not a doubly negated wff (i.e., a wff of the form ~~ϕ). We must detach '~~Q' from step 1 before we can apply ~E. Negation elimination allows us to remove two negation signs only if they are the leftmost symbols of a wff and all the rest of the wff is included in their scope.

The next two rules, conjunction introduction and conjunction elimination, are both extremely simple and obviously valid:

Conjunction Introduction (&I):   From any wffs ϕ and ψ, we may infer the conjunction ϕ & ψ.

Conjunction Elimination (&E):   From a conjunction, we may infer either of its conjuncts.

Some authors call &I *conjunction*, and &E *simplification*. Both are used in the following problems.

SOLVED PROBLEMS

4.3   Prove:

$$P \& Q \vdash Q \& P$$

Solution

1	$P \& Q$	A
2	P	1 &E
3	Q	1 &E
4	$Q \& P$	2, 3 &I

4.4 Prove:

$$P \to (Q \, \& \, R), P \vdash P \, \& \, Q$$

Solution

1	$P \to (Q \, \& \, R)$	A
2	P	A
3	$Q \, \& \, R$	$1, 2 \to$E
4	Q	$3 \,\&$E
5	$P \, \& \, Q$	$2, 4 \,\&$I

4.5 Prove:

$$(P \, \& \, Q) \to (R \, \& \, S), \sim\sim P, Q \vdash S$$

Solution

1	$(P \, \& \, Q) \to (R \, \& \, S)$	A
2	$\sim\sim P$	A
3	Q	A
4	P	$2 \sim$E
5	$P \, \& \, Q$	$3, 4 \,\&$I
6	$R \, \& \, S$	$1, 5 \to$E
7	S	$6 \,\&$E

The fact that two different Greek letters, 'ϕ' and 'ψ', are used in the statement of the &I rule does not imply that the wffs designated by these letters need be distinct. Thus the use of &I in the following problem, though unusual, is perfectly correct:

SOLVED PROBLEM

4.6 Prove:

$$P \vdash P \, \& \, P$$

Solution

1	P	A
2	$P \, \& \, P$	$1, 1 \,\&$I

Though the conjunction of a statement with itself is redundant (and hence unlikely to be uttered in practice), it is grammatically well formed, and thus permissible in theory. Line 1 does double duty in this proof, serving as the source of both of the conjuncts of line 2, so we cite it twice at line 2.

The fifth rule is used to prove disjunctive conclusions:

<u>Disjunction Introduction ($\lor$I):</u> From a wff ϕ, we may infer the disjunction of ϕ with any wff. (ϕ may be either the first or second disjunct of this disjunction.)

Disjunction introduction is sometimes called *addition*. Its validity is an immediate consequence of the truth table for '$\lor$'. If either or both of the disjuncts of a disjunction are true, then the whole disjunction is true. Thus, for example, if the statement 'Today is Tuesday' is true, then the disjunction 'Today is either Tuesday or Wednesday' must also be true (as well as 'Today is either Wednesday or Tuesday', in which 'Today' is the second disjunct). Indeed, if today is Tuesday, then the disjunction of 'Today is Tuesday' with *any* statement (including itself) is true. Disjunction introduction is illustrated in the following proofs.

SOLVED PROBLEMS

4.7 Prove:

$$P \vdash (P \lor Q) \,\&\, (P \lor R)$$

Solution

1	P	A
2	$P \lor Q$	1 $\lor$I
3	$P \lor R$	1 $\lor$I
4	$(P \lor Q) \,\&\, (P \lor R)$	2, 3 &I

4.8 Prove:

$$P, \,\sim\sim(P \to Q) \vdash (R \,\&\, S) \lor Q$$

Solution

1	P	A
2	$\sim\sim(P \to Q)$	A
3	$P \to Q$	2 $\sim$E
4	Q	1, 3 $\to$E
5	$(R \,\&\, S) \lor Q$	4 $\lor$I

4.9 Prove:

$$P \vdash P \lor P$$

Solution

1	P	A
2	$P \lor P$	1 $\lor$I

It is instructive to compare this problem with Problem 4.6. Note that in the second step we cite line 1 only once, since $\lor$I is a rule that applies to only one premise.

The disjunction elimination rule is a bit more complicated than $\lor$I, but on reflection its validity, too, should be apparent:

<u>Disjunction Elimination ($\lor$E):</u>   From wffs of the forms $\phi \lor \psi$, $\phi \to \chi$, and $\psi \to \chi$, we may infer the wff χ.[2]

This rule is sometimes called *constructive dilemma*. To illustrate, assume that today is either Saturday or Sunday (a disjunction). Moreover, assume that if it is Saturday, then tonight there will be a concert, and if it is Sunday, then again tonight there will be a concert. Then we can infer that tonight there will be a concert even without knowing with certainty whether it is Saturday or Sunday. Our disjunctive assumption covers all relevant cases, and in each case we know there will be a concert.

SOLVED PROBLEMS

4.10 Prove:

$$P \lor Q, \, P \to R, \, Q \to R \vdash R$$

---

[2]It is also possible to formulate this rule without mentioning the two conditionals $\phi \to \chi$ and $\psi \to \chi$. Such a formulation would be more elegant, except that $\lor$E would become a hypothetical rule (see next section). Our treatment is dictated primarily by reasons of simplicity.

Solution

$$
\begin{array}{lll}
1 & P \lor Q & A \\
2 & P \to R & A \\
3 & Q \to R & A \\
4 & R & 1, 2, 3 \lor E
\end{array}
$$

This is the formal counterpart of the concert example. At the right side of line 4 we cite the three assumptions needed to apply the rule.

4.11 Prove:

$$(P \lor Q) \ \& \ (P \lor R), P \to S, Q \to S, P \to T, R \to T \vdash S \ \& \ T$$

Solution

$$
\begin{array}{lll}
1 & (P \lor Q) \ \& \ (P \lor R) & A \\
2 & P \to S & A \\
3 & Q \to S & A \\
4 & P \to T & A \\
5 & R \to T & A \\
6 & P \lor Q & 1 \ \& E \\
7 & S & 2, 3, 6 \lor E \\
8 & P \lor R & 1 \ \& E \\
9 & T & 4, 5, 8 \lor E \\
10 & S \ \& \ T & 7, 9 \ \& I
\end{array}
$$

As usual, the use of three different Greek variables in the statement of the $\lor$E rule indicates only that the wffs they denote *may* be different, not that they have to be. The following proof contains an application of $\lor$E (line 3) in which the variables 'ϕ' and 'ψ' stand for the same formula.

SOLVED PROBLEM

4.12 Prove:

$$P \lor P, P \to (Q \ \& \ R) \vdash R$$

Solution

$$
\begin{array}{lll}
1 & P \lor P & A \\
2 & P \to (Q \ \& \ R) & A \\
3 & Q \ \& \ R & 1, 2, 2 \lor E \\
4 & R & 3 \ \& E
\end{array}
$$

With respect to the statement of the $\lor$E rule, ϕ is 'P', ψ is 'P', and χ is '$(Q \ \& \ R)$'. Thus the formula $\phi \lor \psi$ is '$P \lor P$' and the formulas $\phi \to \chi$ and $\psi \to \chi$ are both '$P \to (Q \lor R)$'. Since line 2 of the proof does double duty, serving both as $\phi \to \chi$ and as $\psi \to \chi$, we cite it twice at the right side of line 3.

Our last two rules in this section concern the biconditional. In discussing this connective (Section 3.2), we noted that statements of the form $\phi \leftrightarrow \psi$ are equivalent to statements of the form $(\phi \to \psi)$ & $(\psi \to \phi)$. In view of this equivalence, the introduction and elimination rules for the biconditional function much like &I and &E.[3]

---

[3]Also in this case, it would be possible to formulate the rules for '$\leftrightarrow$' without explicitly relying on the equivalence between biconditionals and conjunctions of conditionals. Biconditional introduction would then become a hypothetical rule.

Biconditional Introduction (↔I):   From any wffs of the forms $(\phi \rightarrow \psi)$ and $(\psi \rightarrow \phi)$, we may infer $\phi \leftrightarrow \psi$.

Biconditional Elimination (↔E):   From any wff of the form $\phi \leftrightarrow \psi$, we may infer either $\phi \rightarrow \psi$ or $\psi \rightarrow \phi$.

The following proofs illustrate the application of these rules:

SOLVED PROBLEMS

4.13   Prove:

$$P \leftrightarrow (Q \vee R), R \vdash P$$

Solution

1	$P \leftrightarrow (Q \vee R)$	A
2	R	A
3	$(Q \vee R) \rightarrow P$	1 ↔E
4	$Q \vee R$	2 ∨I
5	P	3, 4 →E

4.14   Prove:

$$P \rightarrow Q, (P \rightarrow Q) \rightarrow (Q \rightarrow P) \vdash P \leftrightarrow Q$$

Solution

1	$P \rightarrow Q$	A
2	$(P \rightarrow Q) \rightarrow (Q \rightarrow P)$	A
3	$Q \rightarrow P$	1, 2 →E
4	$P \leftrightarrow Q$	1, 3 ↔I

4.15   Prove:

$$P \leftrightarrow Q \vdash Q \leftrightarrow P$$

Solution

1	$P \leftrightarrow Q$	A
2	$P \rightarrow Q$	1 ↔E
3	$Q \rightarrow P$	1 ↔E
4	$Q \leftrightarrow P$	2, 3 ↔I

The analogy between the biconditional rules and the conjunction rules may be clearly seen by comparing Problem 4.15 with Problem 4.3.

4.3   HYPOTHETICAL RULES

We have now introduced eight of the ten inference rules for the propositional calculus. The remaining two, the introduction rules for conditionals and negation, differ from the others in that they employ *hypothetical reasoning*. Hypothetical reasoning is reasoning based on a *hypothesis*, a temporary assumption made "for the sake of argument" in order to show that a particular conclusion would follow. Unlike the other assumptions of a proof, hypotheses are not asserted to be true. They are not premises. They are "logical fantasies," which we entertain temporarily as a special kind of proof strategy.

Suppose that a runner has injured her ankle a week before a big race and that we would like to persuade her to stop running for a few days in order to let the ankle heal. We assert the conditional 'If you keep running now, you won't be able to run the race'. Her response: 'Prove it.'

Now the most widely applicable way to prove a conditional is to hypothesize its antecedent (i.e., to assume it for the sake of argument) and then show that its consequent must follow. To do that, we might reason this way:

> Look, suppose you do keep running now. Your ankle is badly swollen. If it is badly swollen and you keep running, it will not heal in a week. If it does not heal in a week, then you won't be able to run the race. So you won't be able to run the race.

This is a hypothetical argument. The word 'suppose' is a hypothesis indicator, signaling that the statement 'You do keep running' is a hypothesis. From this hypothesis, the conclusion 'You won't be able to run the race' is shown to follow. The argument employs three assumptions:

> Your ankle is badly swollen.
> If your ankle is badly swollen and you keep running, then it will not heal in a week.
> If your ankle does not heal in a week, then you won't be able to run the race.

These are assumptions which feature as premises in our argument. We assert them to be true—unlike the hypothesis, which we assume only for the sake of argument and which in this case we are in fact trying to keep from coming true. Now, these assumptions granted, the hypothetical argument shows that *if* the hypothesis is true, then the conclusion must also be true. But this means that there is no logically possible situation in which the statement 'You do keep running' is true while the statement 'You won't be able to run the race' is false. Hence the conditional

> If you keep running now, then you won't be able to run the race.

must be true, which is just what we set out to prove. We have deduced this conditional by hypothesizing its antecedent and showing that its consequent follows from that hypothesis, together with the assumptions.

Of course, the soundness of the argument depends on the truth of the assumptions—and in a real case their truth may be in doubt. But what is novel here is that it does not depend on the truth of the hypothesis. The hypothesis is introduced only to show that, given the assumptions, it implies the conclusion 'You won't be able to run the race'. Once that has been shown, the hypothesis is discharged or abandoned, and the conditional conclusion is asserted on the strength of the assumptions alone.

Let's now see how all this can be formalized. The three assumptions may be expressed respectively as

$$S$$
$$(S \mathbin{\&} K) \rightarrow \sim H$$
$$\sim H \rightarrow \sim A$$

and the conclusion as

$$K \rightarrow \sim A$$

To derive the conclusion, we hypothesize its antecedent 'K' (just as we did above in the informal version of the argument) and deduce its consequent '$\sim A$' from the three assumptions with the help of this hypothesis. The complete derivation looks like this:

1	S	A
2	$(S \mathbin{\&} K) \rightarrow \sim H$	A
3	$\sim H \rightarrow \sim A$	A
4	$\quad K$	H
5	$\quad S \mathbin{\&} K$	1, 4 &I
6	$\quad \sim H$	2, 5 →E
7	$\quad \sim A$	3, 6 →E
8	$K \rightarrow \sim A$	4–7 →I

The assumptions, as always, are listed first. The hypothesis 'K' is introduced at step 4 and labeled 'H' to indicate that it is a hypothesis. To its left we begin a vertical line which extends downward to indicate the duration of the hypothetical argument, i.e., the part of the proof in which the hypothesis is assumed. The hypothesis is said to be *in effect* for the portion of the proof to the right of this line. Using the assumptions, we derive '$\sim A$' from the hypothesis at step 7. The hypothetical derivation (steps 4 to 7) shows that, given our assumptions, '$\sim A$' follows from 'K', so that *if* '$\sim A$' is true, then 'K' is true as well. Thus at line 8 we can discharge the hypothesis (and indicate we have done so by terminating the vertical line) and assert '$K \rightarrow \sim A$' on the strength of the assumptions alone.

At this final step we cite lines 4 to 7 (the hypothetical derivation) and the rule $\rightarrow$I, which is our conditional introduction rule. Some authors call this rule *conditional proof*. Its formal statement is as follows:

> <u>Conditional Introduction ($\rightarrow$I):</u>   Given a derivation of a wff ψ with the help of a hypothesis ϕ, we may discharge the hypothesis and infer that $\phi \rightarrow \psi$.

With respect to the example above, ϕ is 'K' and ψ is '$\sim A$'.

SOLVED PROBLEMS

4.16   Prove:

$$P \rightarrow Q, \; Q \rightarrow R \vdash P \rightarrow R$$

Solution

1	$P \rightarrow Q$	A
2	$Q \rightarrow R$	A
3	$\mid P$	H
4	$\mid Q$	1, 3 $\rightarrow$E
5	$\mid R$	2, 4 $\rightarrow$E
6	$P \rightarrow R$	3–5 $\rightarrow$I

The conclusion is the conditional '$P \rightarrow R$', We hypothesize its antecedent at line 3, from which we derive its consequent at line 5. This enables us to discharge the hypothesis and infer the conditional by $\rightarrow$I at 6.

4.17   Prove:

$$P \vdash (P \rightarrow Q) \rightarrow Q$$

Solution

1	P	A
2	$\mid P \rightarrow Q$	H
3	$\mid Q$	1, 2 $\rightarrow$E
4	$(P \rightarrow Q) \rightarrow Q$	2–3 $\rightarrow$I

Here the conclusion is a conditional whose antecedent is itself a conditional. We hypothesize its antecedent '$P \rightarrow Q$' at line 2 and derive its consequent 'Q' immediately at line 3, which enables us to obtain the conclusion itself by $\rightarrow$I at line 4.

As with other rules, it should be noted that the wffs designated by the two Greek letters 'ϕ' and 'ψ' in the statement of the $\rightarrow$I rule need not be distinct. This means that we understand the locution 'with the help of a hypothesis' in such a way that every wff can be derived with the help of itself. This is legitimate, since every argument whose conclusion coincides with a premise is obviously valid. Accordingly, the following is an unusual but perfectly licit derivation in which $\rightarrow$I is applied to obtain a conditional whose antecedent and consequent are one and the same formula.

SOLVED PROBLEM

4.18 Prove:

$$Q \vdash P \rightarrow P$$

Solution

1	Q	A
2	$\vert P$	H
3	$P \rightarrow P$	2–2 →I

Here line 2 serves as the entire hypothetical derivation, of which it is both hypothesis and conclusion. At line 3 the double role of 'P' becomes manifest as we introduce the conditional '$P \rightarrow P$' by →I. At the right we write '2–2' to indicate that the hypothetical derivation begins and ends at the same line. Note that the proof is correct even though the premise 'Q' is not actually used to infer the conclusion. This is a case in which the argument is valid but commits a fallacy of relevance (compare Problem 2.19).

Hypothetical rules can be used more than once in the course of a derivation, exactly like nonhypothetical rules. Thus, typically one needs a separate application of →I for each conditional that must be proved (unless some simpler strategy is immediately apparent).

SOLVED PROBLEM

4.19 Prove:

$$P \rightarrow R, \, Q \rightarrow S \vdash (P \rightarrow (R \vee S)) \, \& \, (Q \rightarrow (R \vee S))$$

Solution

1	$P \rightarrow R$	A
2	$Q \rightarrow S$	A
3	$\vert P$	H
4	$\vert R$	1, 3 →E
5	$\vert R \vee S$	4 ∨I
6	$P \rightarrow (R \vee S)$	3–5 →I
7	$\vert Q$	H
8	$\vert S$	2, 7 →E
9	$\vert R \vee S$	8 ∨I
10	$Q \rightarrow (R \vee S)$	7–9 →I
11	$(P \rightarrow (R \vee S)) \, \& \, (Q \rightarrow (R \vee S))$	6, 10 &I

Here we derive a conjunction conclusion by deriving each conjunct and then applying &I. Each derivation involves an application of →I.

One may also need to embed a hypothetical argument within another. This is often the case when a conclusion is a conditional with a conditional consequent. To prove such a conclusion, one begins as with other conditional conclusions: hypothesize the antecedent and derive the consequent from the hypothesis for →I. But since the consequent is itself a conditional, in order to derive it from the antecedent, we must hypothesize its antecedent for a subsidiary step of →I. The result is a nesting of one →I strategy inside another.

SOLVED PROBLEM

4.20   Prove:

$$(P \,\&\, Q) \to R \vdash P \to (Q \to R)$$

Solution

1	$(P \,\&\, Q) \to R$	A
2	P	H
3	Q	H
4	$P \,\&\, Q$	2, 3 &I
5	R	1, 4 →E
6	$Q \to R$	3–5 →I
7	$P \to (Q \to R)$	2–6 →I

We hypothesize the antecedent 'P' of the conclusion at line 2. To derive the consequent '$Q \to R$' from step 2, we hypothesize its antecedent 'Q' at line 3. Since this is a new hypothesis, it requires a new vertical line. Two hypotheses are now in effect. We derive 'R' from 'Q' at line 5. This enables us to discharge the hypothesis 'Q' and infer '$Q \to R$' by →I at line 6. We now have shown that '$Q \to R$' follows from our original hypothesis 'P'. This original hypothesis remains in effect until we discharge it and infer the desired conclusion by →I at line 7.

Since →I is the primary strategy for proving conditionals and application of $\lor$E requires two conditional premises, →I is frequently used as a preliminary to $\lor$E. A common strategy when a disjunctive premise is present is to prove two conditionals, each linking one of the two disjuncts to the desired conclusion, by two separate applications of →I. The conclusion may then be reached by a single step of $\lor$E.

SOLVED PROBLEMS

4.21   Prove:

$$P \lor Q \vdash Q \lor P$$

Solution

1	$P \lor Q$	A
2	P	H
3	$Q \lor P$	2 $\lor$I
4	$P \to (Q \lor P)$	2–3 →I
5	Q	H
6	$Q \lor P$	5 $\lor$I
7	$Q \to (Q \lor P)$	5–6 →I
8	$Q \lor P$	1, 4, 7 $\lor$E

To get from the disjunctive premise '$P \lor Q$' to the conclusion '$Q \lor P$', we need the two conditionals '$P \to (Q \lor P)$' and '$Q \to (Q \lor P)$'. These are established by separate →I proofs (steps 2 to 4 and 5 to 7, respectively). The conclusion then follows by $\lor$E.

4.22   Prove:

$$(P \,\&\, Q) \lor (P \,\&\, R) \vdash P \,\&\, (Q \lor R)$$

Solution

1	$(P \& Q) \vee (P \& R)$	A
2	$P \& Q$	H
3	P	2 &E
4	Q	2 &E
5	$Q \vee R$	4 $\vee$I
6	$P \& (Q \vee R)$	3, 5 &I
7	$(P \& Q) \rightarrow (P \& (Q \vee R))$	2–6 →I
8	$P \& R$	H
9	P	8 &E
10	R	8 &E
11	$Q \vee R$	10 $\vee$I
12	$P \& (Q \vee R)$	9, 11 &I
13	$(P \& R) \rightarrow (P \& (Q \vee R))$	8–12 →I
14	$P \& (Q \vee R)$	1, 7, 13 $\vee$E

The strategy here is essentially the same as that of the previous problem. We need the two conditionals on lines 7 and 13 to get from the disjunctive premise to the conclusion. These are obtained by separate →I proofs (lines 2 to 7 and 8 to 14).

Several important guidelines are to be observed when we use hypothetical reasoning:

(1) *Each hypothesis introduced into a proof begins a new vertical line.* This line continues downward until the hypothesis is discharged by application either of →I or of ~I (the last rule we shall introduce).

(2) *No occurrence of a formula to the right of a vertical line may be cited in any rule applied after that vertical line has ended.* This is to ensure that a formula derived under a hypothesis is not surreptitiously used after that hypothesis has been discharged. Thus, for example, though 'P' appears at line 3 in Problem 4.22, we must derive it anew at line 9 to show that it follows from the second hypothesis, '$P \& R$'. We could not use it (or, indeed, any of lines 2 to 6) in proving '$(P \& R) \rightarrow (P \& (Q \vee R))$', since these lines represent a "logical fantasy" based on the hypothesis of line 2, which is no longer in effect after line 6.

(3) *If two or more hypotheses are in effect simultaneously, then the order in which they are discharged must be the reverse of the order in which they are introduced.* Thus in Problem 4.20, the hypothesis of line 3 must be discharged before the hypothesis of line 2 is.

(4) *A proof is not complete until all hypotheses have been discharged.* Hypotheses which are not discharged are in effect additional assumptions, introduced illicitly into the proof.

Our second hypothetical rule is negation introduction, also known as *reductio ad absurdum* or *indirect proof.* To prove a negated conclusion by negation introduction, we hypothesize the conclusion without its negation sign and derive from it an "absurdity." This shows the hypothesis to be false, whence it follows that the negated conclusion is true.

"Absurdity" in this context means a contradiction of the form $\phi \& \sim\phi$, i.e., a conjunction whose second conjunct is the negation of the first. The chief distinction of contradictions is that they cannot be true (see Section 3.5). So suppose that, given certain assumptions, we can validly derive a contradiction from a hypothesis. Since the derivation is valid and yet it has produced a false conclusion, at least one of its assumptions must be false; for if they were all true, the conclusion (by the definition of validity) would have to be true. Thus if the given assumptions are true, it is the hypothesis that must be false. That is, the falsehood of the hypothesis validly follows from the given assumptions. That is the basis of the rule:

Negation Introduction (~I):   Given a derivation of an absurdity from a hypothesis ϕ, we may discharge the hypothesis and infer $\sim\phi$.

SOLVED PROBLEMS

4.23 Prove:

$$P \to Q, \sim Q \vdash \sim P$$

Solution

1	$P \to Q$	A
2	$\sim Q$	A
3	P	H
4	Q	$1, 3 \to E$
5	$Q \,\&\, \sim Q$	$2, 4 \,\&I$
6	$\sim P$	$3\text{-}5 \sim I$

The conclusion is the negated formula '$\sim P$', so we hypothesize 'P' at line 3. In effect, line 3 says, "Let's pretend that 'P' is true (and we'll show that an absurdity follows)." We succeed in obtaining a contradiction at line 5. This allows us to apply $\sim I$ at 6, discharging the hypothesis and inferring its negation.

4.24 Prove:

$$P \leftrightarrow \sim Q \vdash \sim(P \,\&\, Q)$$

Solution

1	$P \leftrightarrow \sim Q$	A
2	$P \,\&\, Q$	H
3	P	$2 \,\&E$
4	Q	$2 \,\&E$
5	$P \to \sim Q$	$1 \leftrightarrow E$
6	$\sim Q$	$3, 5 \to E$
7	$Q \,\&\, \sim Q$	$4, 6 \,\&I$
8	$\sim(P \,\&\, Q)$	$2\text{-}7 \sim I$

Again the conclusion is negated, and so again we hypothesize the result of removing its negation sign and work toward a contradiction. Deduction of the contradiction at 7 permits us to discharge the hypothesis and infer the conclusion.

$\sim I$ can be used together with $\sim E$ to derive conclusions which are not negated. The strategy in this case is to hypothesize the negation of the conclusion and derive a contradiction from this hypothesis. We can then deduce the double negation of the conclusion by $\sim I$ and remove the two negation signs by $\sim E$. This strategy is very powerful; it often works on problems for which simpler strategies fail. The next two problems are instances.

SOLVED PROBLEMS

4.25 Prove:

$$\sim P \to P \vdash P$$

Solution

1	$\sim P \to P$	A
2	$\sim P$	H
3	P	$1, 2 \to E$
4	$P \,\&\, \sim P$	$2, 3 \,\&I$
5	$\sim\sim P$	$2\text{-}4 \sim I$
6	P	$5 \sim E$

4.26   Prove:

$$\sim(\sim P \mathbin{\&} \sim Q), \sim P \vdash Q$$

Solution

1	$\sim(\sim P \mathbin{\&} \sim Q)$	A
2	$\sim P$	A
3	$\sim Q$	H
4	$\sim P \mathbin{\&} \sim Q$	2, 3 &I
5	$(\sim P \mathbin{\&} \sim Q) \mathbin{\&} \sim(\sim P \mathbin{\&} \sim Q)$	1, 4 &I
6	$\sim\sim Q$	3–5 ~I
7	Q	6 ~E

Where several hypotheses are in effect at once, especially if some are for ~I and some for →I, it may make the proof easier to read if we label them either '(for ~I)' or '(for →I)'. We do this in some of the problems below.

4.27   Prove:

$$\sim P \lor \sim Q \vdash \sim(P \mathbin{\&} Q)$$

Solution

1	$\sim P \lor \sim Q$	A
2	$\sim P$	H (for →I)
3	$P \mathbin{\&} Q$	H (for ~I)
4	P	3 &E
5	$P \mathbin{\&} \sim P$	2, 4 &I
6	$\sim(P \mathbin{\&} Q)$	3–5 ~I
7	$\sim P \to \sim(P \mathbin{\&} Q)$	2–6 →I
8	$\sim Q$	H (for →I)
9	$P \mathbin{\&} Q$	H (for ~I)
10	Q	9 &E
11	$Q \mathbin{\&} \sim Q$	8, 10 &I
12	$\sim(P \mathbin{\&} Q)$	9–11 ~I
13	$\sim Q \to \sim(P \mathbin{\&} Q)$	8–12 →I
14	$\sim(P \mathbin{\&} Q)$	1, 7, 13 $\lor$E

To use the disjunctive premise, we employ the rule $\lor$E. But to do this, we need the conditionals listed on lines 7 and 13. To get them, we hypothesize their antecedents (lines 2 and 8) and seek in each case to prove the consequent, '$\sim(P \mathbin{\&} Q)$', for →I. This wff, however, is negated and is therefore most efficiently proved by ~I. So after introducing our →I hypotheses, we also hypothesize '$P \mathbin{\&} Q$' for ~I (lines 3 and 9). This enables us to prove the desired conditionals and hence to obtain the conclusion by $\lor$E at 14.

Negation introduction may be used in rather complex and surprising ways, as the following problems illustrate.

4.28   Prove:

$$P \to Q \vdash \sim P \lor Q$$

Solution

1	$P \rightarrow Q$	A
2	$\sim(\sim P \vee Q)$	H (for $\sim$I)
3	P	H (for $\sim$I)
4	Q	1, 3 $\rightarrow$E
5	$\sim P \vee Q$	4 $\vee$I
6	$(\sim P \vee Q) \mathbin{\&} \sim(\sim P \vee Q)$	2, 5 &I
7	$\sim P$	3–6 $\sim$I
8	$\sim P \vee Q$	7 $\vee$I
9	$(\sim P \vee Q) \mathbin{\&} \sim(\sim P \vee Q)$	2, 8 &I
10	$\sim\sim(\sim P \vee Q)$	2–9 $\sim$I
11	$\sim P \vee Q$	10 $\sim$E

There is no direct way to derive the conclusion '$\sim P \vee Q$' from the premise '$P \rightarrow Q$', but here we offer an indirect proof, hypothesizing the negation of the conclusion at 2 and working toward a contradiction. It is not possible to derive a contradiction straightforwardly from 2, so our approach must be more subtle. We add another hypothesis 'P', hoping to obtain a contradiction from it. The strategy is then to discharge this extra hypothesis as quickly as possible by $\sim$I to obtain its negation, which will then be used to derive a contradiction from 2. 'P' is shown to lead to a contradiction at 6, which gives us '$\sim P$' by $\sim$I at 7. A further contradiction follows quickly at 9, which allows us to discharge our original hypothesis and obtain '$\sim\sim(\sim P \vee Q)$' at 10. The desired conclusion then follows by a step of $\sim$E.

4.29 Prove:

$$\sim(P \mathbin{\&} Q) \vdash \sim P \vee \sim Q$$

Solution

1	$\sim(P \mathbin{\&} Q)$	A
2	$\sim(\sim P \vee \sim Q)$	H (for $\sim$I)
3	$\sim P$	H (for $\sim$I)
4	$\sim P \vee \sim Q$	3 $\vee$I
5	$(\sim P \vee \sim Q) \mathbin{\&} \sim(\sim P \vee \sim Q)$	2, 4 &I
6	$\sim\sim P$	3–5 $\sim$I
7	P	6 $\sim$E
8	$\sim Q$	H (for $\sim$I)
9	$\sim P \vee \sim Q$	8 $\vee$I
10	$(\sim P \vee \sim Q) \mathbin{\&} \sim(\sim P \vee \sim Q)$	2, 9 &I
11	$\sim\sim Q$	8–10 $\sim$I
12	Q	11 $\sim$E
13	$P \mathbin{\&} Q$	7, 12 &I
14	$(P \mathbin{\&} Q) \mathbin{\&} \sim(P \mathbin{\&} Q)$	1, 13 &I
15	$\sim\sim(\sim P \vee \sim Q)$	2–14 $\sim$I
16	$\sim P \vee \sim Q$	15 $\sim$E

This problem is similar in overall strategy to Problem 4.28. We begin by hypothesizing the negation of the conclusion for $\sim$I. However, no direct route to a contradiction is evident, so we add an extra hypothesis '$\sim P$', seeking to derive a contradiction from it and thus obtain 'P', which will be used in deriving a contradiction from 2. '$\sim P$' yields a contradiction at 5, which allows us to infer 'P' by $\sim$I and $\sim$E at 7. We obtain 'Q' by a similarly devious strategy (lines 8 to 12). Conjoining 'P' and 'Q' at 13 allows us finally to derive a contradiction from our original hypothesis '$\sim(\sim P \vee \sim Q)$' at 14. We then infer its negation by $\sim$I at 15 and obtain the conclusion at 16 by $\sim$E.

There is no one correct way to construct a proof. If a form can be proved at all (i.e., if it is valid), it can be proved by many different permutations of the rules. The shortest, simplest, and easiest proofs, however, are usually obtained by a strategy based on the structure of the conclusion. The suggestions of Table 4-1 are useful guides for planning such a strategy. They usually lead to reasonably efficient proofs, though some problems still require ingenuity and inventiveness.

Table 4-1   Proof Strategies

If the conclusion is a(n):	Then do this:
Atomic formula	If no other strategy is immediately apparent, hypothesize the negation of the conclusion for ~I. If this is successful, then the conclusion can be obtained after the ~I by ~E.
Negated formula	Hypothesize the conclusion without its negation sign for ~I. If a contradiction follows, the conclusion can be obtained by ~I.
Conjunction	Prove each of the conjuncts separately and then conjoin them with &I.
Disjunction	Sometimes (though not often) a disjunctive conclusion can be proved directly simply by proving one of its disjuncts and applying ∨I. Otherwise, hypothesize the negation of the conclusion and try ~I.
Conditional	Hypothesize its antecedent and derive its consequent by →I.
Biconditional	Use →I twice to prove the two conditionals needed to obtain the conclusion by ↔I.

If the strategies of Table 4-1 fail, then try the following. If a disjunctive premise is present, try proving the conditionals needed to get the conclusion by ∨E. If not, add an extra hypothesis whose negation would be useful as an additional premise in the proof. Then, as in Problems 4.28 and 4.29, discharge this extra hypothesis as quickly as possible by ~I to obtain its negation. If even this fails, try the same thing with a different hypothesis. Eventually (if the form you are trying to prove is in fact valid) you should hit on a hypothesis or series of hypotheses that will do the trick.

Various substrategies may develop if the conclusion is more complex. For example, if the conclusion is '~P & ~Q', a conjunction of two negated statements, the typical strategy is to prove each of its conjuncts separately and then join them by &I. Since each conjunct is negated, the substrategy for proving each is to hypothesize it without its negation sign and use ~I. Thus the proof consists of two hypothetical derivations for ~I followed by a step of &I, as in the following problem.

SOLVED PROBLEM

4.30   Prove:

$$\sim(P \lor Q) \vdash \sim P \,\&\, \sim Q$$

Solution

1	$\sim(P \lor Q)$	A
2	P	H (for ~I)
3	$P \lor Q$	2 ∨I
4	$(P \lor Q) \,\&\, \sim(P \lor Q)$	1, 3 &I
5	$\sim P$	2–4 ~I
6	Q	H (for ~I)
7	$P \lor Q$	6 ∨I
8	$(P \lor Q) \,\&\, \sim(P \lor Q)$	1, 7 &I
9	$\sim Q$	6–8 ~I
10	$\sim P \,\&\, \sim Q$	5, 9 &I

The ten basic inference rules presented thus far are *complete* in the sense that they generate a proof for each of the infinitely many valid forms expressible in the language of propositional logic. They are also *valid* in the sense that when applied as stated they generate only valid forms—never invalid ones. In spite of these facts, it is useful to have other rules besides the basic ones. These new rules, the subject matter of the next section, do not enable us to prove anything not provable by the ten basic rules alone, but they help to shorten and simplify proofs and thus save work.

4.4 DERIVED RULES

Every instance of a valid argument form is valid. Thus in proving the form '$P \to Q, \sim Q \vdash \sim P$' (Problem 4.23), we in effect established the validity of any argument which results from that form by replacing 'P' and 'Q' with sentences (no matter how complex these sentences are), for instance:

> If it is raining or snowing, then the sky is not clear.
> It is not the case that the sky is not clear.
> ∴ It is not either raining or snowing.

(See also Problem 3.2.) In this case 'P' stands for 'It is raining or snowing' and 'Q' for 'The sky is not clear'. But notice that if we formalize this argument so as to reveal all of its structure, we obtain not the original form, but:

$$(R \lor S) \to \sim C, \sim\sim C \vdash \sim(R \lor S)$$

This is what is called a *substitution instance* of the original form. A substitution instance of a wff or an argument form is the result of replacing zero or more of its sentence letters by wffs, each occurrence of the same sentence letter being replaced by the same wff. (We say "zero" instead of "one" to allow each form to count as a substitution instance of itself.) The second form above is a substitution instance of the first, because it results from the first by replacing each occurrence of 'P' by '$(R \lor S)$' and each occurrence of 'Q' by '$\sim C$'. In proving the validity of a form, we prove the validity of all its substitution instances. Thus, having proved Problem 4.23, if we encounter premises '$(R \lor S) \to \sim C$' and '$\sim\sim C$' in a proof, we know that '$\sim(R \lor S)$' validly follows (and similarly for any other substitution instance of Problem 4.23). This means that we can legitimately regard the statement of Problem 4.23 (and indeed any previously proved argument form) as a valid inference rule. For each previously proved form, the *associated inference rule* is this:

> From the premises of any substitution instance of the form, we may validly infer the conclusion of that substitution instance.

Inference rules which are derived in this way from previously proved forms are called *derived rules*.[4] Many derived rules have names. The rule associated with Problem 4.23 is called *modus tollens* (MT). It is used with the substitution instance mentioned above in the following proof:

SOLVED PROBLEM

4.31 Prove:

$$(R \lor S) \to \sim C \vdash \sim\sim C \to \sim(R \lor S)$$

---

[4]Which rules count as derived and which as basic is a matter of convention that varies from text to text. Some texts treat all the rules named in this chapter as basic and thus need no distinction between basic and derived rules. Others have fewer basic rules and hence more derived ones.

Solution

$$
\begin{array}{lll}
1 & (R \vee S) \to \sim C & A \\
2 & \quad \sim\sim C & H \text{ (for } \to I) \\
3 & \quad \sim(R \vee S) & 1, 2 \text{ MT} \\
4 & \sim\sim C \to \sim(R \vee S) & 2\text{--}3 \to I
\end{array}
$$

To see the point of introducing derived rules such as MT, contrast the proof of Problem 4.31 with the following problem, which gives the simplest proof of the same form that does not employ derived rules.

SOLVED PROBLEM

4.32 Prove the argument form stated in Problem 4.31 using only the ten basic rules.

Solution

$$
\begin{array}{lll}
1 & (R \vee S) \to \sim C & A \\
2 & \quad \sim\sim C & H \text{ (for } \to I) \\
3 & \quad\quad R \vee S & H \text{ (for } \sim I) \\
4 & \quad\quad \sim C & 1, 3 \to E \\
5 & \quad\quad \sim C \,\&\, \sim\sim C & 2, 4 \,\&I \\
6 & \quad \sim(R \vee S) & 3\text{--}5 \sim I \\
7 & \sim\sim C \to \sim(R \vee S) & 2\text{--}6 \to I
\end{array}
$$

From now on, any previously proved form may be used as a derived rule in a proof. As justification, we cite the lines used as premises and the name of the derived rule (if it has one); otherwise, the problem number at which the associated form was proved. Among the derived rules already proved, one of the most useful is *hypothetical syllogism* (HS), the rule associated with the form '$P \to Q, Q \to R \vdash P \to R$' (Problem 4.16). Also useful are the rules *absorption* (ABS) and *constructive dilemma* (CD).

SOLVED PROBLEMS

4.33 Prove the derived rule ABS, i.e.:

$$P \to Q \vdash P \to (P \,\&\, Q)$$

Solution

$$
\begin{array}{lll}
1 & P \to Q & A \\
2 & \quad P & H \\
3 & \quad Q & 1, 2 \to E \\
4 & \quad P \,\&\, Q & 2, 3 \,\&I \\
5 & P \to (P \,\&\, Q) & 2\text{--}4 \to I
\end{array}
$$

4.34 Prove the derived rule CD, i.e.:

$$P \vee Q, P \to R, Q \to S \vdash R \vee S$$

Solution

$$
\begin{array}{lll}
1 & P \vee Q & A \\
2 & P \to R & A \\
3 & Q \to S & A \\
4 & \quad P & H \text{ (for } \to I) \\
5 & \quad R & 2, 4 \to E \\
6 & \quad R \vee S & 5 \vee I
\end{array}
$$

$$
\begin{array}{lll}
7 & P \rightarrow (R \lor S) & 4\text{--}6 \rightarrow I \\
8 & \quad Q & H \,(\text{for} \rightarrow I) \\
9 & \quad S & 3, 8 \rightarrow E \\
10 & \quad R \lor S & 9 \lor I \\
11 & Q \rightarrow (R \lor S) & 8\text{--}10 \rightarrow I \\
12 & R \lor S & 1, 7, 11 \lor E
\end{array}
$$

We next consider two peculiar derived rules. The first, *repeat* or *reiteration* (RE), allows us to derive in the context of a hypothetical derivation any wff that already occurs in the main derivation, provided that it is not part of a hypothetical derivation whose hypothesis has been discharged. This rule is just an abbreviation of two successive applications of &I and &E, as the following problem illustrates.

SOLVED PROBLEM

4.35 Prove:

$$P \vdash Q \rightarrow P$$

Solution

$$
\begin{array}{lll}
1 & P & A \\
2 & \quad Q & H \,(\text{for} \rightarrow I) \\
3 & \quad P \,\&\, P & 1, 1 \,\&I \\
4 & \quad P & 3 \,\&E \\
5 & Q \rightarrow P & 2\text{--}4 \rightarrow I
\end{array}
$$

Here we derive '*P*' at line 4 in order to conclude '$Q \rightarrow P$' by an application of $\rightarrow$I. Steps 3 and 4 could be abbreviated by simply entering *P* and citing at right the previous line in the argument from which it is reiterated. The abbreviated proof can be written as follows:

$$
\begin{array}{lll}
1 & P & A \\
2 & \quad Q & H \,(\text{for} \rightarrow I) \\
3 & \quad P & 1 \,RE \\
4 & Q \rightarrow P & 2\text{--}3 \rightarrow I
\end{array}
$$

The other peculiar derived rule is called *contradiction* (CON). It allows us to infer any wff from a wff together with its negation: if the premises are inconsistent, the argument is certainly valid, though it cannot have sound instances (see Section 2.4).

SOLVED PROBLEM

4.36 Prove the derived rule CON, i.e.:

$$P, \sim\!P \vdash Q$$

Solution

$$
\begin{array}{lll}
1 & P & A \\
2 & \sim\!P & A \\
3 & \quad \sim\!Q & H \\
4 & \quad P \,\&\, \sim\!P & 1, 2 \,\&I \\
5 & \sim\!\sim\!Q & 3\text{--}4 \,\sim I \\
6 & Q & 5 \,\sim E
\end{array}
$$

This proof is correct, even though we do not actually infer the contradiction at line 4 from '$\sim\!Q$'. The idea is this: Since '$P \,\&\, \sim\!P$' certainly follows validly from 1 and 2, it also follows validly from 1 and 2 together with the hypothesis '$\sim\!Q$'. (That is what we establish at line 4.) Since this contradiction is necessarily false, either the hypothesis or one of the assumptions must be false.

So, once the assumptions are granted, it must be the hypothesis which is false (this is what we conclude at line 5). Of course, we know that at least one assumption must be false, since the assumptions are inconsistent. But that only shows that any argument of this form is unsound. All such arguments are nevertheless valid, as our proof shows (compare Problems 2.21 and 3.22).

CON is useful in the proof of another important rule, *disjunctive syllogism* (DS), which we discussed in Section 3.1.

SOLVED PROBLEM

4.37 Prove the derived rule DS, i.e.:

$$P \vee Q, \sim P \vdash Q$$

Solution

1	$P \vee Q$	A
2	$\sim P$	A
3	$\quad P$	H (for $\rightarrow$I)
4	$\quad Q$	2, 3 CON
5	$P \rightarrow Q$	3–4 $\rightarrow$I
6	$\quad Q$	H (for $\rightarrow$I)
7	$Q \rightarrow Q$	6–6 $\rightarrow$I
8	Q	1, 5, 7 $\vee$E

The overall strategy is to prove the conditionals needed to get the conclusion from the disjunctive premise '$P \vee Q$' by $\vee$E. This involves two conditional proofs (lines 3 to 5 and 6 to 7). The first employs CON. The second is a one-step derivation like the one of Problem 4.18. It is a useful exercise to prove DS using only the ten basic rules.

The following problems illustrate typical uses of some of the derived rules.

SOLVED PROBLEMS

4.38 Prove:

$$\sim P \rightarrow Q, R \rightarrow S, \sim P \vee R, \sim Q \vdash S$$

Solution

1	$\sim P \rightarrow Q$	A
2	$R \rightarrow S$	A
3	$\sim P \vee R$	A
4	$\sim Q$	A
5	$Q \vee S$	1, 2, 3 CD
6	S	4, 5 DS

4.39 Prove:

$$P \rightarrow Q, (P \,\&\, Q) \rightarrow R, \sim R \vdash \sim P$$

Solution

1	$P \rightarrow Q$	A
2	$(P \,\&\, Q) \rightarrow R$	A
3	$\sim R$	A
4	$P \rightarrow (P \,\&\, Q)$	1 ABS
5	$P \rightarrow R$	2, 4 HS
6	$\sim P$	3, 5 MT

4.5 THEOREMS

Some wffs are provable without making any nonhypothetical assumptions. These are the *theorems* or *laws* of the propositional calculus. (Semantically, they are just the tautologies, i.e., those wffs all of whose instances are logically necessary.) To indicate that a wff is a theorem, we write the symbol '⊢' in front of it. As before, this symbol asserts that the formula on its right is provable using only the formulas on its left as assumptions; hence when there are no formulas on its left, it asserts that the formula on its right is a theorem. The proof of a theorem typically begins with one or more hypotheses, which are later discharged by →I or ~I.

SOLVED PROBLEMS

4.40 Prove the theorem:

$$\vdash \sim(P \& \sim P)$$

Solution

1	$P \& \sim P$	H
2	$\sim(P \& \sim P)$	1–1 ~I

This is the simplest possible *reductio ad absurdum* proof. Line 1 serves as the entire hypothetical derivation, of which '$P \& \sim P$' is both the hypothesis and the conclusion.

4.41 Prove the theorem:

$$\vdash P \to (P \lor Q)$$

Solution

1	P	H
2	$P \lor Q$	1 ∨I
3	$P \to (P \lor Q)$	1–2 →I

4.42 Prove the theorem:

$$\vdash P \to ((P \to Q) \to Q)$$

Solution

1	P	H (for →I)
2	$P \to Q$	H (for →I)
3	Q	1, 2 →E
4	$(P \to Q) \to Q$	2–3 →I
5	$P \to ((P \to Q) \to Q)$	1–4 →I

4.43 Prove the theorem:

$$\vdash P \leftrightarrow \sim\sim P$$

Solution

1	P	H (for →I)
2	$\sim P$	H (for ~I)
3	$P \& \sim P$	1, 2 &I
4	$\sim\sim P$	2–3 ~I
5	$P \to \sim\sim P$	1–4 →I
6	$\sim\sim P$	H (for →I)
7	P	6 ~E
8	$\sim\sim P \to P$	6–7 →I
9	$P \leftrightarrow \sim\sim P$	5, 8 ↔I

4.44   Prove the theorem:

$$\vdash P \vee \sim P$$

Solution

1	$\sim (P \vee \sim P)$	H (for $\sim$I)
2	P	H (for $\sim$I)
3	$P \vee \sim P$	2 $\vee$I
4	$(P \vee \sim P)\ \&\ \sim (P \vee \sim P)$	1, 3 &I
5	$\sim P$	2–4 $\sim$I
6	$P \vee \sim P$	5 $\vee$I
7	$(P \vee \sim P)\ \&\ \sim (P \vee \sim P)$	1, 6 &I
8	$\sim\sim (P \vee \sim P)$	1–7 $\sim$I
9	$P \vee \sim P$	8 $\sim$E

Every substitution instance of a theorem is provable under any set of assumptions whatsoever. Therefore we may legitimately introduce a theorem or any of its substitution instances at any line of a proof. The derived rule which enables us to do this is called *theorem introduction* (TI). As with the other derived rules, the purpose of TI is to make proofs shorter and more efficient. When using TI, we cite the number of the problem at which the relevant theorem was proved. TI is used three times in the following proof.

SOLVED PROBLEM

4.45   Prove the theorem:

$$\vdash (P \vee Q) \vee (\sim P \vee \sim Q)$$

Solution

1	$P \vee \sim P$	TI 4.44
2	$P \rightarrow (P \vee Q)$	TI 4.41
3	$\sim P \rightarrow (\sim P \vee \sim Q)$	TI 4.41
4	$(P \vee Q) \vee (\sim P \vee \sim Q)$	1, 2, 3 CD

Notice that we cite no previous lines in using TI, since the instance of the theorem is not inferred from any previous premise. It is instructive to prove this theorem without using TI.

4.6   EQUIVALENCES

We shall call a biconditional which is a theorem an *equivalence*. If $\phi \leftrightarrow \psi$ is an equivalence, then ϕ and ψ validly imply one another and are said to be *interderivable*. Thus 'P' and '$\sim\sim P$' are interderivable in view of the equivalence proved in Problem 4.43. To prove an equivalence, we follow the usual strategy for proving biconditionals: prove the two conditionals needed for $\leftrightarrow$I by two separate conditional proofs.

SOLVED PROBLEM

4.46   Prove the equivalence:

$$\vdash (P \rightarrow Q) \leftrightarrow \sim (P\ \&\ \sim Q)$$

Solution

1	$P \to Q$	H (for $\to$I)
2	$P \,\&\, {\sim}Q$	H (for ${\sim}$I)
3	P	2 &E
4	Q	1, 3 $\to$E
5	${\sim}Q$	2 &E
6	$Q \,\&\, {\sim}Q$	4, 5 &I
7	${\sim}(P \,\&\, {\sim}Q)$	2–6 ${\sim}$I
8	$(P \to Q) \to {\sim}(P \,\&\, {\sim}Q)$	1–7 $\to$I
9	${\sim}(P \,\&\, {\sim}Q)$	H (for $\to$I)
10	P	H (for $\to$I)
11	${\sim}Q$	H (for ${\sim}$I)
12	$P \,\&\, {\sim}Q$	10, 11 &I
13	$(P \,\&\, {\sim}Q) \,\&\, {\sim}(P \,\&\, {\sim}Q)$	9, 12 &I
14	${\sim}{\sim}Q$	11–13 ${\sim}$I
15	Q	14 ${\sim}$E
16	$P \to Q$	10–15 $\to$I
17	${\sim}(P \,\&\, {\sim}Q) \to (P \to Q)$	9–16 $\to$I
18	$(P \to Q) \leftrightarrow {\sim}(P \,\&\, {\sim}Q)$	8, 17 $\leftrightarrow$I

The first of the two conditionals is proved at lines 1 to 8. Since its consequent '${\sim}(P \,\&\, {\sim}Q)$' is negated, this first $\to$I proof employs a *reductio* substrategy. The second conditional is proved at lines 9 to 17. Its consequent '$P \to Q$' is itself a conditional, so the substrategy here is $\to$I. We hypothesize the antecedent 'P' at line 10 and then use a *reductio* argument (steps 11 to 14) to get '${\sim}{\sim}Q$' and hence the consequent 'Q'. This enables us to derive '$P \to Q$' by one step of $\to$I at line 16 and then '${\sim}(P \,\&\, Q) \to (P \to Q)$' by another at line 17.

Like derived rules, many equivalences have names. Table 4-2 lists some of the most important equivalences, together with previously solved problems that exhibit some of the substrategies of their proofs. The actual proofs (except for DN, which was proved in Problem 4.43) are left as exercises for the reader.

Table 4-2   Equivalences

Equivalence	*Name*	*See Problem(s)*
${\sim}(P \,\&\, Q) \leftrightarrow ({\sim}P \lor {\sim}Q)$	De Morgan's law (DM)	4.27, 4.29
${\sim}(P \lor Q) \leftrightarrow ({\sim}P \,\&\, {\sim}Q)$	De Morgan's law (DM)	4.30
$(P \lor Q) \leftrightarrow (Q \lor P)$	Commutation (COM)	4.21
$(P \,\&\, Q) \leftrightarrow (Q \,\&\, P)$	Commutation (COM)	4.3
$(P \lor (Q \lor R)) \leftrightarrow ((P \lor Q) \lor R)$	Association (ASSOC)	
$(P \,\&\, (Q \,\&\, R)) \leftrightarrow ((P \,\&\, Q) \,\&\, R)$	Association (ASSOC)	
$(P \,\&\, (Q \lor R)) \leftrightarrow ((P \,\&\, Q) \lor (P \,\&\, R))$	Distribution (DIST)	4.22
$(P \lor (Q \,\&\, R)) \leftrightarrow ((P \lor Q) \,\&\, (P \lor R))$	Distribution (DIST)	
$P \leftrightarrow {\sim}{\sim}P$	Double negation (DN)	4.43
$(P \to Q) \leftrightarrow ({\sim}Q \to {\sim}P)$	Transposition (TRANS)	
$(P \to Q) \leftrightarrow ({\sim}P \lor Q)$	Material implication (MI)	4.28
$((P \,\&\, Q) \to R) \leftrightarrow (P \to (Q \to R))$	Exportation (EXP)	4.20
$P \leftrightarrow (P \,\&\, P)$	Tautology (TAUT)	4.6
$P \leftrightarrow (P \lor P)$	Tautology (TAUT)	4.9

Equivalences play a special role in proofs. If two formulas are interderivable, one may validly be substituted for *any* occurrence of the other, *as either a whole formula or a subwff of some larger wff.* If one formula is obtained from another by such a substitution, then it is possible to derive each from the other using only the ten basic rules of inference.[5] Thus, for example, since DN establishes the interderivability of 'P' and '$\sim\sim P$', it also guarantees that '$(Q \to \sim\sim P)$' is provable from '$(Q \to P)$'. Each equivalence, then, may be treated as a derived rule of inference that allows us to replace either of the two interderivable wffs by the other, either as an entire wff, or as a subwff of some larger wff. More precisely:

If ϕ and ψ are interderivable and ϕ is a subwff of some wff χ, from χ we may infer the result of replacing one or more occurrences of ϕ in χ by ψ.

As justification, we cite the line on which χ occurs and the name of the equivalence or (if it has no name) the problem number at which it is proved.

SOLVED PROBLEM

4.47 Prove:

$$\vdash \sim(P \,\&\, Q) \to (Q \to \sim P)$$

Solution

1	$(Q \to \sim P) \leftrightarrow \sim(Q \,\&\, \sim\sim P)$	TI 4.46
2	$(Q \to \sim P) \leftrightarrow \sim(Q \,\&\, P)$	1 DN
3	$\sim(Q \,\&\, P) \to (Q \to \sim P)$	4 $\leftrightarrow$E
4	$\sim(P \,\&\, Q) \to (Q \to \sim P)$	3 COM

We begin with an instance of the theorem proved in Problem 4.46, using TI. This instance contains a doubly negated component, so at line 2 we eliminate the double negation by DN. At line 3 we simplify the biconditional to a conditional by $\leftrightarrow$E. Lastly we apply COM to the negated part of the antecedent to obtain line 4.

Proof of an equivalence also establishes the derivability of all substitution instances of that equivalence. Thus DN asserts not only the interderivability of 'P' and '$\sim\sim P$', but also of 'Q' and '$\sim\sim Q$', of '$S \,\&\, \sim R$' and '$\sim\sim(S \,\&\, \sim R)$', and so on. Hence DN may be used to justify substituting any member of one of these pairs for the other.

SOLVED PROBLEM

4.48 Prove:

$$P \leftrightarrow Q \vdash \sim((P \to Q) \to \sim(Q \to P))$$

Solution

1	$P \leftrightarrow Q$	A
2	$P \to Q$	1 $\leftrightarrow$E
3	$Q \to P$	1 $\leftrightarrow$E
4	$(P \to Q) \,\&\, (Q \to P)$	3 &I
5	$\sim\sim(P \to Q) \,\&\, (Q \to P)$	4 DN
6	$\sim(\sim(P \to Q) \lor \sim(Q \to P))$	5 DM
7	$\sim((P \to Q) \to \sim(Q \to P))$	6 MI

[5]This can be proved, but once again the proof is beyond the scope of this book.

Like the other derived rules, equivalences enable us to prove no more than do the ten basic rules. Their virtue lies in their ability to shorten and simplify proofs. For the reader's convenience, the ten basic rules are listed in Table 4-3, and the most important of the derived rules are summarized in Table 4-4.

Table 4-3　The Ten Basic Rules

Negation introduction (~I)	Given a derivation of an absurdity from a hypothesis ϕ, discharge the hypothesis and infer $\sim\phi$.
Negation elimination (~E)	From a wff of the form $\sim\sim\phi$, infer ϕ.
Conditional introduction ($\rightarrow$I)	Given a derivation of a wff ψ from a hypothesis ϕ, discharge the hypothesis and infer $\phi \rightarrow \psi$.
Conditional elimination ($\rightarrow$E)	From a conditional and its antecedent, infer its consequent.
Conjunction introduction (&I)	From any wffs ϕ and ψ, infer the conjunction $\phi \& \psi$.
Conjunction elimination (&E)	From a conjunction, infer either of its conjuncts.
Disjunction introduction ($\lor$I)	From a wff ϕ, infer the disjunction of ϕ with any wff.
Disjunction elimination ($\lor$E)	From wffs of the forms $\phi \lor \psi$, $\phi \rightarrow \chi$, and $\psi \rightarrow \chi$, infer χ.
Biconditional introduction ($\leftrightarrow$I)	From wffs of the forms $\phi \rightarrow \psi$ and $\psi \rightarrow \phi$, infer $\phi \leftrightarrow \psi$.
Biconditional elimination ($\leftrightarrow$E)	From a wff of the form $\phi \leftrightarrow \psi$, infer either $\phi \rightarrow \psi$ or $\psi \rightarrow \phi$.

Table 4-4　Important Derived Rules

Modus tollens (MT)	From wffs of the forms $\phi \rightarrow \psi$ and $\sim\psi$, infer $\sim\phi$.
Hypothetical syllogism (HS)	From wffs of the forms $\phi \rightarrow \psi$ and $\psi \rightarrow \chi$, infer $\phi \rightarrow \chi$.
Absorption (ABS)	From a wff of the form $\phi \rightarrow \psi$, infer $\phi \rightarrow (\phi \& \psi)$.
Constructive dilemma (CD)	From wffs of the forms $\phi \lor \psi$, $\phi \rightarrow \chi$, and $\psi \rightarrow \omega$, infer $\chi \lor \omega$.
Repeat (RE)	From any wff ϕ, infer ϕ.
Contradiction (CON)	From wffs of the forms ϕ and $\sim\phi$, infer any wff.
Disjunctive syllogism (DS)	From wffs of the forms $\phi \lor \psi$ and $\sim\phi$, infer ψ.
Theorem introduction (TI)	Any substitution instance of a theorem may be introduced at any line of a proof.
Equivalence introduction (Abbreviation used depends on equivalence used; see Table 4-2, the table of equivalences.)	If ϕ and ψ are interderivable and ϕ is a subwff of some wff χ, from χ infer the result of replacing one or more occurrences of ϕ in χ by ψ.

The following problems illustrate typical uses of equivalences and some of the other derived rules.

SOLVED PROBLEMS

4.49 Prove:

$$P \lor Q, \sim Q \vdash P$$

Solution

1	$P \lor Q$	A
2	$\sim Q$	A
3	$Q \lor P$	2 COM
4	P	2, 3 DS

Notice that this form is not DS, since (in contrast to DS) the second premise is the negation of the second conjunct of the first premise. Thus to use DS in the proof, we must first apply COM at line 3.

4.50   Prove:

$$(P \lor Q) \mathbin{\&} (P \lor R) \vdash \sim P \to (Q \mathbin{\&} R)$$

Solution

1	$(P \lor Q) \mathbin{\&} (P \lor R)$	A
2	$P \lor (Q \mathbin{\&} R)$	1 DIST
3	$\sim\sim P \lor (Q \mathbin{\&} R)$	2 DN
4	$\sim P \to (Q \mathbin{\&} R)$	3 MI

4.51   Prove:

$$(\sim P \lor Q) \lor R, (Q \lor R) \to S \vdash P \to S$$

Solution

1	$(\sim P \lor Q) \lor R$	A
2	$(Q \lor R) \to S$	A
3	$\sim P \lor (Q \lor R)$	1 ASSOC
4	$P \to (Q \lor R)$	3 MI
5	$P \to S$	2, 4 HS

4.52   Prove:

$$P \to (Q \to \sim R), R \vdash \sim P \lor \sim Q$$

Solution

1	$P \to (Q \to \sim R)$	A
2	R	A
3	$(P \mathbin{\&} Q) \to \sim R$	1 EXP
4	$\sim\sim R$	2 DN
5	$\sim(P \mathbin{\&} Q)$	3, 4 MT
6	$\sim P \lor \sim Q$	5 DM

4.53   Prove:

$$\sim P \to P \vdash P$$

Solution

1	$\sim P \to P$	A
2	$\sim\sim P \lor P$	1 MI
3	$P \lor P$	2 DN
4	P	3 TAUT

4.54   Prove:

$$\sim P \lor \sim Q, R \to P, \sim\sim Q \lor \sim S \vdash \sim S \lor \sim R$$

Solution

$$
\begin{array}{lll}
1 & \sim P \vee \sim Q & A \\
2 & R \to P & A \\
3 & \sim\sim Q \vee \sim S & A \\
4 & \sim P \to \sim R & 2 \text{ TRANS} \\
5 & \sim Q \to \sim S & 3 \text{ MI} \\
6 & \sim R \vee \sim S & 1, 4, 5 \text{ CD} \\
7 & \sim S \vee \sim R & 6 \text{ COM}
\end{array}
$$

Supplementary Problems

I   The following arguments are all valid. Formalize each, using the interpretation indicated below, and prove the validity of the resulting form, using only the ten introduction and elimination rules.

Sentence Letter	Interpretation
C	The conclusion of this argument is true.
P	The premises of this argument are all true.
S	This argument is sound.
V	This argument is valid.

(1)   This argument is not unsound. Therefore, this argument is sound.

(2)   This argument is sound. Therefore, this argument is not unsound.

(3)   If this argument is sound, then it is valid. It is not valid; therefore, it is not sound.

(4)   If this argument is sound, then it is not invalid. It is sound. Hence it is valid.

(5)   If this argument is sound, then it is not invalid. So if it is invalid, then it is unsound.

(6)   This argument is both sound and valid. Therefore, either it is sound or it is invalid.

(7)   This argument is not both sound and invalid. It is sound. Therefore, it is valid.

(8)   This argument is sound only if all its premises are true. But not all its premises are true. Hence it is unsound.

(9)   If this argument's conclusion is untrue, then this argument is unsound. So it is not the case that this argument is sound and has an untrue conclusion.

(10)   If this argument is unsound and valid, then not all its premises are true. All its premises are true. It is valid. Therefore, it is sound.

(11)   If this argument is valid and all its premises are true, then it is sound. If it is sound, then its conclusion is true. All its premises are true. Therefore, if this argument is valid, then its conclusion is true.

(12)   Either this argument is unsound, or else it is valid and all its premises are true. Hence it is either unsound or valid.

(13)   This argument is sound if and only if it is valid and all its premises are true. Not all its premises are true. Hence it is unsound.

(14)   This argument is sound if and only if it is valid and all its premises are true. Hence if it is valid, then it is sound if all its premises are true.

(15)   This argument is unsound only if either not all its premises are true or it is invalid. But it is valid and all its premises are true. Therefore, it is sound.

II   Using only the ten introduction and elimination rules, provide proofs for the equivalences listed in Table 4-2 (except for DN, which is proved in Problem 4.43).

III Using basic or derived rules, prove the validity of the following argument forms.

(1) $P \leftrightarrow Q, Q \leftrightarrow R \vdash P \leftrightarrow R$

(2) $P \leftrightarrow Q \vdash \sim P \leftrightarrow \sim Q$

(3) $\sim P \vee Q \vdash \sim(P \mathbin{\&} \sim Q)$

(4) $P \rightarrow Q, P \rightarrow \sim Q \vdash \sim P$

(5) $(P \rightarrow Q) \mathbin{\&} (P \rightarrow R) \vdash P \rightarrow (Q \mathbin{\&} R)$

(6) $P \rightarrow Q \vdash (P \mathbin{\&} R) \rightarrow (Q \mathbin{\&} R)$

(7) $P \rightarrow Q \vdash (P \vee R) \rightarrow (Q \vee R)$

(8) $\sim P \rightarrow P \vdash P$

(9) $\sim P \vdash P \rightarrow Q$

(10) $P \mathbin{\&} Q \vdash P \rightarrow Q$

IV Using basic or derived rules, prove the following theorems.

(1) $\vdash P \rightarrow P$

(2) $\vdash P \rightarrow (Q \rightarrow (P \mathbin{\&} Q))$

(3) $\vdash \sim(P \leftrightarrow \sim P)$

(4) $\vdash (P \rightarrow Q) \rightarrow (\sim Q \rightarrow \sim P)$

(5) $\vdash (P \mathbin{\&} Q) \vee (\sim P \vee \sim Q)$

(6) $\vdash Q \rightarrow (P \vee \sim P)$

(7) $\vdash (P \mathbin{\&} \sim P) \rightarrow Q$

(8) $\vdash P \vee (P \rightarrow Q)$

(9) $\vdash \sim P \vee (Q \rightarrow P)$

(10) $\vdash (P \rightarrow Q) \vee (Q \rightarrow P)$

V Using basic or derived rules, prove the following equivalences.

(1) $\vdash (P \mathbin{\&} Q) \leftrightarrow \sim(\sim P \vee \sim Q)$

(2) $\vdash (P \vee Q) \leftrightarrow \sim(\sim P \mathbin{\&} \sim Q)$

(3) $\vdash (P \mathbin{\&} Q) \leftrightarrow \sim(P \rightarrow \sim Q)$

(4) $\vdash (P \vee Q) \leftrightarrow \sim P \rightarrow Q$

(5) $\vdash P \leftrightarrow ((P \mathbin{\&} Q) \vee (P \mathbin{\&} \sim Q))$

(6) $\vdash \sim(P \rightarrow Q) \leftrightarrow (P \mathbin{\&} \sim Q)$

(7) $\vdash (P \leftrightarrow Q) \leftrightarrow ((P \mathbin{\&} Q) \vee (\sim P \mathbin{\&} \sim Q))$

(8) $\vdash \sim(P \leftrightarrow Q) \leftrightarrow ((\sim P \mathbin{\&} Q) \vee (P \mathbin{\&} \sim Q))$

(9) $\vdash (P \mathbin{\&} \sim P) \leftrightarrow (Q \mathbin{\&} \sim Q)$

(10) $\vdash (P \vee \sim P) \leftrightarrow (Q \vee \sim Q)$

Answers to Selected Supplementary Problems

I   (5) $S \rightarrow \sim\sim V \vdash \sim V \rightarrow \sim S$

1	$S \rightarrow \sim\sim V$	A
2	$\sim V$	H (for $\rightarrow$I)
3	S	H (for $\sim$I)
4	$\sim\sim V$	1, 3 $\rightarrow$E
5	$\sim V \,\&\, \sim\sim V$	2, 4 &I
6	$\sim S$	3–5 $\sim$I
7	$\sim V \rightarrow \sim S$	2–6 $\rightarrow$I

(10) $(\sim S \,\&\, V) \rightarrow \sim P, P, V \vdash S$

1	$(\sim S \,\&\, V) \rightarrow \sim P$	A
2	P	A
3	V	A
4	$\sim S$	H (for $\sim$I)
5	$\sim S \,\&\, V$	3, 4 &I
6	$\sim P$	1, 5 $\rightarrow$E
7	$P \,\&\, \sim P$	2, 6 &I
8	$\sim\sim S$	4–7 $\sim$I
9	S	8 $\sim$E

(15) $\sim S \leftrightarrow (\sim P \vee \sim V), V \,\&\, P \vdash S$

1	$\sim S \leftrightarrow (\sim P \vee \sim V)$	A
2	$V \,\&\, P$	A
3	$\sim S$	H (for $\sim$I)
4	$\sim S \rightarrow (\sim P \vee \sim V)$	1 $\leftrightarrow$E
5	$\sim P \vee \sim V$	3, 4 $\rightarrow$E
6	$\sim P$	H (for $\rightarrow$I)
7	$V \,\&\, P$	H (for $\sim$I)
8	P	7 &E
9	$P \,\&\, \sim P$	6, 8 &I
10	$\sim(V \,\&\, P)$	7–9 $\sim$I
11	$\sim P \rightarrow \sim(V \,\&\, P)$	6–10 $\rightarrow$I
12	$\sim V$	H (for $\rightarrow$I)
13	$V \,\&\, P$	H (for $\sim$I)
14	V	13 &E
15	$V \,\&\, \sim V$	12, 14 &I
16	$\sim(V \,\&\, P)$	13–15 $\sim$I
17	$\sim V \rightarrow \sim(V \,\&\, P)$	12–16 $\rightarrow$I
18	$\sim(V \,\&\, P)$	5, 11, 17 $\vee$E
19	$(V \,\&\, P) \,\&\, \sim(V \,\&\, P)$	2, 18 &I
20	$\sim\sim S$	3–19 $\sim$I
21	S	20 $\sim$E

III   (5) $(P \rightarrow Q) \,\&\, (P \rightarrow R) \vdash P \rightarrow (Q \,\&\, R)$

1	$(P \rightarrow Q) \,\&\, (P \rightarrow R)$	A
2	$(\sim P \vee Q) \,\&\, (P \rightarrow R)$	1 MI
3	$(\sim P \vee Q) \,\&\, (\sim P \vee R)$	2 MI
4	$\sim P \vee (Q \,\&\, R)$	3 DIST
5	$P \rightarrow (Q \,\&\, R)$	4 MI

(10) $P \& Q \vdash P \to Q$

    1 $P \& Q$ A
    2 | P H (for $\to$I)
    3 | Q 1 &E
    4 $P \to Q$ 2–3 $\to$I

IV   (5) $\vdash (P \& Q) \vee (\sim P \vee \sim Q)$

    1 $(P \& Q) \vee \sim (P \& Q)$ TI 3.45
    2 $(P \& Q) \vee (\sim P \vee \sim Q)$ 1 DM

   (10) $\vdash (P \to Q) \vee (Q \to P)$

    1 $P \vee \sim P$ TI 4.44
    2 $(P \vee \sim P) \vee Q$ 1 $\vee$I
    3 $P \vee (\sim P \vee Q)$ 2 ASSOC
    4 $\sim Q \vee (P \vee (\sim P \vee Q))$ 3 $\vee$I
    5 $(\sim Q \vee P) \vee (\sim P \vee Q)$ 4 ASSOC
    6 $(\sim P \vee Q) \vee (\sim Q \vee P)$ 5 COM
    7 $(P \to Q) \vee (\sim Q \vee P)$ 6 MI
    8 $(P \to Q) \vee (Q \to P)$ 7 MI

V   (5) $\vdash P \leftrightarrow ((P \& Q) \vee (P \& \sim Q))$

    1 | P H (for $\to$I)
    2 | $Q \vee \sim Q$ TI 4.44
    3 | $P \& (Q \vee \sim Q)$ 1, 2 &I
    4 | $(P \& Q) \vee (P \& \sim Q)$ 3 DIST
    5 $P \to ((P \& Q) \vee (P \& \sim Q))$ 1–4 $\to$I
    6 | $(P \& Q) \vee (P \& \sim Q)$ H (for $\to$I)
    7 | $P \& (Q \vee \sim Q)$ 6 DIST
    8 | P 7 &E
    9 $((P \& Q) \vee (P \& \sim Q)) \to P$ 6–8 $\to$I
   10 $P \leftrightarrow ((P \& Q) \vee (P \& \sim P))$ 5, 9 $\leftrightarrow$I

   (10) $\vdash (P \vee \sim P) \leftrightarrow (Q \vee \sim Q)$

    1 | $P \vee \sim P$ H (for $\to$I)
    2 | $Q \vee \sim Q$ TI 4.44
    3 $(P \vee \sim P) \to (Q \vee \sim Q)$ 1–2 $\to$I
    4 | $Q \vee \sim Q$ H (for $\to$I)
    5 | $P \vee \sim P$ TI 4.44
    6 $(Q \vee \sim Q) \to (P \vee \sim P)$ 4–5 $\to$I
    7 $(P \vee \sim P) \leftrightarrow (Q \vee \sim Q)$ 3, 6 $\leftrightarrow$I

Chapter 5

The Logic of Categorical Statements

5.1 CATEGORICAL STATEMENTS

The propositional logic of Chapters 3 and 4 concerns only logical relations generated by truth-functional operators, expressions like 'not', 'and', 'or', 'if ... then', and 'if and only if'. These relations are fundamental, but they are still only a small part of the subject matter of logic. This chapter takes a preliminary look at another part, the logical relations generated by such expressions as 'all', 'some', and 'no'. Then, in Chapter 6, we combine the material of Chapters 3, 4, and 5 and add some new elements, thereby attaining a perspective from which we can see how all these studies are interrelated.

The main reason to go beyond propositional logic is that there are valid arguments whose validity does not depend solely on truth-functional operators. For example, the following is a valid argument which cannot be understood by propositional logic alone:

> Some four-legged creatures are gnus.
> All gnus are herbivores.
> ∴ Some four-legged creatures are herbivores.

Because none of the statements in the argument is truth-functionally compound, *from the viewpoint of propositional logic* these statements lack internal structure. If we try to formalize the argument in propositional logic, the best we can do is something like

$P, Q \vdash R$

But this form is clearly invalid, since any three statements P, Q, and R such that P and Q are true and R is false constitute a counterexample.

Yet from a more discriminating perspective, the statements of the above example do have internal structures, and their structures constitute a valid form. These structures consist, not of truth-functional relations among statements, as in propositional logic, but of relations between terms occurring within the statements themselves. To see this clearly, let us represent the form of the argument as follows:

> Some F are G.
> All G are H.
> ∴ Some F are H.

Here the letters 'F', 'G', and 'H' are placeholders, not for sentences, as in propositional logic, but for *class terms*, such as 'gnu', 'herbivore', and 'four-legged creature'. Class terms (also called *predicates*) denote classes (sets) of objects. The term 'gnu', for example, denotes the class of all gnus, and the term 'herbivore', the class of all herbivores. Every substitution of class terms for these letters (replacing each occurrence of the same letter by the same class term) produces a valid argument.

SOLVED PROBLEM

5.1 Represent the following argument in schematic form:

> All logicians are philosophers.
> Some logicians are mathematicians.
> ∴ Some mathematicians are philosophers.

111

Solution

Using '*L*' for 'logicians', '*P*' for 'philosophers', and '*M*' for 'mathematicians', the form of the argument can be represented thus:

> All *L* are *P*.
> Some *L* are *M*.
> ∴ Some *M* are *P*.

The class terms just mentioned are all simple common nouns; but noun *phrases*, such as 'blue thing', 'ugly bug-eyed toad', or 'friend of mine', are also class terms. Adjectives and adjective phrases may function as class terms as well. Thus the adjectives 'old', 'circular', and 'vicious', for example, denote respectively the sets of all old, circular, and vicious things; and the adjective phrase 'in the northern hemisphere' denotes the set of things located in the northern hemisphere. In addition, verbs and verb phrases may be regarded as class terms: 'move', 'love Bill', and 'have had a car wreck' denote, respectively, the sets of all things that move, love Bill, and have had a car wreck.

To facilitate substitution and comparison of class terms, we shall rewrite them as common nouns and noun phrases if they are not already in that form. That is, we shall regard nouns and noun phrases as the standard form for class terms. A class term expressed as an adjective, adjective phrase, verb, or verb phrase can usually be converted into a noun phrase by adding the word 'things'. Thus 'old' becomes 'old things', 'in the northern hemisphere' becomes 'things in the northern hemisphere', 'love Bill' becomes 'things that love Bill', and so on.

SOLVED PROBLEM

5.2 Represent the following argument in schematic form:

> All cups on the coffee table have been hand-painted by my art teacher.
> All things that have been hand-painted by my art teacher are beautiful.
> ∴ All cups on the coffee table are beautiful.

Solution

This argument is of the form:

> All *C* are *H*.
> All *H* are *B*.
> ∴ All *C* are *B*.

where '*C*' is replaced by the noun phrase 'cups on the coffee table', '*H*' by the noun phrase 'things that have been hand-painted by my art teacher' (obtained by converting the verb phrase 'have been painted by my art teacher' into standard form), and '*B*' by the noun phrase 'beautiful things' (obtained by converting the adjective 'beautiful' into standard form).

In statements, class terms are often related to one another by the expressions 'all' and 'some', which are called *quantifiers*. Quantifiers, like truth-functional operators, are logical operators; but instead of indicating relationships among sentences, they express relationships among the sets designated by class terms. Statements of the form 'All *A* are *B*', for example, assert that the set *A* is a subset of the set *B*; that is, all the members of *A* are also members of *B*. And, by a convention universally employed in logic, statements of the form 'Some *A* are *B*' assert that the set *A* shares at least one member with the set *B*.

This is a departure from ordinary usage, according to which 'Some *A* are *B*' typically signifies that the set *A* shares *more than one* member with the set *B*. It also departs from ordinary usage in another way. Typically, when we say that *some A* are *B*, we presuppose that not all *A* are *B*. Suppose, for example, that someone says, "Some friends of mine are angry with me." (This statement is of the form 'Some *A* are *B*', where '*A*' is replaced by the noun phrase 'friends of mine' and '*B*' by the adjective phrase 'angry with me'—or, in standard form, 'things (or persons) that are angry with me'.) Normally, this would be understood as presupposing or suggesting that not all this person's friends are angry with him or her. This presupposition is absent from the logical notion of 'some'. In the logical sense of 'some', it is perfectly correct to say that some friends of mine are angry with me, even when all of them are.

In addition to the quantifiers 'all' and 'some', we shall also consider the negative quantifier 'no'. Statements of the form 'No *A* are *B*' assert that the sets *A* and *B* are disjoint, i.e., share no members.

The only other expression occurring in the three arguments considered so far is 'are'. This word (together with its grammatical variants 'is', 'am', 'was', 'were', etc.) is called the *copula*, because it couples or joins subject and predicate. Each statement in the three arguments is therefore made up of a quantifier followed by a class term, a copula, and, finally, another class term. Statements so constructed are called *categorical statements*. The study of these statements is the oldest branch of western logic and goes back to the work of Aristotle in the fourth century B.C. (It underwent important revisions in the nineteenth century—a point to which we will return in Section 5.3.) In addition, we shall find it convenient to regard negations of categorical statements as categorical statements themselves, though some authors do not.

The first class term in a categorical statement is the *subject term*; the second is the *predicate term*. Unnegated categorical statements come in four distinct forms, each of which is traditionally designated by a vowel:

Designation	Form
A	All *S* are *P*.
E	No *S* are *P*.
I	Some *S* are *P*.
O	Some *S* are not *P*.

Here '*S*' stands for the subject term and '*P*' for the predicate term.

Note that *O*-form statements contain the expression 'not'. 'Not' plays two very different roles in categorical statements. When applied to an entire sentence, it expresses truth-functional negation—the operation, familiar from Chapters 3 and 4, that makes true statements false and false ones true. But when applied to class terms, as in *O*-form sentences, 'not' has a different function. In 'Some trees are not oaks', for example, 'not' modifies only the class term 'oaks', not the whole sentence. The result is a new class term 'not oaks' (or, more perspicuously, 'non-oaks'), which designates the set of all things which are not oak trees. (This is a motley set, containing everything from quarks to quasars—but not oak trees.) In general, the set of all things which are not members of a given set *S* is called the *complement* of *S*, and when 'not' is applied to class terms, it is said to express *complementation* rather than negation. Thus 'Some trees are not oaks', where 'not' expresses complementation, asserts that the set of trees shares at least one member with the complement of the set of oaks, i.e., that there are trees that are not oaks. It is quite different in meaning from 'It is not the case that some trees are oaks', where 'It is not the case that' expresses truth-functional negation. 'Some trees are not oaks' is true (there are pines and maples, for example). But 'It is not the case that some trees are oaks' is false, since 'Some trees are oaks' is true.

Because of its dual role in the logic of categorical statements, 'not' is ambiguous and should be treated with caution. To minimize misunderstanding, in the text of this chapter we will use 'It is not the case that' or the symbol '~' to express negation and 'not' or 'non-' exclusively to express complementation. In the exercises, however, 'not' may be used in either sense; it is up to the reader to determine which, by deciding whether it applies to the entire sentence or just to a class term within the sentence.

SOLVED PROBLEM

5.3 Analyze the logical form of the following categorical statement:

>   All men are not rational.

Solution

>   This statement is ambiguous, depending upon whether the 'not' is read as negation or as complementation. If it is read as negation, then (using 'M' for 'men' and 'R' for 'rational things') its form is '$\sim$(All M are R)', a negated A-form statement, which asserts that not all men are rational but leaves open the possibility that some are. If 'not' is read as complementation, then its form is 'All M are non-R', an A-form statement with a complemented predicate term. This latter statement says that all men are nonrational, i.e., that no men are rational.

The prefix 'non-' expresses complementation unambiguously in categorical statements. Other prefixes, such as 'un-', 'im-', 'in-', and 'ir-', may also express complementation, but they require care, since they can express other forms of opposition as well. The set of impossible things, for example, is the complement of the set of possible things. A thing is either possible or impossible; there is no third option. But the set of unhappy things is not the complement of the set of happy things. Many things (such as rocks, or people in a neutral state of mind) are neither happy nor unhappy. The complement of the set of happy things is the set of nonhappy things (which includes those things that are unhappy as well as things that are neither happy nor unhappy).

The class terms of a categorical statement may occur without complementation operators or with one or more. Thus, for example, 'All nonvertebrates are nonmammals' is an A-form statement with complemented subject and predicate terms. Statements of the form 'Some S are not P' have two forms. They are O-form statements with the predicate term 'P', but they may also be regarded as I-form statements with the complemented predicate term 'not P'. (It is customary to write 'Some S are not P' when the O form is to be emphasized and 'Some S are non-P' when the I form is, but these are precisely the same statement, since in these contexts both 'non' and 'not' express complementation.)

Sometimes more than one complementation operator is applied to a single class term. Double complementation is like double negation, in that the two negatives "cancel out." Thus 'All men are not nonmortal', an A-form statement with a doubly complemented predicate term, is logically equivalent to the simpler A-form statement 'All men are mortal'. It is customary in the logic of categorical statements to perform the "cancellation" without comment and to regard a statement involving double complementation as identical to the statement which results from removing it. On the other hand, note that a statement like 'It is not the case that all men are not mortal' involves one negation and one complementation, and the two cannot be "canceled" without altering the meaning of the statement.

SOLVED PROBLEM

5.4 Some of the following sentences express categorical statements; others do not. Formalize those which do, and indicate which of the four forms (or their negations) they exemplify.

(*a*)   All embezzlers are wicked.

(*b*)   Not all embezzlers are wicked.

(*c*)   All embezzlers are nonwicked.

(*d*)   Some embezzlers are not wicked.

(*e*)   If Jack is an embezzler, then Jack is wicked.

(*f*)   Nobody in this room is leaving.

(*g*)   Nobody is in this room.

(h)   Some of these towels are wet and some aren't.

(i)   Diamonds are expensive.

(j)   A few of the miners were nonsmokers.

(k)   Socrates is mortal.

(l)   Anything fun is illegal.

(m)   It is not true that anything fun is illegal.

(n)   Death is ubiquitous.

(o)   It's raining.

(p)   There are mice in the attic.

(q)   It isn't true that no skeletons are buried in the yard.

(r)   Not all nondrinkers are nonsmokers.

Solution

(a)   All *E* are *W* (using '*E*' for 'embezzlers' and '*W*' for 'wicked' or, in standard form, 'wicked things'); form *A*.

(b)   ~(All *E* are *W*); negation of form *A*. Notice that in this context, 'not' expresses negation; what is being said is that it is not the case that all embezzlers are wicked.

(c)   All *E* are non-*W*; form *A* with complemented predicate term.

(d)   Some *E* are not *W*; form *O*. (Can also be regarded as the *I*-form statement 'Some *E* are non-*W*'.)

(e)   This is a conditional, not a categorical statement.

(f)   No *P* are *L* (using '*P*' for 'people in this room' and '*L*' for 'things that are leaving'); form *E*.

(g)   No *P* are *I* (similar to part *f*, but here '*P*' stands for 'people' and '*I*' for 'things in this room'); form *E*.

(h)   This is a conjunction of two categorical statements, the first *I*-form and the second *O*-form, but it is not itself a categorical statement.

(i)   This can be rendered with some distortion as the *A*-form categorical statement 'All *D* are *E*', where '*D*' stands for 'diamonds' and '*E*' for 'expensive things'. In most contexts, however, it would be taken to mean not that all diamonds without exception are expensive, but, more reasonably, that diamonds are typically expensive. Hence the categorical rendering is not wholly accurate.

(j)   This can be regarded, again with some distortion, as the *O*-form statement 'Some *M* were not *S*' (which can also be seen as an *I*-form statement with a complemented predicate term). The distortion results from the fact that the phrase 'a few' indicates that the number was small, a connotation lost in the formalization.

(k)   This can be read as 'All things that are Socrates are mortal things', i.e., as the *A*-form statement 'All *S* are *M*', but most logicians prefer to treat 'Socrates' as a proper name (see Section 6.2).

(l)   All *F* are non-*L* (where '*F*' stands for 'fun things' and '*L*' for 'legal things'); form *A*.

(m)   ~(All *F* are non-*L*); negation of form *A*.

(n)   Cannot be rendered into categorical form without considerable distortion.

(o)   Not categorical.

(p)   Some *M* are *I* (where '*M*' stands for 'mice' and '*I*' for 'things in the attic'); form *I*.

(q)   ~(No *S* are *B*) (where '*S*' stands for 'skeletons' and '*B*' for 'things buried in the yard'); negation of form *E*.

(r)   ~(All non-*D* are non-*S*), negation of form *A* with complemented subject and predicate terms.

5.2 VENN DIAGRAMS

In working with categorical statements, it often helps to visualize relationships among sets. The most vivid way to do this is by *Venn diagrams*, an invention of the nineteenth-century mathematician John Venn. A categorical statement is represented in a Venn diagram by two overlapping circles, representing the sets designated by the statement's two terms (subject and predicate). The area inside a circle represents the contents of the set; the area outside it represents the contents of its complement. The area of overlap between the two circles stands for the members (if any) which the corresponding sets share. To show that a set or part of a set has no members, we block out the part of the diagram which represents it.

For example, form *E*, 'No *S* are *P*', asserts that the sets *S* and *P* share no members. Its Venn diagram therefore consists of two overlapping circles (one for *S* and one for *P*), with the overlapping part blocked out (see Fig. 5-1). Empty (unblocked) areas of the diagram are areas about which we have no information; *if an area is empty, we must not assume that the corresponding set has members*. Thus the diagram indicates that there may or may not be objects in *S* and there may or may not be objects in *P*, but there are no objects in both.

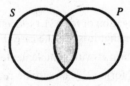

Fig. 5-1

Form *A*, 'All *S* are *P*', says that *S* is a subset of *P*, or (what comes to the same thing) that *S* has no members which are not also in *P*. Thus in its diagram the part of the *S* circle outside the *P* circle is blocked out (see Fig. 5-2).

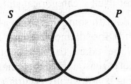

Fig. 5-2

To show that a set or part of a set has at least one member, we put an *X* in the corresponding part of the diagram. Thus form *I*, 'Some *S* are *P*', which says that *S* and *P* share at least one member, is represented as shown in Fig. 5-3. And the form *O*, 'Some *S* are not *P*', which asserts that *S* has at least one member not in *P*, is as shown in Fig. 5-4.

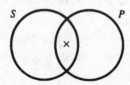

Fig. 5-3

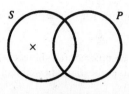

Fig. 5-4

Once again, areas in the diagram of a statement which are not blocked out and do not contain an X are areas about which the statement gives no information. They may contain members or they may not; the statement does not say.

For statements involving complementation it is useful to frame the diagram. The area inside the frame represents the set of all things—loosely speaking, the universe. The members of a set's complement are then represented by the area outside the circle for that set but inside the frame.

SOLVED PROBLEMS

5.5 Draw a Venn diagram for 'All non-S are P'.

Solution

This form asserts that the complement of the set S is a subset of the set P. This means that any part of the complement of S which lies outside P is empty. On the diagram, the complement of S is represented by the area that lies outside the S circle but inside the frame. To show that the part of this area outside the P circle is empty, we blocked it out as in Fig. 5-5. The result correctly diagrams 'All non-S are P'.

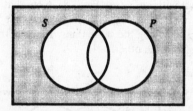

Fig. 5-5

5.6 Draw a Venn diagram for 'Some non-S are not P'.

Solution

This form says that some members of the complement of S are also members of the complement of P. The complement of a set is represented by the area outside its circle but inside the frame. Thus to diagram this form we place an X outside both circles but inside the frame (see Fig. 5-6).

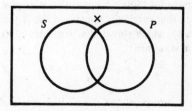

Fig. 5-6

5.7   Draw a Venn diagram for 'No non-*S* is *P*'.

Solution

This is form *E* with a complemented subject term. It asserts that the complement of *S* shares no members with *P*, and so in the diagram we block out the area in which *P* and the complement of *S* overlap (see Fig. 5-7).

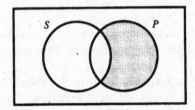

Fig. 5-7

Negation is represented on a Venn diagram by switching blocked areas and X's, i.e., blocking out any area in which there is an *X* and replacing any blocked out area with an *X*.[1]

SOLVED PROBLEMS

5.8   Draw a Venn diagram for '~(All *S* are *P*)'.

Solution

The diagram for 'All *S* are *P*' was given in Fig. 5-2. To diagram its negation, we remove the shading and replace it with an *X* (see Fig. 5-8). Note that the result is the same as the diagram for form *I*, 'Some *S* are not *P*' (Fig. 5-4). This shows that '~(All *S* are *P*)' and 'Some *S* are not *P*' are different ways of saying the same thing. If one is true, the other must be as well; that is, each validly follows from the other.

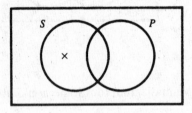

Fig. 5-8

5.9   Draw a Venn diagram for '~(Some *S* are not *P*)'.

Solution

Again, the diagram for 'Some *S* are not *P*' is as given in Fig. 5-4. To diagram its negation, we block out the area that originally contained the *X* (see Fig. 5-9). This is just the diagram for 'All *S* are *P*' (Fig. 5-2), which shows that statements of the form '~(Some *S* are not *P*)' are equivalent to *A*-form statements.

---

[1]This simple way of representing negation is correct only for the modern (Boolean) interpretation of categorical statements; complications arise when it is applied to the traditional (Aristotelian) interpretation, which is not discussed here. (See Section 5.3 for a brief account of the differences between the two interpretations.)

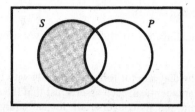

Fig. 5-9

5.10   Draw a Venn diagram for '~(No S are P)'.

Solution

The diagram for 'No S are P' was given in Fig. 5-1. Now to diagram its negation we remove the shading and replace it with an X (see Fig. 5-10). The result is the same as the diagram for 'Some S are P' given in Fig. 5-3, indicating that statements of these two forms say the same thing and hence validly imply one another.

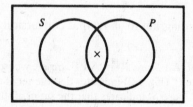

Fig. 5-10

5.3   IMMEDIATE INFERENCES

Inferences from one categorical statement to another are called *immediate inferences*. As some of the problems of the preceding section suggest, we can test the validity of immediate inference forms by using Venn diagrams. The test is very simple; to check the validity of an inference form, diagram its premise. If in doing so you have also diagramed its conclusion, then the form is valid. If not, it is invalid. If two statement forms have exactly the same diagram, then in diagraming either, we automatically diagram the other. They are therefore logically equivalent; i.e., any instance of one validly follows from any instance of the other. This is the case, for example, with '~(All S are P)' and 'Some S are not P', as Problem 5.8 illustrates, and with 'All S are P' and '~(Some S are not P)', as shown in Problem 5.9. Taken together, these two results say that any pair of A- and O-form statements having the same subject and predicate terms are *contradictories*; that is, each validly implies the negation of the other, and the negation of each validly implies the other. Any two I- and E-form statements with the same subject and predicate terms are also contradictories, as the reader can verify with the help of Problem 5.10.

Inferences from one of a pair of contradictories to the negation of the other, or vice versa, are one kind of immediate inference, but there are a variety of others. Two categorical statement forms are said to be *converses* if one results from the other by exchanging subject and predicate terms. Thus 'Some P are S' is the converse of 'Some S are P', 'All P are S' is the converse of 'All S are P', and so on. An inference from a categorical statement to its converse is called *conversion*. Conversion is a valid immediate inference for forms E and I, but it is invalid for forms A and O.[2]

---

[2]Recall, however, that invalid forms generally have some valid instances.

SOLVED PROBLEMS

5.11   Use a Venn diagram to show that conversion is valid for form *E*.

Solution

We compare the diagram for form *E* (Fig. 5-1) with the diagram for its converse (Fig. 5-11) and note that they are identical; each asserts just that the sets *S* and *P* share no members. Hence an *E*-form statement and its converse are logical equivalents; each validly follows from the other.

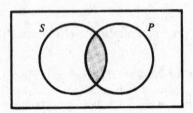

Fig. 5-11

5.12   Use a Venn diagram to show that conversion is not valid for form *A*.

Solution

The converse of 'All *S* are *P*' is 'All *P* are *S*', whose diagram is shown in Fig. 5-12. The diagram for 'All *S* are *P*' (Fig. 5-2) indicates that the set of things which are *S* and not *P* is empty; the diagram for 'All *P* are *S*' indicates that the set of all things which are *P* and not *S* is empty. Clearly, the first set might be empty while the second is not, so that 'All *S* are *P*' is true while 'All *P* are *S*' is false.

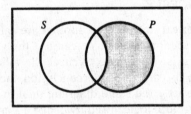

Fig. 5-12

For example, let *S* be the set of salesmen and *P* the set of people. Then the premise 'All salesmen are people' is true (the set of things which are salesmen and not people is empty). But the conclusion 'All people are salesmen' is false (the set of things which are people and not salesmen is not empty!). Hence the form

       All *S* are *P*
∴ All *P* are *S*.

is invalid. Notice that this same example, with '*P*' standing for 'salesmen' and '*S*' for 'people', also suffices to show that the form

       All *P* are *S*.
∴ All *S* are *P*.

is invalid. (This form is just a notational variant of the first.) Thus conversion is an invalid form in both directions.

The reader can similarly verify that conversion is valid for form *I* but not for form *O*.

Two categorical statements are *contrapositives* if one results from the other when we replace its subject term with the complement of its predicate term and its predicate term with the complement of its subject term. Thus, for example, 'All *S* are *P*' and 'All non-*P* are non-*S*' are contrapositives. An inference from one of a pair of contrapositives to the other is called *contraposition*. Contraposition is valid for forms *A* and *O*, but invalid for forms *E* and *I*. We demonstrate its invalidity for form *E* and leave the other cases to the reader.

SOLVED PROBLEM

5.13   Use a Venn diagram to show that contraposition is invalid for form *E*.

Solution

The contrapositive of the *E*-form statement 'No *S* are *P*' is 'No non-*P* are non-*S*', which asserts that the complement of *P* shares no members with the complement of *S*. To diagram this statement, we block out the area on the diagram that represents things which are both non-*S* and non-*P* (Fig. 5-13). 'No *S* are *P*', as its diagram shows (Fig. 5-1), asserts the emptiness of a wholly different subset of the universe. One of these subsets could be empty while the other is not; in diagraming neither premise do we automatically diagram the other; hence neither form implies the other.

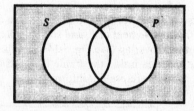

Fig. 5-13

To illustrate, Let *S* be the set of snakes and *P* the set of people. Then the premise 'No snakes are people' is surely true; but its contrapositive, 'No nonpeople are nonsnakes', is just as surely false: bees, for example, are both nonpeople and nonsnakes. Thus we see that the form

No *S* are *P*.
∴ No non-*P* are non-*S*.

is invalid. To see by example that the form

No non-*P* are non-*S*.
∴ No *S* are *P*.

is likewise invalid, suppose that everything that exists is either physical or spiritual and that some things, say people, are both. (This is logically possible, even if it isn't actually so.) Then the premise 'No nonphysical things are nonspiritual things' would be true; but the conclusion 'No spiritual things are physical things' would be false, people being an exception. Hence it is logically possible for this premise to be true while the conclusion is false. Contraposition for *E*-form propositions is thus invalid in both directions.

The four basic categorical forms may be classified by *quality* and *quantity*. There are two qualities, *affirmative* and *negative*, and two quantities, *universal* and *particular*. Each of the four basic forms of categorical statements has a characteristic combination of quality and quantity, as the following table indicates:

		QUALITY	
		Affirmative	Negative
QUANTITY	Universal	A	E
	Particular	I	O

A third type of immediate inference is one in which we change the quality of a categorical statement (while keeping its quantity the same) and replace the predicate term by its complement. Such an inference is called *obversion*; statements obtainable from one another by obversion are *obverses*.

The obverse of 'No S are P', for example, is 'All S are non-P'. The obverse of 'Some S are not P' is itself (changing the quality and canceling out the double complementation gives 'Some S are P' and then replacing the predicate term 'P' by its complement 'not P' returns us to 'Some S are not P'). Obverse categorical statements are always logically equivalent; their Venn diagrams are always identical.

SOLVED PROBLEM

5.14 Use a Venn diagram to show that obversion is valid for form A:

Solution

The obverse of the A-form 'All S are P' is 'No S are non-P'. This asserts that S shares no members with the complement of P, and so in the diagram we block out the area in which S and the complement of P overlap (see Fig. 5-14). But this diagram is equivalent to that for form A (Fig. 5-2), since the area inside the frame and outside the two circles is not blocked out and contains no X, and is therefore not informative.

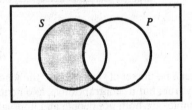

Fig. 5-14

The logic originally developed by Aristotle for categorical statements recognized as valid a number of immediate inferences that are not considered valid in modern logic. This discrepancy can be traced to the presupposition in Aristotelian logic that all subject and predicate terms designate nonempty sets, i.e., sets with at least one member. Modern logic assumes no such thing.

To understand the effects of this presupposition, consider a pair of I- and O-form statements with the same subject and predicate terms, i.e., 'Some S are P' and 'Some S are not P'. Now assume (as Aristotelian logic does) that the terms 'S' and 'P' designate nonempty sets. Then the set S has members; if none of these members is P then some (indeed all of them) must be not-P, and if none is not-P then some (indeed all) must be P. Hence 'Some S are P' and 'Some S are not P' cannot both be false. From the negation of either, we can validly infer the other. Two statements so related were called *subcontraries* in Aristotelian logic, and the relationship of subcontrariety was held to be the basis for valid immediate inferences. However, if we remove the presupposition that the subject term is nonempty, such inferences may become invalid. Suppose, for example, that 'S' is the term 'submarine over a mile long'. (There are, of course, no such things.) Then, no matter what 'P' is, the statements 'Some S are P' and 'Some S are not P' are both false. It is false, for example, both that some submarines over a mile long are pink and that some submarines over a mile long are not pink—simply because there aren't any submarines over a mile long. We cannot validly infer either of these statements from the negation of the other.

The presupposition of nonemptiness limited the applicability of Aristotelian logic. Moreover, once this presupposition was abandoned, logic was greatly simplified. Thus Aristotelian logic was for the most part abandoned, and the extra inferences it recognized (such as those involving subcontraries) were banished from subsequent logical systems.

The acceptance of empty terms by modern logic does, however, create a problem: What is the truth value of *A*-form statements with empty subject terms? For example, is the statement 'All submarines over a mile long are pink' true or false? In a sense it doesn't matter, since there aren't any such submarines; but the demands of generality and completeness require a decision. Modern logic achieves an important gain in simplicity and systematicity by the stipulation that *all A*-form statements with empty subject terms are true. After all, if the set corresponding to the subject term is empty, then it is certainly included in the set corresponding to the predicate term, since the empty set is included in *every* set. Thus the statement 'All submarines over a mile long are pink' is true.

Note that, by the same pattern, the statement 'All submarines over a mile long are not pink' is also true. This appears contradictory, but the appearance is illusory. 'Not' in this statement expresses complementation, not negation. Where *S* is empty, 'All *S* are *P*' says merely that the empty set is a subset of the set *P* and 'All *S* are not *P*' says merely that the empty set is a subset of the complement of *P*. This is not a contradiction.

5.4 CATEGORICAL SYLLOGISMS

Immediate inferences are one-premise arguments whose premise and conclusion are both categorical statements. *Categorical syllogisms* are two-premise arguments consisting entirely of categorical statements. To qualify as a categorical syllogism, such an argument must contain exactly three class terms: the subject and predicate term of the conclusion (these are, respectively, the *minor* and *major* terms of the syllogism) and a third term (the *middle* term) which occurs in both premises. In addition, the major and minor terms must each occur once in a premise.

We owe the concept of a categorical syllogism to Aristotle, and this class of arguments is singled out for study primarily in deference to tradition. The argument with which this chapter began

> Some four-legged creatures are gnus.
> All gnus are herbivores.
> ∴ Some four-legged creatures are herbivores.

is a categorical syllogism. Its major term is 'herbivores', its minor term is 'four-legged creatures', and its middle term is 'gnus'.

Venn diagrams provide a quick and effective test for the validity of the forms of categorical syllogisms. To diagram a syllogistic form, we draw three overlapping circles to represent the three terms in the premises. The circles are then labeled (in any order) with letters standing for these three terms. For the form of the argument above, we will use the letters '*F*', '*G*', and '*H*' for 'four-legged creatures', 'gnus', and 'herbivores', respectively (Fig. 5-15). It is crucial that the drawing include a three-cornered middle region, which represents the things common to all three sets *F*, *G*, and *H*.

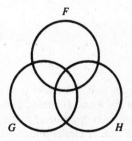

Fig. 5-15

We diagram each of the premises in turn. The resulting diagram may then be used to test the form for validity in just the way that we tested immediate inferences; if in diagraming the premises we have thereby diagramed the conclusion as well, then the form is valid; if not, then it is invalid.

SOLVED PROBLEMS

5.15 Use a Venn diagram to determine whether the argument above is valid.

Solution

The first premise asserts that the set of four-legged creatures shares at least one member with the set of gnus, so we need to put an X in the region of overlap between circles F and G. This region, however, is divided into two parts by the H circle. Which of these two parts should contain the X? We know, of course, that gnus are herbivores, so it might seem that we should put the X in the three-cornered middle region. But this would be a mistake. We are diagraming only the information contained in the first premise; to bring in additional knowledge is "cheating." The first premise does not indicate whether four-legged gnus are herbivores or not. We show this by putting the X on the border between herbivores and nonherbivores (Fig. 5-16). The second premise says that the set of gnus is a subset of the set of herbivores. To diagram it, we block out the region representing gnus which are not herbivores (Fig. 5-17). Notice that in diagraming the second premise we eliminate the possibility that the X represents a nonherbivore. Thus we now see that the X (which represents at least one four-legged gnu) must be contained in the three-cornered middle region. The premises require, in other words, that at least one four-legged gnu be a herbivore. But that is just what the conclusion says. Hence in diagraming just the premises, we have diagramed the conclusion. This shows that if the premises are true, then the conclusion must be true as well, i.e., that the argument is valid. Indeed, it shows that any argument of this form is valid, since it makes no difference to the diagram whether 'F', 'G', and 'H' stand for 'four-legged creatures', 'gnus', and 'herbivores' or some other class terms.

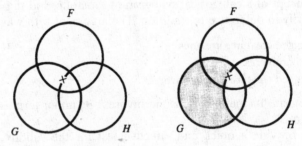

Fig. 5-16   **Fig. 5-17**

5.16 Construct a Venn diagram to test the following form for validity:

No F are G.
All G are H.
∴ No F are H.

Solution

The first premise asserts that sets F and G share no members. We therefore block out the lens-shaped area between the F and G circles. The second premise says that G is a subset of H, and so we block out the crescent opposite H in the G circle (Fig. 5-18). This still leaves a region of overlap between the F and H circles about which we have no information. It is, in other words, consistent with the premises that there are F's which are H but not G. Now if there are

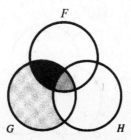

Fig. 5-18

such F's, then the conclusion 'No F are H' is false. Thus this conclusion may be false while the premises are true; the form is invalid.

5.17   Construct a Venn diagram to test the following form for validity:

> All F are G.
> No G are H.
> ∴ No F are H.

Solution

To diagram the first premise, we block out the crescent part of the F circle opposite G. To diagram the second, we block out the lens-shaped area shared by the G and H circles. This automatically blocks out the area shared by the F and H circles, showing that the conclusion must be true, given the premises (Fig. 5-19). The form is therefore valid.

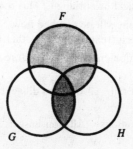

Fig. 5-19

5.18   Construct a Venn diagram to test the following form for validity:

> Some F are G.
> Some G are H.
> ∴ Some F are H.

Solution

The premises are diagramed by placing X's in the crescent areas shared by the F and G and by the G and H circles, respectively. These X's must be placed on the lines within the crescents, since neither premise tells us to which part of the relevant crescent the X belongs. Thus the diagram is as shown in Fig. 5-20. Now it is perfectly consistent with this diagram that both of these X's fall outside the three-cornered middle region, in which case F and H need not have any members in common. But if F and H need not have any members in common, then the conclusion 'Some F are H' can be false. Hence the premises do not require the truth of the conclusion. The form is invalid.

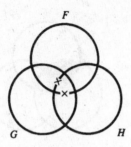

Fig. 5-20

Supplementary Problems

Some of the following arguments can be formalized as immediate inferences involving a pair of categorical statements or as categorical syllogisms; others cannot. Formalize those which can and test the resulting forms for validity with Venn diagrams.

(1)   No one has conquered the world. Therefore, it is not true that someone has conquered the world.

(2)   No one has conquered the world. Therefore, not everyone has conquered the world.

(3)   No one has conquered the world. Therefore, everyone is not a world conqueror.

(4)   No one has conquered the world. Therefore, someone has not conquered the world.

(5)   Every self-contradictory truth is surprising. Therefore, there are self-contradictory truths which are surprising.

(6)   All square numbers are nonprime. So all prime numbers are nonsquare.

(7)   Miracles are impossible. Therefore it is not the case that some miracles are possible.

(8)   Either everything is physical or nothing is. This rock is physical. Hence everything is physical.

(9)   No arms control agreement will be reached now. So no arms control agreement will be reached at any other time. (*Hint:* Treat 'reached at any other time' as the complement of 'reached now'.)

(10) Everyone who is incompetent fails. Therefore it is not true that some who succeed are incompetent.

(11) All who are incompetent fail. All who are diligent succeed. Therefore, none who are incompetent is diligent.

(12) Everything he says is nonsense. All nonsense is contemptible. Therefore, everything he says is contemptible.

(13) If Jean is sick, she won't come to work. If she doesn't come to work, none of us will have anything to do. So if Jean is sick, none of us will have anything to do.

(14) Some people are nonsmokers. Some are nondrinkers. Therefore, some nonsmokers are nondrinkers.

(15) Anything suitable for the job must be able to withstand that strain. No metal is strong enough to withstand that strain. Thus nothing suitable for the job is a metal.

(16) None of the players was injured. Some of the players missed practice. So no one who missed practice was injured.

(17) All good things must pass. No dictatorship is a good thing. Consequently, some dictatorships do not pass.

(18) Every electron has a negative charge. No positron has a negative charge. Therefore, no positron is an electron.

(19) Every electron has a negative charge. No positron has a negative charge. Therefore some positrons are not electrons.

(20) Each fatty meal he eats is another nail in his coffin. And one of the nails in his coffin is going to kill him. It follows that one of those fatty meals he eats is going to kill him.

Answers to Selected Supplementary Problems

(1)   No *P* are *C*.

       ∴ ~(Some *P* are *C*.)

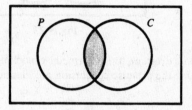

Fig. 5-21

('*P*' stands for 'people' and '*C*' for 'things that have conquered the world'.) We have diagramed the premise (Fig. 5-21), which asserts that the sets *P* and *C* share no members. But in doing so we have diagramed the conclusion as well. For the diagram of 'Some *P* are *C*' is just an *X* in the lens-shaped area; and to diagram its negation, '~(Some *P* are *C*)', we remove this *X* and block out the area which contained it. The result is just the diagram for 'No *P* are *C*'. Hence the premise and conclusion are logically equivalent, and the inference form is valid.

(4)   No *P* are *C*.

       ∴ Some *P* are not *C*.

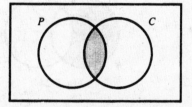

Fig. 5-22

(Same symbols as in Problem 1.) We have diagramed the premise (Fig. 5-22), but in doing so we have not diagramed the conclusion (compare Fig. 5-4). Hence this form, somewhat surprisingly, is invalid. This is because the premise does not guarantee that there are any people: the crescent area inside the *P* circle is empty, indicating that the premise provides no information about this region. Note that we have treated the 'not' in the conclusion as expressing complementation; if one treats it as expressing negation then the conclusion becomes

       ~(Some *P* are *C*.)

which under the current interpretation means "it is not the case that some people have conquered the world." This form is the same as in Problem 1 and is therefore valid.

(7)   All *M* are non-*P*.

       ∴ ~(Some *M* are *P*.)

(We use '*M*' for 'miracles' and '*P*' for 'possible', or, in standard form, 'possible things'.) The premise asserts that the set *M* is a subset of the complement of the set *P*, and hence that the lens-shaped area between *M* and *P* is empty. The conclusion has the same form as the conclusion

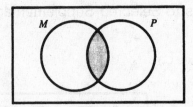

Fig. 5-23

of Problem 1; hence its diagram, too, is formed by blocking out this lens-shaped area (Fig. 5-23). Since the diagrams for the premise conclusion are identical, the inference is valid.

(8)   The statements of this argument (which, incidentally, is valid) are not categorical.

(12)   All *S* are *N*.
            All *N* are *C*.
       ∴ All *S* are *C*.

(We use '*S*' for 'things said by him', '*N*' for 'nonsensical things', and '*C*' for 'contemptible things'.) This is a valid categorical syllogism. (See Fig. 5-24.)

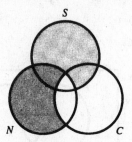

Fig. 5-24

(13)   The statements of this argument are not categorical, but the argument is valid, since it is an instance of hypothetical syllogism (Section 4.4).

(16)   No *P* were *I*.
            Some *P* were *M*.
       ∴ No *M* were *I*.

(We use '*P*' for 'players', '*I*' for 'things (or persons) that were injured', and '*M*' for 'things (or persons) that missed practice'.) This is an invalid categorical syllogism. (See Fig. 5-25.)

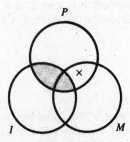

Fig. 5-25

(20)   All *F* are *N*.
              Some *N* are *K*.
         ∴ Some *F* are *K*.

(We use '*F*' for 'fatty meals he eats', '*N*' for 'nails in his coffin', and '*K*' for 'things that are going to kill him'.) On our formalization, this is an invalid categorical syllogism (Fig. 5-26). However, the formalization introduces some distortion, since the term 'one' in the conclusion of the argument may well mean "exactly one," whereas the term 'some' with which we have replaced it means (in its logical sense) "at least one."

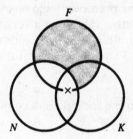

Fig. 5-26

Chapter 6

Predicate Logic

6.1 QUANTIFIERS AND VARIABLES

Chapter 5 introduced the logical concept of a quantifier ('all', 'some', 'no') in the narrow context of the logic of categorical statements. In this chapter we combine the quantifiers with the operators of propositional logic (the truth-functional connectives) to produce a more encompassing logical apparatus. This apparatus—known as predicate logic or quantification theory—is very powerful and proves adequate to testing the deductive validity of a very wide range of argument forms.

As a first step, we note that reformulation of categorical statements reveals the presence of some of the logical operators of Chapters 3 and 4. Consider, for example, the A-form statement

       All S are P.

Using the letter 'x' as a variable standing for individual objects, we may express the same statement as follows:

       For all x, if x is S then x is P.

This reformulation has the same truth conditions as the original formulation: it asserts that any object that is contained in the set S is also contained in the set P, which is just another way of saying that S is included in P. However, unlike the original formulation, the new formulation contains an occurrence of the conditional 'if . . . then'. Likewise, the E-form statement

       No S are P.

can be expressed as follows:

       For all x, if x is S then it is not the case that x is P.

This asserts that any object that is contained in the set S is not contained in the set P, i.e., that S and P share no members, which is exactly the condition expressed by the E-form statement. Again, the new formulation contains a conditional; in addition it contains the negation connective 'it is not the case that'.

Let us adopt the symbol '∀' to mean "for all," "for any," and the like. This symbol is called a *universal quantifier*. Moreover, instead of writing 'x is S', let us simply write 'Sx'—and treat 'x is P' similarly. Finally, let us use the symbolic notation for the truth-functional operators 'if . . . then', 'it is not the case that', etc., familiar from Chapter 3. Then our reformulations of the A-form statement 'All S are P' and the E-form statement 'No S are P' can be symbolized as follows, respectively:

       $\forall x(Sx \rightarrow Px)$
       $\forall x(Sx \rightarrow \sim Px)$

Note that there is no need to have a special symbol for the negative quantifier 'no', since we can express it with the help of '∀' together with the negation connective '∼'. To express I and O statements, however, it is convenient to have a second kind of quantifier. The I proposition

       Some S are P.

may be expressed as

       For some x, x is S and x is P.

which contains a conjunction. We adopt the symbol '∃' to mean "for some," or, more precisely, "for at least one." (As in the previous chapter, we shall understand "some" so as to include the limit cases "exactly one" and "all". See Section 5.1.) We can then rewrite the *I* proposition as

$\exists x(Sx \ \& \ Px)$

An equivalent way of expressing the meaning of '∃' in English is "there exists . . . such that." Hence the formula above can also be read as

There exists an *x* such that *x* is *S* and *x* is *P*.

Accordingly, '∃' is called the *existential quantifier*.
    The *O* proposition

Some *S* are not *P*.

in this notation is

$\exists x(Sx \ \& \ {\sim}Px)$

which means

For some *x*, *x* is *S* and it is not the case that *x* is *P*.

or, equivalently,

There exists an *x* such that *x* is *S* and *x* is not *P*.

    Notice that it is wrong to render the *I* proposition 'Some *S* are *P*' as

$\exists x(Sx \rightarrow Px)$

which contains a conditional rather than a conjunction. This formula means "there exists an *x* such that, *if x* is *S*, then *x* is *P*." It can therefore be true even if nothing is *S*. It is true, for instance, that for at least one *x* (let *x* be any person you like), *if x* has seven heads, then *x* is peculiar. But it is false that some seven-headed people are peculiar, since there aren't any seven-headed people; thus these two statements are not equivalent. It is similarly incorrect to use a conditional in the symbolization of an *O* proposition.

    A and *E* propositions, on the other hand, must always be formalized by conditionals, rather than conjunctions. It is wrong, for example, to formalize the *A* proposition 'All *S* is *P*' as:

$\forall x(Sx \ \& \ Px)$

This means, "For all *x*, *x* is *both S* and *P*." But clearly the statement 'All sharks are predators' is not the equivalent of 'For all *x*, *x* is *both* a shark and a predator', which means that everything is a predatory shark.

SOLVED PROBLEM

6.1 Interpreting the letters '*F*' and '*G*' as the predicates 'is a frog' and 'is green', respectively, formalize each of the following sentences:

(*a*) Frogs are green.

(*b*) There is at least one green frog.

(*c*) Some frogs are not green.

(*d*) There aren't any green frogs.

(*e*) No frogs are green.

(*f*) Frogs are not green.

(*g*) Not everything that is a frog is green.

Solution

(a) $\forall x(Fx \rightarrow Gx)$. This formalization treats the plural noun 'frogs' as expressing a universal quantification, i.e., as meaning 'all frogs'. This is customary in predicate logic, and we shall follow this practice here. However, in some contexts it might be more appropriate to regard a plural as expressing a form of generality that allows for some exceptions. (See Problem 5.4(i).)

(b) $\exists x(Fx \& Gx)$

(c) $\exists x(Fx \& \sim Gx)$. This formalization treats 'not' as expressing negation. If one read it as expressing complementation instead, one would need a separate predicate letter, say 'H', to represent the predicate 'non-green'; the sentence could then be formalized as an ordinary I-form statement, '$\exists x(Fx \& Hx)$.

(d) $\sim \exists x(Fx \& Gx)$

(e) $\forall x(Fx \rightarrow \sim Gx)$. This statement says the same thing as (d), hence it can also be formalized as '$\sim \exists x(Fx \& Gx)$'.

(f)   This expresses the same proposition as (e), so it can be formalized as $\forall x(Fx \rightarrow \sim Gx)$ or, equivalently, $\sim \exists x(Fx \& Gx)$. As in (c), 'not' is treated as expressing negation. If one read it as expressing complementation (see Section 5.1), the sentence could then be formalized as an ordinary A-form statement, '$\forall x(Fx \rightarrow Hx)$' (where '$H$' is interpreted as 'non-green').

(g) $\sim \forall x(Fx \rightarrow Gx)$

The new notation reveals previously unnoticed structure in categorical propositions. But its chief advantage is that it allows us to combine the concepts of propositional and categorical logic to express a rich variety of logical structures.

SOLVED PROBLEM

6.2 Interpreting the letter 'R' as the sentence 'It is raining' and the letters 'F', 'G', 'H', and 'I' as the predicates 'is a frog', 'is green', 'is hopping', and 'is iridescent', respectively, formalize each of the following sentences:

(a)   Everything is a frog.

(b)   Something is a frog.

(c)   Not everything is a frog.

(d)   Nothing is a frog.

(e)   Green frogs exist.

(f)   Everything is either green or iridescent.

(g)   Everything is a green frog.

(h)   It is raining and some frogs are hopping.

(i)   If it is raining, then all frogs are hopping.

(j)   Some things are green and some are not.

(k)   Some things are both green and iridescent.

(l)   Either everything is a frog or nothing is a frog.

(m)   Everything is either a frog or not a frog.

(n)   All frogs are frogs.

(o)   Only frogs are green.

(p)   Iridescent frogs do not exist.

(q)   All green frogs are hopping.

(*r*)   Some green frogs are not hopping.

(*s*)   It is not true that some green frogs are hopping.

(*t*)   If nothing is green, then green frogs do not exist.

(*u*)   Green frogs hop if and only if it isn't raining.

Solution

(*a*) $\forall x Fx$

(*b*) $\exists x Fx$

(*c*) $\sim\forall x Fx$

(*d*) $\forall x \sim Fx$ (This can also be expressed as '$\sim\exists x Fx$'.)

(*e*) $\exists x(Gx \& Fx)$

(*f*) $\forall x(Gx \vee Ix)$

(*g*) $\forall x(Gx \& Fx)$

(*h*) $R \& \exists x(Fx \& Hx)$

(*i*) $R \rightarrow \forall x(Fx \rightarrow Hx)$

(*j*) $\exists x Gx \& \exists x \sim Gx$

(*k*) $\exists x(Gx \& Ix)$

(*l*) $\forall x Fx \vee \forall x \sim Fx$

(*m*) $\forall x(Fx \vee \sim Fx)$

(*n*) $\forall x(Fx \rightarrow Fx)$

(*o*) $\forall x(Gx \rightarrow Fx)$

(*p*) $\sim\exists x(Ix \& Fx)$

(*q*) $\forall x((Gx \& Fx) \rightarrow Hx)$

(*r*) $\exists x((Gx \& Fx) \& \sim Hx)$

(*s*) $\sim\exists x((Gx \& Fx) \& Hx)$

(*t*) $\forall x \sim Gx \rightarrow \sim\exists x(Gx \& Fx)$

(*u*) $\forall x((Gx \& Fx) \rightarrow (Hx \leftrightarrow \sim R))$

One should be very careful in symbolizing sentences with quantifiers, as there is no straightforward way to match English phrases with the symbols '∀' and '∃'. For instance, typically a phrase such as 'anything' stands for the universal quantifier and a phrase such as 'something' stands for the existential quantifier, but the overall structure of the sentence may indicate a different interpretation. Thus, in 'If something is cheap, then it is not of good quality', the word 'something' suggests the presence of an existential quantifier. But on closer inspection it is clear that the sentence expresses a universal quantification: it is a general statement to the effect that cheap things are not of good quality. Also, in English often the quantifier is not expressed by an explicit quantifier phrase, but by an adverb (such as 'always' or 'sometimes') or even by a connective. For instance, the statement 'Only expensive things are nice' means that the nice things can be found only among the expensive things, and is therefore a categorical statement of the form 'All *N* are *E*'.

SOLVED PROBLEM

6.3 Interpreting the letters '*F*' and '*G*' as the predicates 'is a frog' and 'is green', respectively, formalize each of the following sentences:

(*a*)   If something is a frog, then it is green.

(*b*)   If anything at all is a frog, then something is green.

- (c) Anything that is a frog is green.
- (d) If anything is green, then frogs are green.
- (e) If everything is green, then frogs are green.
- (f) Invariably, frogs are green.
- (g) Occasionally, frogs are green.
- (h) A frog is green.
- (i) A frog is always green.
- (j) Only frogs are green.

Solution

- (a) $\forall x(Fx \rightarrow Gx)$
- (b) $\exists xFx \rightarrow \exists xGx$. The antecedent features the word 'anything'; this is typically a sign of universal quantification, but in the present context it clearly expresses an existential condition. (The condition could be reformulated as 'If there are any frogs at all'.)
- (c) $\forall x(Fx \rightarrow Gx)$
- (d) $\exists xGx \rightarrow \forall x(Fx \rightarrow Gx)$. This case is similar to (b), except that the consequent is a general claim about the color of frogs and is therefore formalized as an A-form statement.
- (e) $\forall xGx \rightarrow \forall x(Fx \rightarrow Gx)$
- (f) $\forall x(Fx \rightarrow Gx)$. Here the adverb 'Invariably' reinforces the reading of the plural 'frogs' as expressing universal quantification. (Compare Problem 6.1(a).)
- (g) $\exists x(Fx \,\&\, Gx)$. In this case, the adverb 'Occasionally' weakens the reading of the plural, leaving room for frogs that are not green. In fact, the adverb suggests that only a few frogs are green, a connotation that is lost in the formalization. To capture this connotation, the sentence can be rendered as a conjunction $\exists x(Fx \,\&\, Gx) \,\&\, \exists x(Fx \,\&\, {\sim}Gx)$.
- (h) This sentence is ambiguous. It may be interpreted as a general statement about frogs (as in (a) and (c)), or as an existential statement of the form $\exists x(Fx \,\&\, Gx)$.
- (i) $\forall x(Fx \rightarrow Gx)$. In this case the ambiguity is resolved by the adverb 'always'.
- (j) $\forall x(Gx \rightarrow Fx)$

6.2 PREDICATES AND NAMES

Not all statements contain quantifiers. There are, for example, simple subject-predicate statements which attribute a property to an individual person or thing. We shall interpret lowercase letters from the beginning and middle of the alphabet as proper names of individuals, and we shall adopt the convention (standard in logic but contrary to English grammar) or writing the subject after the predicate. Thus the sentence

> Jones is a thief.

may be formalized as

> Tj

where 'j' is interpreted as the proper name 'Jones' and 'T' as the predicate 'is a thief'.

So far we have considered only predicates that are class terms, i.e., nouns, adjectives, or verbs which designate classes of objects. (See Section 5.1.) The result of combining such a predicate with a proper name is always a sentence, and the sentence will be true or false depending on whether or not the object designated by the name is a member of the class designated by the predicate. (For instance, the sentence 'Jones is a thief' will be true if Jones is indeed a thief, and false otherwise.) There are, however, many predicates that must be combined with two or more proper names to make a sentence. In English, for example, this is true of such transitive verbs as 'hits', 'loves', or 'defies', which take both a subject and an object. These are usually written in logical notation in the order predicate-subject-object. Thus the statement

　　　　Bob loves Cathy.

is formalized as

　　　　　Lbc

and the sentence

　　　　Cathy loves Bob.

is formalized as:

　　　　　Lcb

Transitive verbs are just one subclass of the broader class of *relational predicates*, predicates which designate *relations* between two or more objects (rather than *classes* of objects) and which combine two or more proper names to make a sentence. Here are some other examples of relational predicates that take two names to make a sentence: 'is next to', 'thumbs his nose at', 'is taller than', 'is less than', 'is a subset of'. The expressions 'gave ... to' and 'is between ... and' are examples of predicates that take three names. And 'is closer to ... than ... is to' takes four.

A predicate which takes only one name (a class term) is called a *nonrelational* or *one-place* predicate. A relational predicate which takes two names is called a *two-place* predicate; one which takes three names a *three-place* predicate; and so on. When sentences containing such predicates are formalized, the letters representing names are written after the predicate letter in the order in which they occur in English, unless some other order is specified. Thus the statement

　　　　Cathy gave Fido to Bob.

is formalized as

　　　　　Gcfb

SOLVED PROBLEM

6.4 Formalize the following statements, interpreting the letters '*a*', '*b*', and '*c*' as the proper names 'Alex', 'Bob', and 'Cathy'; '*M*' and '*N*' as the one-place predicates 'is a mechanic' and 'is a nurse'; '*L*' and '*T*' as the two-place predicates 'likes' and 'is taller than'; and '*I*' as the three-place predicate 'introduced ... to'.

(*a*)　Cathy is a mechanic.

(*b*)　Bob is a mechanic.

(*c*)　Cathy and Bob are mechanics.

(*d*)　Either Cathy or Bob is a mechanic.

(*e*)　Cathy is either a mechanic or a nurse (or both).

(*f*)　If Cathy is a mechanic, then she isn't a nurse.

(*g*)　Cathy is taller than Bob.

(*h*)　Bob likes Cathy.

(*i*)　Bob likes himself.

(*j*)　Cathy likes either Bob or Alex.

(*k*)　Alex introduced Cathy to Bob.

(*l*)　Cathy introduced herself to Bob but not to Alex.

Solution

(a) Mc

(b) Mb

(c) $Mc \,\&\, Mb$

(d) $Mc \lor Mb$

(e) $Mc \lor Nc$

(f) $Mc \rightarrow {\sim}Nc$

(g) Tcb

(h) Lbc

(i) Lbb

(j) $Lcb \lor Lca$

(k) $Iacb$

(l) $Iccb \,\&\, {\sim}Icca$

We have seen that some subject-predicate or relational statements involve quantifiers (like 'Frogs are green') and others do not ('Jones is a thief'). In English there are also mixed statements that involve both quantifiers and proper names. For example, the sentence

> Jones like everything.

is obtained by combining the predicate 'likes' with the name 'Jones' and the quantifier 'everything'. However, in formalizing this sentence we cannot treat the quantifier as a name. (There is no object corresponding to 'everything'.) Rather, the quantifier acts as a logical operator: it indicates that every individual object is such that Jones likes it. This becomes clearer if we reformulate the sentence as follows:

> For anything x, John likes x.

Accordingly, the sentence is formalized as

> $\forall x Ljx$.

Sentences involving more than one quantifier are treated in a similar fashion. For instance, the sentence

> Something likes everything.

involves two quantifiers, as is clear from the following reformulation:

> There exists an x such that, for all y, x likes y.

In our notation, this becomes

> $\exists x \forall y Lxy$.

SOLVED PROBLEM

6.5 Formalize the following statements using the same interpretation as in Problem 6.4.

(a)   Bob likes nothing.

(b)   Nothing likes Bob.

(c)   Something likes itself.

(*d*)   There is something which Cathy does not like.

(*e*)   Cathy likes something which Bob likes.

(*f*)   There is something which both Bob and Cathy like.

(*g*)   There is something which Bob likes and something which Cathy likes.

(*h*)   If Bob likes himself, then he likes something.

(*i*)   If Bob doesn't like himself, then he likes nothing.

(*j*)   If Bob likes something, then he likes everything.

(*k*)   Everything likes everything.

(*l*)   There is some one thing which is liked by everything.

(*m*)   Everything likes at least one thing.

Solution

(*a*) $\forall x \sim Lbx$

(*b*) $\forall x \sim Lxb$

(*c*) $\exists x Lxx$

(*d*) $\exists x \sim Lcx$

(*e*) $\exists x (Lcx \ \& \ Lbx)$

(*f*) $\exists x (Lbx \ \& \ Lcx)$

(*g*) $\exists x Lbx \ \& \ \exists x Lcx$

(*h*) $Lbb \rightarrow \exists x Lbx$

(*i*) $\sim Lbb \rightarrow \forall x \sim Lbx$

(*j*) $\exists x Lbx \rightarrow \forall x Lbx$

(*k*) $\forall x \forall y Lxy$

(*l*) $\exists x \forall y Lyx$

(*m*) $\forall x \exists y Lxy$

   There are also English sentences that are obtained by combining a predicate with one or more common nouns, as in

   Jones likes a mechanic.

Common nouns are class terms, so their formalization requires a quantifier. In our example this becomes more perspicuous if we reformulate the sentence as

   There exists an *x* such that *x* is a mechanic and Jones likes *x*.

which may be formalized as follows:

   $\exists x (Mx \ \& \ Ljx)$.

   If the sentence involves more than one common noun, then its formalization may involve a corresponding number of variables and quantifiers. For example, the sentence

   A nurse likes a mechanic.

means that there exist a nurse *x* and a mechanic *y* such that *x* likes *y*. In our notation, this may be expressed as

   $\exists x \exists y ((Nx \ \& \ My) \ \& \ Lxy)$.

SOLVED PROBLEM

6.6 Formalize the following statements using the same interpretation as in Problem 6.4.

(*a*)   A mechanic likes Bob.

(*b*)   A mechanic likes herself.

(*c*)   Every mechanic likes Bob.

(*d*)   Bob likes a nurse.

(*e*)   Some mechanic likes every nurse.

(*f*)   There is a mechanic who is liked by every nurse.

(*g*)   Bob introduced a mechanic to Cathy.

(*h*)   A mechanic introduced Bob to Alex.

(*i*)   A mechanic introduced herself to Bob and Alex.

(*j*)   Cathy introduced a mechanic and a nurse to Bob.

(*k*)   Cathy introduced a mechanic to a nurse.

(*l*)   A mechanic introduced a nurse to Cathy.

(*m*)   Some mechanic introduced a nurse to a mechanic.

Solution

(*a*) $\exists x(Mx \mathbin{\&} Lxb)$

(*b*) $\exists x(Mx \mathbin{\&} Lxx)$

(*c*) $\forall x(Mx \rightarrow Lxb)$

(*d*) $\exists x(Nx \mathbin{\&} Lbx)$

(*e*) $\exists x(Mx \mathbin{\&} \forall y(Ny \rightarrow Lxy))$

(*f*) $\exists x(Mx \mathbin{\&} \forall y(Ny \rightarrow Lyx))$

(*g*) $\exists x(Mx \mathbin{\&} Ibxc)$

(*h*) $\exists x(Mx \mathbin{\&} Ixba)$

(*i*) $\exists x(Mx \mathbin{\&} (Ixxb \mathbin{\&} Ixxa))$

(*j*) $\exists x(Mx \mathbin{\&} Icxb) \mathbin{\&} \exists x(Nx \mathbin{\&} Icxb)$

(*k*) $\exists x\exists y((Mx \mathbin{\&} Ny) \mathbin{\&} Icxy)$

(*l*) $\exists x\exists y((Mx \mathbin{\&} Ny) \mathbin{\&} Ixyc)$

(*m*) $\exists x\exists y\exists z((Mx \mathbin{\&} Ny) \mathbin{\&} (Mz \mathbin{\&} Ixyz))$

You should keep the following points in mind when formalizing statements:

(1)   *Different variables do not necessarily designate different objects.* Thus, for example, the formula '$\forall x\forall yLxy$' (Problem 6.5(*k*)) asserts that for *any x* and *any* (not necessarily distinct) *y*, *x* likes *y*. It asserts, in other words, not only that every thing likes every *other* thing, but also that everything likes itself.

(2)   *Choice of variables makes no difference to meaning.* Thus Problem 6.5(*c*), for example, is symbolized just as correctly by '$\exists yLyy$' or '$\exists zLzz$' as by '$\exists xLxx$'. However, when two or more quantifiers govern overlapping parts of the same formula, as in Problem 6.5(*m*), a different variable must be used with each. This formula could be written equivalently as '$\forall y\exists xLyx$' or '$\forall x\exists zLxz$', for example, but not as '$\forall x\exists xLxx$'. In this last case we cannot tell which of the two

final occurrences of 'x' is governed by which quantifier. We shall regard such formulas as ill formed, i.e., ungrammatical.[1]

(3) *The same variables used with two different quantifiers does not necessarily designate the same object in each case.* Thus '$\exists x Lbx$ & $\exists x Lcx$' is a correct formalization of 'There is something which Bob likes and something which Cathy likes' (Problem 6.5(*g*)), a statement which neither affirms nor denies that each likes the same thing. But some people find it more natural to write '$\exists x Lbx$ & $\exists y Lcy$', which is an equivalent and equally correct formalization of the same statement. It is legitimate to use the same variable with each quantifier in formalizing this statement because the quantifiers govern nonoverlapping parts of the formula (each applies only to the conjunct of which it is a part).

(4) *Many, if not most, English sentences which mix universal and existential quantifiers are ambiguous.* Thus, the seemingly simple sentence 'Something is liked by everything' can mean either 'There is some one thing which is liked by everything' (Problem 6.5(*l*)) or 'Everything likes at least one thing (not necessarily the same thing or things in each case)' (Problem 6.5(*m*)). These more awkward but less ambiguous sentences are not equivalent. Indeed, suppose there is one thing that likes everything while there are also things that don't like anything; then the first sentence is true whereas the second is false. This difference in meaning is revealed formally by the difference in order of the quantifiers. The formalizations of these sentences are wholly unambiguous.

(5) *The order of consecutive quantifiers affects meaning only when universal and existential quantifiers are mixed.* Consecutive occurrences of universal quantifiers can be permuted without change of meaning, as can consecutive occurrences of existential quantifiers. Thus, whereas '$\exists x \forall y Lyx$' (Problem 6.5(*l*)) and '$\forall y \exists x Lyx$' (a variant of Problem 6.5(*m*)) have different meanings, '$\exists x \exists y Lxy$' and '$\exists y \exists x Lxy$' both mean simply and unambiguously "something likes something."

(6) *Nested quantifiers may combine with the truth-functional operators in many equivalent ways.* For example, the sentence 'Cathy introduced a mechanic to a nurse' (Problem 6.6(*k*)) can be rendered as '$\exists x \exists y((Mx \& Ny) \& Icxy)$' but also as '$\exists x(Mx \& \exists y(Ny \& Icxy))$'. This and similar equivalences may be verified intuitively. However, a more precise account of the syntax and semantics of the symbolic language is in order before we can deal with logical equivalences in general.

6.3 FORMATION RULES

We now define more precisely the formal language exemplified in the previous sections. As with the language of propositional logic, we divide the vocabulary of this language into two parts, the *logical symbols* (whose interpretation or function remains fixed in all contexts) and the *nonlogical symbols* (whose interpretation varies from problem to problem):

Logical Symbols

Logical operators: '$\sim$', '&', '$\lor$', '$\rightarrow$', '$\leftrightarrow$'

Quantifiers:[2] '$\forall$', '$\exists$'

Variables: lowercase letters from 'u' through 'z'

Brackets: '(', ')'

---

[1] In some versions of predicate logic, such formulas are counted as wffs, under the convention that if a variable occurs within the scope of two or more quantifiers on that variable the quantifier nearest the variable is the one that governs it.

[2] Quantifier notation varies. Some authors omit the symbol for the universal quantifier and write 'for all x' as '(x)'. The existential quantifier is then written as '$(\exists x)$'. Sometimes both quantifier symbols and brackets are used. Occasionally the symbol '$\land$' is used instead of '$\forall$' and '$\lor$' instead of '$\exists$'.

Nonlogical Symbols

Name letters: lowercase letters from '*a*' through '*t*'

Predicate letters: uppercase letters

To ensure that we have enough nonlogical symbols for any task, we permit the addition of numerical subscripts to the symbols we already have. Thus 'a_3' counts as a name letter and 'P_{147}' as a predicate letter. Subscripted nonlogical symbols are seldom needed, however.

We define a *formula* in our language as any finite sequence of elements of the vocabulary (logical or nonlogical symbols).

An *atomic formula* is a predicate letter followed by zero or more name letters. Atomic formulas consisting of a single predicate letter followed by no name letters are just the sentence letters (atomic formulas) of the propositional calculus. They are used to symbolize whole sentences when there is no need to represent the internal structure of these sentences. If the number of name letters following a predicate letter in an atomic formula is *n*, then that letter represents an *n*-place predicate. (Sentence letters may be thought of as zero-place predicates.)

The concept of a *well-formed formula* (wff) of the language of predicate logic is defined by the following formation rules (we use Greek letters to stand for expressions in our language, as indicated):

(1)   Any atomic formula is a wff.

(2)   If ϕ is a wff, so is $\sim\!\phi$.

(3)   If ϕ and ψ are wffs, so are $(\phi\ \&\ \psi)$, $(\phi \lor \psi)$, $(\phi \to \psi)$, and $(\phi \leftrightarrow \psi)$.

(4)   If ϕ is a wff containing a name letter α, then any formula of the form $\forall\beta\phi^\beta_\alpha$ or $\exists\beta\phi^\beta_\alpha$ is a wff, where ϕ^β_α is the result of replacing one or more of the occurrences of α in ϕ by some variable β not already in ϕ.

Only formulas obtainable by (possibly) repeated application of these rules are wffs.[3]

Rules 2 and 3 are the same as for the language of propositional logic (see Section 3.3). Rule 1 is broader; in view of the definition of 'atomic formula' given above, it allows more kinds of atomic formulas. Rule 4 is completely new. To see how it works, let's use it to generate some quantified formulas from a specific unquantified wff ϕ. Let ϕ be the wff '$(Fa\ \&\ Gab)$'. (We can tell that this formula is a wff, because 'Fa' and 'Gab' are wffs, by rule 1, so that '$(Fa\ \&\ Gab)$' is a wff, by rule 3.) Now ϕ contains two name letters, 'a' and 'b'. Either can serve as what the rule calls α. Let's use 'a'. The rule tells us to choose some variable β which is not already in ϕ. Any variable will do, since ϕ contains no variables; so let β be the variable 'x'. Then there are three formulas of the form ϕ^β_α which result from replacing one or more occurrences of α (that is, 'a') in ϕ (that is, '$Fa\ \&\ Gab$') by β (that is, 'x'):

$(Fx\ \&\ Gxb)$	(Both occurrences of 'a' replaced by 'x')
$(Fx\ \&\ Gab)$	(Only first occurrence of 'a' replaced)
$(Fa\ \&\ Gxb)$	(Only second occurrence of 'a' replaced)

These formulas themselves are not wffs, but rule 4 stipulates that the result of prefixing any of them with a universal quantifier followed by 'x', that is, $\forall\beta\phi^\beta_\alpha$, is a wff. Thus

$\forall x(Fx\ \&\ Gxb)$

$\forall x(Fx\ \&\ Gab)$

$\forall x(Fa\ \&\ Gxb)$

---

[3]Some authors use somewhat different formation rules. (Some common points of variation were mentioned in notes 1 and 2.) Readers should compare formation rules carefully when using different texts.

are all wffs by rule 4. The rule also says that the result of prefixing any of them with an existential quantifier followed by '*x*', that is, ∃β*φ*β̸, is a wff. Hence

> ∃*x(Fx & Gxb)*
> ∃*x(Fx & Gab)*
> ∃*x(Fa & Gxb)*

are all wffs as well. We have thus generated six quantified wffs from the unquantified wff '*(Fa & Gab)*' by single applications of rule 4. We could have generated others by using the name '*b*' for α or variables other than '*x*' for β. Rule 4 can be reapplied as well. For example, now that we know that '∀*x(Fx & Gxb)*' is a wff, we can apply rule 4 a second time, using '*b*' for α and '*y*' for β to obtain the following two wffs:

> ∀*y*∀*x(Fx & Gxy)*
> ∃*y*∀*x(Fx & Gxy)*

Notice that rule 4 is the only rule which allows us to introduce variables into a wff, and we can introduce a variable only by prefixing the formula with a quantifier for that variable. Thus any formula containing a variable without a corresponding quantifier (for example, '*Fx*') is not a wff. Similarly, any formula containing a quantifier-plus-variable without a further occurrence of that variable (for example, '∃*xPa*') is not a wff.

The purpose of the phrase 'by some variable β not already in φ' in rule 4 is to ensure that quantifiers using the same variable never apply to overlapping parts of the same formula. Thus, for instance, though '∃*xLxa*' is a wff by rules 1 and 4, this clause prevents us from adding another quantifier on *x* to obtain, for example, '∀*x*∃*xLxx*', which, as we noted earlier (comment 2, Section 6.2), is ill formed. Notice that this clause also prevents a formula such as '∀*x(Fx & ∃xGx)*' from being well formed, since this could be obtained by rule 4 only from something like the wff '*(Fa & ∃xGx)*' which already contains '*x*'. But it does not prohibit such formulas as '(∃*xFx & ∃xGx)*', where the quantifiers apply to nonoverlapping parts of the formula. This formula is indeed a wff, since it is obtained by rule 3 from '∃*xFx*' and '∃*xGx*', which are themselves wffs, by rules 1 and 4 (see comment 3, Section 6.2).

In general, to show that a given formula is a wff, we show that it can be constructed strictly by the formation rules.

SOLVED PROBLEMS

6.7   Show that '~∃*x(~Fx & ∀zGzx)*' is a wff.

Solution

> By rule 1, '*Fa*' and '*Gba*' are wffs. Hence, by rules 2 and 4, respectively, '~*Fa*' and '∀*zGza*' are wffs. Application of rule 3 to these two formulas shows that '(~*Fa & ∀zGza*)' is a wff. So, by rule 4, '∃*x(Fx & ∀zGzx)*' is a wff, whence it follows by rule 2 that '~∃*x(Fx & ∀zGzx)*' is a wff.

6.8   Explain why the following formulas are not wffs.

(*a*)   ∀*xLxz*

(*b*)   (*Fa*)

(*c*)   (∃*xFx & Gx*)

(*d*)   ∀*x(Fx)*

(*e*)   (∀*xFx*)

(*f*)   ∃*x*∀*yFx*

(*g*)   ∃*x*∃*x(Fx & ~Gx)*

(*h*)   ∃*xFx & ∃xGx*

Solution

(*a*)   The variable '*z*' lacks a quantifier.

(*b*)   Atomic wffs are not enclosed in brackets (see rule 1).

(*c*)   The final occurrence of '*x*' is not quantified. Note, however, that '∃*x*(*Fx* & *Gx*)' *is* a wff.

(*d*)   Unnecessary brackets.

(*e*)   Unnecessary brackets.

(*f*)   '∀*y*' requires a second occurrence of '*y*' (see rule 4).

(*g*)   Quantifiers apply to overlapping parts of the formula.

(*h*)   Outer brackets missing (see rule 3).

As in the propositional calculus, we adopt the informal convention of dropping paired outer brackets. Thus, although the formula in Problem 6.8(*h*) is not officially a wff, from now on we will permit the use of such formulas as abbreviations for wffs where convenient. It is not permissible, however, to drop brackets when they are not both the outermost symbols. Thus, for example, we may not drop brackets from '∀*x*(*Fx* → *Gx*)'.

6.4 MODELS

We now define more precisely the semantics of our language, which we have implicitly presupposed in the previous sections. Relative to the semantics of propositional logic, there are two sources of complexity. One is that some atomic formulas of the language of predicate logic are treated as compound expressions, whereas all atomic formulas of the language of propositional logic were unanalyzed atomic units (sentence letters). These new atomic formulas (such as '*Fa*') cannot be assigned a truth value directly: their truth value will have to depend on the interpretation of the predicate and name letters occurring therein (in our example, the letters '*F*' and '*a*'). The second source of complexity is that the new language contains logical operators—the quantifiers—which were absent in the language of propositional logic. The method of Venn diagrams introduced in Section 5.2 is of some help in visualizing the intended interpretation of these operators. But predicate logic is not just a reformulation of the logic of categorical statements. Its language includes wffs in which a quantifier combines with other quantifiers or with the truth-functional operators in ways that do not correspond to any categorical statements, and Venn diagrams are not adequate for dealing with such a variety of linguistic structures.

The main concept needed to deal with the first difficulty is that of a *model*, or *interpretation structure*. A model supplies an interpretation (or semantic value) for the nonlogical symbols occurring in a given wff or set of wffs. The nature of the interpretation depends on the type of symbol:

Symbol	*Interpretation*
name letter	individual object (e.g., the Moon)
zero-place predicate letter (sentence letter)	truth value (T or F)
one-place predicate letter	class of objects (e.g., the class of people)
n-place predicate letter ($n > 1$)	relation between *n* objects (e.g., the relation that holds between a pair of objects just in case the first is bigger than the second)

It is not assumed that different symbols of the same type have different interpretations. As with English names and predicates, distinct name letters can designate the same object and distinct predicate letters can stand for the same class or relation (think of the English predicates 'person' and 'human being', for example). On the other hand, if the same predicate letter is used to represent predicates with a different number of places, as in '*Fa & Fbc*', then we shall assume that the model interprets the letter in different ways, one for each distinct use. This amounts to saying that a model will treat different uses of a predicate letter as distinct predicates, just as an English dictionary treats a word with multiple meanings as distinct words (for example, the noun 'ties' and the transitive verb 'ties'). In practice, however, this convention will rarely be necessary, since it is generally advisable to translate different predicates by distinct letters, or by the same letter with distinct subscripts.

In addition to specifying an interpretation for the nonlogical symbols, a model for a wff or set of wffs specifies a *domain* or *universe* of interpretation: this is simply the class of objects relative to which the name and predicate letters are interpreted. For example, if the domain of interpretation is the class of living beings and '*F*' is interpreted as the class of human beings, then all of us will be included in the interpretation; if the domain is a class of mythological creatures, however, the interpretation of '*F*' may contain all sorts of animals, but none of us. There is no restriction on the sorts of object that may be included in a model's universe, as long as the universe is not empty.

A model, then, specifies a domain or universe together with a corresponding interpretation of the nonlogical symbols occurring in some given wffs. Given a model, every atomic formula ϕ built up from those symbols is assigned a truth value according to the following rules:

(1)   If ϕ consists of a single sentence letter, then its truth value is the one specified directly by the model.

(2)   If ϕ consists of a predicate letter followed by a single name letter, then ϕ is assigned the value T if the object designated by the name letter is a member of the class designated by the predicate letter; otherwise ϕ is assigned the value F.

(3)   If ϕ consists of a predicate letter followed by two or more name letters, then ϕ is assigned the value T if the objects designated by the name letters stand in the relation designated by the predicate letter; otherwise ϕ is assigned the value F.

A sentence that is assigned the value T in a model is also said to be *true* in the model, and a sentence that is assigned the value F is also said to be *false* in the model.

SOLVED PROBLEM

6.9 Evaluate the following atomic formulas in the model given below:

(*a*)   *Pc*

(*b*)   *Pe*

(*c*)   *Se*

(*d*)   *Tce*

(*e*)   *Tec*

(*f*)   *Tee*

Universe: every object in the world, including people and animals

c : Bill Clinton

e : the Empire State Building

P : the class of all people

S : the class of all saxophone players

T : the relation that holds between two things x and y just in the case that x is taller than y

Solution

(*a*) T

(*b*) F

(*c*) F

(*d*) F

(*e*) T

(*f*) F

There is of course a close relationship between the notion of a model and the intuitive notion of an interpretation involved in formalizing English sentences. When we specify an interpretation in order to formalize certain sentences, we establish a correspondence between the symbols of the formal language and certain names and predicates of English. (See Problems 6.1 to 6.6.) When we specify a model, we establish a correspondence between the symbols of the formal language and certain objects, classes, or relations that we *describe* in English. However, note that there are in principle many different ways of describing such objects, classes, or relations. For instance, in the model of Problem 6.9 we could have specified the interpretation of '*c*' with the description 'the 42nd President of the United States of America' rather than with the name 'Bill Clinton'; and we could have specified the interpretation of the predicate letter '*P*' by listing the names of all people: although practically very unfeasible (if not impossible), such a description would in principle be perfectly legitimate.

Once we know how to evaluate atomic formulas, the truth values of compound wffs that do not contain quantifiers are determined as in propositional logic: we rely on the conditions specified by the truth tables for the truth-functional operators. (See Section 3.4.)

SOLVED PROBLEM

6.10 Evaluate the following wff relative to the model of Problem 6.9:

$$(Pc \mathbin{\&} Tce) \lor (\sim Pc \mathbin{\&} Tec)$$

Solution

This wff is a disjunction. The first disjunct is a conjunction, and this conjunction is false in the model since its second conjunct, '*Tce*', is false. The second disjunct is also a conjunction, and this conjunction is also false in the model since its first conjunct, '$\sim Pc$', is the negation of a true atomic formula, and is therefore false. Thus, both disjuncts are false. Hence the disjunction is false, i.e., the model assigns it the value F.

We now consider the truth conditions for quantified wffs (the second difficulty mentioned at the beginning of this section).[4] Suppose we have to evaluate the wff '$\forall x Sx$' in the model of Problem 6.9, where '*S*' is interpreted as the class of all saxophone players relative to the class of all existing things. Since this wff says "Everything is a saxophone player", it is clearly false in the model: we can easily produce an object, say the Empire State Building, which is not a saxophone player, i.e., is not a member of the interpretation of '*S*' in the model. We can express this more precisely by saying that there exists an *instance* of the quantified wff '$\forall x Sx$', say '*Se*', which is false in the model. This instance is obtained by dropping the quantifier '$\forall x$' and replacing the variable '*x*' in the formula that follows by a name letter, '*e*', yielding an atomic formula which can be evaluated by rule (2). On the other hand, the wff '$\exists x Sx$' is true in the model, since we can easily find an object in the universe, say Bill Clinton, who is a saxophone player. That is, we can find an instance of '$\exists x Sx$', say '*Sc*', which is true in the model. This

---

[4] There are many slightly different ways of defining the truth conditions of quantified wffs. The method followed here is close to the one presented in Benson Mates, *Elementary Logic*, 2d edn, New York, Oxford University Press, 1972.

procedure can easily be generalized to quantified wffs whose instances are not atomic, as our next examples illustrate.

SOLVED PROBLEMS

6.11 Evaluate the following wff in the model given below:

$$\forall x(Px \rightarrow Sx)$$

Universe: the class of all people, past and present

b : George Bush

c : Bill Clinton

P : the class of all twentieth-century US Presidents

S : the class of saxophone players

Solution

The instance '$Pc \rightarrow Sc$' is true in the model: it is a conditional whose antecedent is true by rule (2) (Bill Clinton is a member of the class of twentieth-century US Presidents) and whose consequent is also true by rule (2) (Bill Clinton is a saxophone player). However, the instance '$Pb \rightarrow Sb$' is false, since its antecedent is true (Bush has been a US President in the twentieth century) but its consequent is false (Bush is not a saxophone player). But a universally quantified wff can be true only if every instance of it is true. Thus the wff '$\forall x(Px \rightarrow Sx)$' is false.

6.12 Evaluate the following wff in the model of Problem 6.11:

$$\exists x(\sim Px \vee Sx)$$

Solution

The instance '$\sim Pb \vee Sb$' is false in this model: it is a disjunction whose disjuncts are both false. The first disjunct is false because it is the negation of the atomic wff 'Pb', which is true by rule (2) (George Bush is a member of the class of twentieth-century US Presidents); the second disjunct is an atomic sentence which is also false by rule (2) (George Bush is not a saxophone player). However, the instance '$\sim Pc \vee Sc$' is true in the model, since its second disjunct is true (Bill Clinton is a member of the class of saxophone players). An existentially quantified wff is true as long as at least one instance of it is true. Thus, the wff '$\exists x(\sim Px \vee Sx)$' is true.

In some cases, it may be necessary to consider a greater number of instances than the model initially allows. For example, to evaluate the wff '$\exists x \sim Px$' in the model of Problem 6.11, it is not enough to consider the instances '$\sim Pb$' and '$\sim Pc$'. Both these instances are false in the model, since both are negations of true atomic formulas. However, intuitively the fact that both Bush and Clinton are twentieth-century US Presidents is not enough to deny the existence of *something* which is not a twentieth-century US President: the quantifier ranges over all objects in the universe, including those which are not designated by any name letters for which the model is defined. To overcome this difficulty, we construct one more substitution instance, say '$\sim Pa$', and we consider its truth values for all the possible interpretations of the new name letter 'a'. If 'a' is interpreted as Bush or as Clinton, then the truth value of '$\sim Pa$' coincides with that of '$\sim Pb$' and '$\sim Pc$'. But if 'a' is interpreted, for instance, as Walt Disney, then '$\sim Pa$' becomes true, since it is the negation of the false atomic formula 'Pa'. Technically, the result of extending the given model by interpreting a new name letter 'a' is a new model called an *a-variant* of the given model. Since there is at least one *a*-variant that assigns the value T to the instance '$\sim Pa$', the existentially quantified wff '$\exists x \sim Px$' is evaluated as true in the model.

SOLVED PROBLEM

6.13 Evaluate the following wff in the model given below:

$$\exists x(\sim Px \,\&\, Mx)$$

Universe: the class of all people, past and present

c : Bill Clinton

M : the class of all male people

P : the class of all twentieth-century US Presidents

Solution

The instance '$\sim Pc \,\&\, Mc$' is false in the given model, but this is not enough to evaluate the existential wff '$\exists x(\sim Px \,\&\, Mx)$': we must consider every object in the universe, not just Bill Clinton. We thus construct the new instance '$\sim Pa \,\&\, Ma$' and consider all possible a-variants of the model, i.e., all possible ways of interpreting the new name letter 'a' on the class of all people. These a-variants fall into two broad categories. On the one hand, there are a-variants that interpret 'a' as an element of the class of twentieth-century US Presidents. Such models assign the value T to the wff 'Pa', and hence the value F to its negation '$\sim Pa$'. Therefore, they all assign the value F to the conjunction '$\sim Pa \,\&\, Ma$'. On the other hand, there are a-variants that interpret 'a' as an element of the complement of the class of twentieth-century US Presidents, i.e., as an individual which is not a twentieth-century US President. Such a-variants assign the value F to the wff 'Pa', and hence the value T to its negation '$\sim Pa$'. Furthermore, among these a-variants there will be some which interpret 'a' as a male person, and some which do not. Those that do interpret 'a' as a male person will assign the value T to the wff 'Ma', and, accordingly, they will assign the value T to the conjunction '$\sim Pa \,\&\, Ma$'. Since there are a-variants of the given model in which the instance '$\sim Pa \,\&\, Ma$' is true, we conclude that the existential wff '$\exists x(\sim Px \,\&\, Mx)$' is true in the model itself.

We may now state the truth conditions for quantified wffs more precisely. Let M be any model and suppose α is a name letter. An α-*variant* of M is defined, quite generally, as any model that results from M by freely interpreting α as an object in the universe of M. If M does not assign any interpretation to α, then an α-variant will be a slightly "richer" model than M: in addition to the interpretations provided by M, it will provide an interpretation for α as well. (This was indeed the case in the a-variants of our examples above.) On the other hand, if M already did assign an interpretation to α, then an α-variant of M will simply represent a new way of interpreting α, keeping everything else exactly as in M. In this case it is not assumed that the new interpretation of α is *different* from its interpretation in M: the interpretation may be the same, in which case the α-variant coincides with M itself. In other words, whenever a model assigns an interpretation to a name letter α, it counts as an α-variant of itself.

Using the notion of an α-variant, the truth conditions for quantified sentences are now defined as follows:

(4)   A universal quantification $\forall\beta\phi$ is true in a model M if the wff ϕ^{α}_{β} is true in every α-variant of M, where α is the first name letter in the alphabetic order not occurring in ϕ and ϕ^{α}_{β} is the result of replacing all occurrences of β in ϕ by α; if ϕ^{α}_{β} is false in some α-variant of M, then $\forall\beta\phi$ is false in M.

(5)   An existential quantification $\exists\beta\phi$ is true in M if the wff ϕ^{α}_{β} is true in some α-variant of M, where α and ϕ^{α}_{β} are as in (4); if ϕ^{α}_{β} is false in every α-variant of M, then $\exists\beta\phi$ is false in M.

(In Problem 6.13, the wff ϕ is '$(\sim Px \,\&\, Mx)$', the name letter α is 'a', the variable β is 'x', and the formula ϕ^{α}_{β} is '$(\sim Pa \,\&\, Ma)$'.)

We assume the alphabetic order to be defined in such a way that the first name letter after 't' is 'a_1', the next one 'a_2', and so on. Since wffs are finite sequences of symbols, we are assured that there is always a first name letter available. The requirement that α be alphabetically the *first* name letter not occurring in α is purely stipulative; any other choice would do, as long as α is a name letter foreign to φ. This proviso, however, is crucial. To see this, suppose φ is the wff '∃xTxc', and suppose we want to evaluate it in the model of Problem 6.9, where 'c' is interpreted as Bill Clinton and 'T' as the relation *taller than*. If we take α to be the name letter 'c', the formula φ%ᵦ becomes 'Tcc'. And since nothing is taller than itself, this formula will come out true in every α-variant of the model. This would result in an assignment of the value F to the wff '∃xTxc'. Yet clearly this wff should be true in the model, for there is some taller object than Bill Clinton (for instance the Empire State Building).

SOLVED PROBLEMS

6.14 Evaluate the following wff in the model M given below:

$$\exists x Gxa$$

Universe: the class of all people

a : Bill Clinton

G : the relation that holds between two persons x and y just in case x is younger than y

Solution

   This is an existentially quantified wff and the first name letter not occurring in it is '*b*'. Thus, by condition (5), the wff is true in M just in case there exists at least one *b*-variant of M in which the wff '*Gba*' is true. There are many such *b*-variants, since there are many people younger than Bill Clinton. Thus, the wff '∃xGxa' is true in M. Note that, in this example, Bill Clinton is the interpretation of the name letter '*a*', not of the name letter '*c*' (which in fact is left uninterpreted by the model). There is no constraint on how a nonlogical symbol may be interpreted by a model as long as names are interpreted as objects, predicates as classes or relations, etc.

6.15 Evaluate the following wff in the model M given below:

$$Pa \rightarrow \exists x(Px \;\&\; Mx)$$

Universe: the class of all people, past and present

a : Bill Clinton

M : the class of all male people

P : the class of all twentieth-century US Presidents

Solution

   This wff is a conditional. The antecedent of the conditional is an atomic formula which is true in M by rule (2) (since the object designated by '*a*' is a twentieth-century US President). The consequent is an existentially quantified wff and by condition (5) it is true in M just in case there exists some *a*-variant of M in which the wff '*Pa & Ma*' is true. (Here '*a*' is the first name letter that has no occurrences in '∃x(Px & Mx)'; the fact that it occurs in '*Pa*' is irrelevant.) The wff '*Pa & Ma*' is a conjunction, and both conjuncts are true in any *a*-variant where '*a*' is interpreted as a male US President. Since there are male US Presidents, such an *a*-variant is sure to exist. (M itself counts as one, since Bill Clinton is a male President.) Hence the wff '∃x(Px & Mx)' is true in M. Since both the antecedent and the consequent of the conditional '*Pa* → ∃x(Px & Mx)' are true in M, we conclude that the conditional is true in M.

6.16 Evaluate the following wff in the model M given below:

$$\exists x Px \;\&\; \exists x Mx$$

Universe: the class of all people, past and present

M : the class of all male people

P : the class of all twentieth-century US Presidents

Solution

The wff is a conjunction (not to be confused with the wff '$\exists x(Px \& Mx)$' which is a quantified wff). So it is true if both conjuncts are true. The first conjunct is an existentially quantified wff '$\exists xPx$' which contains no name letters. So we consider the values of 'Pa' in the a-variants of M. Since the class of twentieth-century US Presidents is nonempty, there certainly exist a-variants that interpret 'a' as a twentieth-century US President. In such models 'Pa' is true. Consequently '$\exists xPx$' is true in M (condition (5)). A similar reasoning shows that the second conjunct, '$\exists xMx$', is also true in M. Hence the wff '$\exists xPx \& \exists xMx$' is true as well. Note that the model of this example is undefined for every name letter.

6.17 Evaluate the following wff in the model M given below:

$$\forall x(Ex \lor Ox)$$

Universe: the class of all positive integers

E : the class of all even integers

O : the class of all odd integers

Solution

'$\forall x(Ex \lor Ox)$' is a universally quantified wff in which no name letter occurs, so by condition (4) it is true in M if and only if the instance '$Ea \lor Oa$' is true in every a-variant of M. There are only two possibilities. Either 'a' is interpreted as an even integer, in which case it is an element of the class designated by 'E', or 'a' is interpreted as an odd integer, in which case it is an element of the class designated by 'O'. In both cases, one of the disjuncts of '$Ea \lor Oa$' will be true, hence the disjunction will be true as well. So in every a-variant of M the wff '$Ea \lor Oa$' is true. Hence '$\forall x(Ex \lor Ox)$' is true. This example also illustrates that a model universe may consist of anything we like—in this case, numbers.

6.18 Evaluate the following wff in the model M given below:

$$\forall x(Wx \rightarrow Bx)$$

Universe: the class of all living creatures

B : the class of all blue things

W: the class of all winged horses

Solution

This wff is true in M iff the conditional '$Wa \rightarrow Ba$' is true in every a-variant of M. No matter how 'a' is interpreted, the antecedent of this conditional is bound to be false, since there are no winged horses. Hence, since a conditional with false antecedent is true, '$Wa \rightarrow Ba$' is indeed true in every a-variant of M. We conclude that '$\forall x(Wx \rightarrow Bx)$' is true in M.

6.19 Evaluate the following wff in the model M of Problem 6.18:

$$\forall x(Wx \rightarrow {\sim}Bx)$$

Solution

This wff is true in M iff the conditional '$Wa \rightarrow {\sim}Ba$' is true in every a-variant of M. The same reasoning used in Problem 6.18 shows that this is indeed the case. Hence '$\forall x(Wx \rightarrow {\sim}Bx)$' is true in M.

The last two examples show that any universally quantified material conditional whose antecedent is empty (i.e., applies to nothing) is true. This is just another way of saying that all A-form propositions with empty subject terms are true (see the end of Section 5.3).

Our next examples show that conditions (4) and (5) can be applied more than once to evaluate wffs which involve nested quantifiers.

SOLVED PROBLEMS

6.20 Evaluate the following wff in the model M given below:

$$\forall x(Ex \rightarrow \forall yGxy)$$

Universe: the class of all positive integers

E : the class of even integers

G : the relation *greater than*

Solution

This wff is true in M iff the conditional '$Ea \rightarrow \forall yGay$' is true in every model that counts as an *a*-variant of M. Let M' be such a model. If '*a*' is interpreted as an odd integer, the antecedent 'Ea' is false in M', hence the conditional will be true regardless of the value of the consequent. However, suppose that M' interprets '*a*' as an even integer. In this case the antecedent is true in M', hence the conditional will be true in M' only if the consequent, '$\forall yGay$', is also true in M'. This is a universally quantified wff, so it will be true in M' just in case the instance 'Gab' (obtained by replacing '*y*' with the first available name letter, '*b*') is true in every *b*-variant of M'. But clearly this is not the case: any *b*-variant of M' in which '*b*' is interpreted as an integer greater than the one designated by '*a*' will make 'Gab' false. Thus, if '*a*' is interpreted by M' as an even integer, the conditional '$Ea \rightarrow \forall yGay$' comes out false in M'. Since M' is an *a*-variant of M, it follows that the wff '$\forall x(Ex \rightarrow \forall yGxy)$' is false in M.

6.21 Evaluate the following wff in the model M of Problem 6.20:

$$\exists y\forall yGxy$$

Solution

This wff is true in M just in case the wff '$\forall yGay$' is true in some *a*-variant of M. Suppose there is such an *a*-variant of M, say M'. If '$\forall yGay$' is true in M', then the wff 'Gab' must be true in every *b*-variant of M'. But this is impossible, since no matter what number is designated by '*a*', we can consider infinitely many *b*-variants in which '*b*' is interpreted as a greater number than '*a*'. Thus the assumption that there is some *a*-variant of M in which '$\forall yGay$' is true leads to a contradiction. Hence we conclude that '$\exists x\forall yGxy$' is false in M. (Intuitively: it is false that some integer is greater than all the integers.)

6.22 Evaluate the following wff in the model M of Problem 6.20:

$$\forall y\exists xGxy$$

Solution

This wff is true in M just in case the wff '$\exists xGxa$' is true in every *a*-variant of M. Let M' be any such *a*-variant. '$\exists xGxa$' is true in M' if the wff 'Gba' is true in some *b*-variant of M'. This is obviously the case, for no matter what number is designated by '*a*', we can always consider a *b*-variant in which '*b*' is interpreted as a greater number than '*a*'. Since we have made no specific assumptions on M', this means that '$\exists xGxa$' is true in every *a*-variant of M. Hence '$\forall y\exists xGxy$' is true in M. (Intuitively: it is true that, given any integer *y*, there exists an integer *x* that is greater than *y*.)

These last two examples should be compared with comment 4 toward the end of Section 6.2: they show that wffs in which mixed consecutive quantifiers are listed in a different order have different truth conditions.

6.5 REFUTATION TREES

The semantic rules of the previous section, together with the semantic rules for the truth-functional operators of Chapter 3, enable us to provide a rigorous characterization of the concept of deductive validity for predicate logic. Recall that an argument form is valid if and only if there is no possible situation in which its premises are true while its conclusion is false. A possible situation (in the relevant sense) is now characterized as a model in which all the nonlogical symbols occurring in the argument receive an interpretation. Thus, an argument form of predicate logic is valid if and only if there is no such model in which its premises are assigned the value true while its conclusion is assigned the value false. For example, we have seen in Problems 6.21 and 6.22 that there is a model in which the wff '$\forall y \exists x Gxy$' is true while the wff '$\exists x \forall y Gxy$' is false. This means that any argument of the form

$$\forall y \exists x Gxy \vdash \exists x \forall y Gxy$$

is invalid. By contrast, the argument in the opposite direction is valid.

SOLVED PROBLEM

6.23 Show the validity of the argument form:

$$\exists x \forall y Gxy \vdash \forall y \exists x Gxy$$

Solution

If M is any model in which the premise is true, then the universe of M contains some one object—call it 'The One'—which bears the relation designated by 'G' to every other object in the universe. (The reader should verify this by a careful application of the semantic rules.) But if this is the case, then every object in the universe is such that there is at least one object (namely The One) which bears that relation to it, and therefore the conclusion is also true in the model.

This characterization of validity is quite general and can be applied to every argument form that can be expressed in the language of predicate logic. However, our examples also show that the work needed to test an argument for validity in this way may be long and burdensome and may call for a lot of ingenuity. As with propositional logic, the characterization of validity in terms of truth-value assignments is significant from a theoretical viewpoint, but not very efficient when it comes to its practical application.

In fact, predicate logic is worse than propositional logic in this regard. The method of truth tables is certainly inefficient for problems involving numerous sentence letters. Nevertheless it is a reliable, algorithmic method: given any argument form of propositional logic (with finitely many premises), its table will contain finitely many rows, so they can in principle be checked in a finite period of time. (We could program a computer to do the checking for us.) By contrast, our characterization of the procedure for testing the validity of an argument form of predicate logic gives no hint of any finite bound on the number of models that must be checked. In particular, there is no limit to the size of a model's universe, hence no limit to the number of variants to be considered in the course of evaluating a quantified wff. And relative to any class of objects, there is no limit to the way a predicate can be interpreted: any different subclass counts as a different interpretation, to which there corresponds a different model. Indeed, it is a fact that in predicate logic, unlike propositional logic or the logic of categorical statements, there is not and cannot in principle be any algorithmic procedure which reliably detects invalidity in every case. Predicate logic is in this sense *undecidable*. (The undecidability of predicate logic can be proved by metalogical reasoning, under a well-confirmed assumption known as Church's thesis.[5])

---

[5]An elegant version of the proof is given in Richard Jeffrey, *Formal Logic: Its Scope and Limits*, 2d edn, New York, McGraw-Hill, 1981, Chap. 6.

There are, however, rule-governed procedures for testing validity which yield an answer after finitely many operations for many, though not all, predicate logic argument forms. One such procedure is a generalization of the refutation tree technique of Chapter 3, and the rest of this section will be devoted to its outline. Like the technique of Chapter 3, this new technique enables us to discover the validity of a valid argument in a finite number of steps (though the number of steps may be very large, and we may not be able to tell in advance whether or not we are going to get an answer). But unlike the technique of Chapter 3, it sometimes fails to reveal the invalidity of invalid forms.

Our generalized tree technique incorporates the refutation tree rules of propositional logic, but has in addition four new rules to deal with sentences containing quantifiers. For some predicate logic trees, just the rules of propositional logic suffice.

SOLVED PROBLEM

6.24   Use the refutation tree rules for propositional logic to determine whether the following argument form is valid:

$$\forall x Fx \rightarrow \forall x Gx, \ {\sim}\forall x Gx \vdash {\sim}\forall x Fx$$

Solution

```
1   √        ∀xFx → ∀xGx
2            ~∀xGx
3            ~~∀xFx

4   ~∀xFx   1→      ∀xGx   1→
5   X       3,4~    X      2,4 ~
```

All paths close, so the form is valid. Only propositional logic rules are needed, because the form is a substitution instance of *modus tollens* (MT), which is valid by propositional logic.

Since the language of predicate logic has two logical symbols not included in the language of propositional logic (namely, '∀' and '∃') and we need two rules for each (to deal with negated and unnegated sentences in which they occur), there are four new rules for refutation trees in predicate logic. The first is the universal quantification rule.

<u>Universal Quantification (∀):</u>   If a wff of the form ∀βφ appears on an open path, then if α is a name letter that occurs in some wff on that path, write φα/β (the result of replacing all occurrences of β in φ by α) at the bottom of the path. If (and only if) no wff containing a name letter appears on the path, then choose some name letter α and write φα/β at the bottom of the path. In either case, do not check ∀βφ.

The intuitive rationale for this rule is this. In a refutation tree, each step consists in the breaking down of a wff into simpler wff(s), all of which must be true if the former is assumed to be true. If the wff in question is a universally quantified formula ∀βφ, *any* wff that can be obtained from φ by replacing each occurrence of the variable β by a name α would have to be true: for if φ is true of everything, then it must be true of the object denoted by α, whatever that might be. In principle there is no limit to the number of names that one may consider in this regard. However, in a refutation tree we are interested in seeing whether every path will eventually close. Accordingly, we may confine ourselves to those names that already appear on some open path, for only a wff involving such a name may be inconsistent with some other wff in the tree, forcing the path to close. This explains the first part of the rule (which applies in most cases). The second part applies in those cases when no name yet occurs in any open path: in such cases we have to introduce some name to start the process, so we choose some name α (any name will do) and write φα/β.

This rule is a formal expression of the fact that whatever is true of everything must be true of any particular individual. It is used, for example, in proving the validity of:

> All frogs are green
> Everything is a frog
> ∴ Alf is green

Using 'F' for 'is a frog', 'G' for 'is green', and 'a' for 'Alf', we may formalize this argument as

$$\forall x(Fx \rightarrow Gx), \forall xFx \vdash Ga$$

SOLVED PROBLEM

6.25 Show the validity of this argument form.

Solution

```
1          ∀x(Fx → Gx)
2          ∀xFx
3          ~Ga
4    √     Fa → Ga          1 ∀
5          Fa               2 ∀

6   ~Fa      4 →      Ga    4 →
7    X      5, 6 ~     X    3, 6 ~
```

Since the name letter 'a' occurs in '$\sim Ga$' (line 3), this is the letter we use in obtaining lines 4 and 5 by the rule for universal quantification. The tree then closes after an application of the conditional rule at 6. Hence the form is valid. (With respect to the rule, here ϕ is '$(Fx \rightarrow Gx)$', the name letter α is 'a', the variable β is 'x', and the formula ϕ^α_β is '$(Fa \rightarrow Ga)$' from which we drop the outer brackets according to our convention.)

Note that the rule does not require that we check $\forall \beta \phi$, since no matter how many wffs we infer from it by $\forall$, we still have not exhausted all its implications. But, though universally quantified wffs are never checked, their trees may close (in which case we know the inference being tested is valid) or may reach a point at which the tree is not closed but no more rules apply (in which case we know that the inference is invalid).

SOLVED PROBLEM

6.26 Construct a refutation tree to decide whether the following form is valid:

$$Fa \rightarrow Gb, \forall x \sim Fx \vdash \sim Gb$$

Solution

```
1    √     Fa → Gb
2          ∀x~Fx
3    √     ~~Gb
4          Gb              3 ~~
5          ~Fa             2 ∀
6          ~Fb             2 ∀

7   ~Fa      1 →      Gb    1 →
```

Applying nonbranching rules first, we check line 3 and obtain 'Gb' at 4 by double negation and derive '$\sim Fa$' and '$\sim Fb$' from line 2 by separate steps of the rule for universal quantification at 5 and 6. Notice that line 2 remains unchecked. At this point $\forall$ can be applied no more, since we have used it with both of the name letters that occur in wffs on the path and ∀ can be used to introduce a new name letter only if no name letters appear in wffs on the path. Hence all that remains is to apply the conditional rule at line 7. This fails to close the tree, and so the form is invalid.

Unlike refutation trees for propositional logic, refutation trees for predicate logic do not generate a complete list, in any sense, of counterexamples for an invalid argument. Each open path of a finished refutation tree for predicate logic can be interpreted as a model universe containing just the objects mentioned by name on the path. The atomic wffs or negations of atomic wffs on the path indicate what is true of these objects in this model.

In Problem 6.26, the two open paths (the one that branches to the right as well as the one that branches to the left) both represent a universe containing just two objects, a and b. In this model universe, b has the property G and both a and b are not F. (The paths do not specify whether or not a is G; this happens to be irrelevant to the case.) Now 'Fa' is false in this model; hence the premise '$Fa \rightarrow Ga$' is true, since a material conditional is true if its antecedent is false. Moreover, since the only objects in this universe are a and b and neither is F, the premise '$\forall x \sim Fx$' is true. But the conclusion '$\sim Gb$' is false. This model therefore shows that there is at least one instance of this form whose conclusion can be false while its premises are true. Hence it shows that the form is invalid.

SOLVED PROBLEM

6.27 Construct a refutation tree to decide whether the following form is valid:

$$\forall x(Fx \rightarrow Gx), \forall xGx \vdash Fa$$

Solution

```
1           ∀x(Fx → Gx)
2           ∀xGx
3           ~Fa
4    √      Fa → Ga      1 ∀
5           Ga           2 ∀
                /     \
6    ~Fa    4 →         Ga    4 →
```

The form is invalid. We apply the rule for universal quantification at 4 and 5 and the conditional rule at 6 as in Problem 6.25, but the tree does not close. No further rules are applicable, even though lines 1 and 2 remain unchecked, since we have applied ∀ to both 1 and 2 for every name letter occurring in a wff on the path ('a' is the only name letter occurring in a wff on the path).

In this problem, both open paths represent a model universe containing only one object, a, which has the property of being G but not F. In this model, the premises '$\forall x(Fx \rightarrow Gx)$' and '$\forall xGx$' are true, but the conclusion 'Fa' is false; hence the form is invalid. It is obvious that '$\forall xGx$' is true in the model, since the model contains only one object, a, and a is G. It is also obvious that 'Fa' is false. And the truth of '$\forall x(Fx \rightarrow Gx)$' follows from the fact that the class designated by 'F' is empty (see Problems 6.18 and 6.19).

The next two quantifier rules are rules for negated wffs. The first expresses the equivalence between a negated existential quantification and universally quantified negation. For instance, we have seen in Problem 6.2(d) that the sentence 'Nothing is a frog' can be formalized as '$\sim \exists xFx$' or, equivalently, as '$\forall x \sim Fx$'. The second rule expresses the matching equivalence between a negated universal quantification and an existentially quantified negation. Both equivalences can be verified semantically, but we shall leave the verification as an exercise for the reader.

Negated Existential Quantification (~∃):   If an unchecked wff of the form ~∃βφ appears on an open path, check it and write ∀β~φ at the bottom of every open path that contains the newly checked wff.

Negated Universal Quantification (~∀):   If an unchecked wff of the form ~∀βφ appears on an open path, check it and write ∃β~φ at the bottom of every open path that contains the newly checked wff.

SOLVED PROBLEMS

6.28   Construct a refutation tree to decide whether the following form is valid:

$$\forall x(Fx \to Gx), \sim\exists xGx \vdash \sim Fa$$

Solution

$$
\begin{array}{llll}
1 & & \forall x(Fx \to Gx) & \\
2 & \checkmark & \sim\exists xGx & \\
3 & & \sim\sim Fa & \\
4 & \checkmark & \forall x\sim Gx & 2\ \sim\exists \\
5 & & \sim Ga & 4\ \forall \\
6 & \checkmark & Fa \to Ga & 1\ \forall \\
\end{array}
$$

$$
\begin{array}{llll}
7 & \sim Fa \quad 6\to & Ga \quad 6\to \\
8 & X \quad\ \ 3,7\sim & X \quad 5,7\sim \\
\end{array}
$$

The form is valid. Notice that there is no need to apply the negated negation rule at line 3, since the tree closes anyway (compare Problem 3.28).

6.29   Construct a refutation tree to decide whether the following form is valid:

$$\sim\exists x(Fx \ \& \ Gx) \vdash \sim Fa$$

Solution

$$
\begin{array}{llll}
1 & \checkmark & \sim\exists x(Fx \ \& \ Gx) & \\
2 & \checkmark & \sim\sim Fa & \\
3 & & \forall x\sim(Fx \ \& \ Gx) & 1\ \sim\exists \\
4 & & Fa & 2\ \sim\sim \\
5 & \checkmark & \sim(Fa \ \& \ Ga) & 3\ \forall \\
\end{array}
$$

$$
\begin{array}{llll}
6 & \sim Fa \quad 5\sim\& & \sim Ga \quad 5\sim\& \\
7 & X \quad\ \ 4,6,\sim & \\
\end{array}
$$

The open path shows that in a universe consisting of one object, *a*, which is *F* but not *G*, the premise is true and the conclusion false. Hence the form is invalid. Note that no further rules can be applied to the unchecked wffs of the open path, so that the tree is complete.

6.30   Construct a refutation tree to decide whether the following form is valid:

$$\forall xFx \to \forall xGx, \sim\exists xGx \vdash \exists x\sim Fx$$

Solution

$$
\begin{array}{llll}
1 & \checkmark & \forall xFx \to \forall xGx & \\
2 & \checkmark & \sim\exists xGx & \\
3 & & \sim\exists x\sim Fx & \\
4 & & \forall x\sim Gx & 2\ \sim\exists \\
\end{array}
$$

$$
\begin{array}{llll}
5 & \checkmark & \sim\forall xFx \quad 1\to & \forall xGx \quad 1\to \\
6 & & \exists x\sim Fx \quad 5\sim\forall & Ga \quad\ \ 5\ \forall \\
7 & & X \quad\quad 3,6\sim & \sim Ga \quad 4\ \forall \\
8 & & & X \quad\quad 6,7\sim \\
\end{array}
$$

The form is valid.

The fourth and last quantifier rule allows us to simplify a wff by dropping an initial existential quantifier:

Existential Quantification (∃):   If an unchecked wff of the form ∃βφ appears on an open path, check it. Then choose a name letter α that does not yet appear anywhere on that path and write φα⁄β (the result of replacing every occurrence of β in φ by α) at the bottom of every open path that contains the newly checked wff.

The intuitive rationale for this rule is that if an existentially quantified wff '∃xφ' is true, then there must exist an object of which φ is true. We may not know what that object is, but we can give it a name. And since we do not want to make any specific assumptions about the object, we choose a name letter that does not yet appear on the path on which '∃xφ' appears.

SOLVED PROBLEMS

6.31   Construct a refutation tree to decide whether the following form is valid:

∃x(Fx & Gx) ⊢ ∃xFx & ∃xGx

Solution

1	√	∃x(Fx & Gx)		
2	√	~(∃xFx & ∃xGx)		
3	√	Fa & Ga	1 ∃	
4		Fa	3 &	
5		Ga	3 &	
6	√ ~∃xFx	2 ~&	√ ~∃xGx	2 ~&
7	∀x~Fx	6 ~∃	∀x~Gx	6 ~∃
8	~Fa	7 ∀	~Ga	7 ∀
9	X	4, 8 ~	X	5, 8 ~

This form is valid. Intuitively, at line 3 we suppose that *a* is one of the objects that satisfy the property of being *F* and *G*, whose existence is guaranteed by premise 1. In fact, *a* may not be such an object, but since we make no specific assumption about *a*, we may as well think of it as a *representative* of such objects. This is enough to show that the argument is valid. The reader may verify that the tree would equally close if we used a different name in place of '*a*' at line 3. (With respect to the rule, here φ is '(Fx & Gx)', the name letter α is '*a*', the variable β is '*x*', and the formula φα⁄β is '(Fa & Ga)', from which we drop the outer brackets.)

6.32   Construct a refutation tree to decide whether the following form is valid:

∃xFx ⊢ ∀xFx

Solution

1	√	∃xFx	
2	√	~∀xFx	
3		Fa	1 ∃
4	√	∃x~Fx	2 ~∀
5		~Fb	4 ∃

We introduce the new name letter '*a*' by ∃ at step 3, replace '~∀xFx' by its existential equivalent at step 4, and introduce yet another name letter, '*b*', with the second application of ∃ at 5. (The existential quantifier rule requires that this second name letter be different from the first.) No further rules apply. The tree is complete yet contains an open path; therefore the form is invalid. The tree represents a universe containing two objects, *a* and *b*, such that *a* is *F* and *b* is not *F*. In such a universe, '∃xFx' is true but '∀xFx' is not.

6.33   Construct a refutation tree to decide whether the following form is valid:

$\exists xFx, \exists xGx \vdash \exists x(Fx \mathbin{\&} Gx)$

Solution

1	√	$\exists xFx$	
2	√	$\exists xGx$	
3	√	$\sim\exists x(Fx \mathbin{\&} Gx)$	
4		Fa	1 $\exists$
5		Gb	2 $\exists$
6		$\forall x\sim(Fx \mathbin{\&} Gx)$	3 $\sim\exists$
7	√	$\sim(Fa \mathbin{\&} Ga)$	6 $\forall$
8	√	$\sim(Fb \mathbin{\&} Gb)$	6 $\forall$

9	$\sim Fa$	7 $\sim\&$			$\sim Ga$	7 $\sim\&$	
10	X	4, 9 $\sim$					
11			$\sim Fb$	8 $\sim\&$	$\sim Gb$	8 $\sim\&$	
12					X	5, 11 $\sim$	

The form is invalid. The open path represents a universe containing two objects, a and b, such that a is F but not G and b is G but not F. Notice that in applying the rule for existential quantification we introduce the new name letter 'a' at line 4 and another new name letter, 'b', in the second application at line 5. Notice too that the tree is complete, since no further rules apply to the open path.

The refutation tree test for validity of a form without premises in predicate logic is the same as for propositional logic: Negate the conclusion and then apply the rules to construct a tree. The form is valid if and only if all paths of the finished tree are closed.

SOLVED PROBLEMS

6.34   Construct a refutation tree to decide whether the following form is valid.

$\vdash \forall xFx \rightarrow Fa$

Solution

1	√	$\sim(\forall xFx \rightarrow Fa)$	
2		$\forall xFx$	1 $\sim\rightarrow$
3		$\sim Fa$	1 $\sim\rightarrow$
4		Fa	2 $\forall$
5		X	3, 4 $\sim$

The form is valid.

6.35   Construct a refutation tree to decide whether the following form is valid:

$\vdash \sim(\exists xFx \mathbin{\&} \forall x\sim Fx)$

Solution

1	√	$\sim\sim(\exists xFx \mathbin{\&} \forall x \sim Fx)$	
2	√	$\exists xFx \mathbin{\&} \forall x\sim Fx$	1 $\sim$
3	√	$\exists xFx$	2 $\&$
4		$\forall x\sim Fx$	2 $\&$
5		Fa	3 $\exists$
6		$\sim Fa$	4 $\forall$
7		X	5, 6 $\sim$

The form is valid.

The quantifier rules work the same way for wffs containing multiple quantifiers as they do for wffs containing single quantifiers.

SOLVED PROBLEMS

6.36   Construct a refutation tree to decide whether the following form is valid:

$$\forall x \forall y (Fxy \to \sim Fyx) \vdash \sim \exists x Fxx$$

Solution

1		$\forall x \forall y (Fxy \to \sim Fyx)$	
2	$\checkmark$	$\sim\sim \exists x Fxx$	
3	$\checkmark$	$\exists x Fxx$	$2 \sim\sim$
4		Faa	$3\,\exists$
5		$\forall y (Fay \to \sim Fya)$	$1\,\forall$
6	$\checkmark$	$Faa \to \sim Faa$	$5\,\forall$

7	$\sim Faa$	$6 \to$		$\sim Faa$	$6 \to$
8	X	$4, 7 \sim$		X	$4, 7 \sim$

The form is valid. Notice that we apply the existential rule at line 3 before applying the universal rule at lines 5 and 6. Doing this as a general policy minimizes the length of trees.

6.37   Construct a refutation tree to decide whether the following form is valid:

$$\exists x \forall y Lxy \vdash \forall x \exists y Lyx$$

Solution

1	$\checkmark$	$\exists x \forall y Lxy$	
2	$\checkmark$	$\sim \forall x \exists y Lyx$	
3		$\forall y Lay$	$1\,\exists$
4	$\checkmark$	$\exists x \sim \exists y Lyx$	$2 \sim\forall$
5	$\checkmark$	$\sim \exists y Lyb$	$4\,\exists$
6		$\forall y \sim Lyb$	$5 \sim\exists$
7		$\sim Lab$	$6\,\forall$
8		Lab	$3\,\forall$
9		X	$7, 8 \sim$

The tree's only path closes, and so the form is valid. (Compare Problem 6.23.)

6.38   Construct a refutation tree to decide whether the following form is valid:

$$\exists x \exists y Lxy \vdash \exists x Lxx$$

Solution

1	$\checkmark$	$\exists x \exists y Lxy$	
2	$\checkmark$	$\sim \exists x Lxx$	
3	$\checkmark$	$\exists y Lay$	$1\,\exists$
4		Lab	$3\,\exists$
5		$\forall x \sim Lxx$	$2 \sim\exists$
6		$\sim Laa$	$5\,\forall$
7		$\sim Lbb$	$6\,\forall$

The existential quantifier rule is used twice (steps 3 and 4), and at each use a new name letter is introduced. At line 6 no further rules apply, but the path remains open; the form is therefore invalid.

6.39 Construct a refutation tree to decide whether the following form is valid:

$\forall x \exists y Lxy \vdash Laa$

Solution

1		$\forall x \exists y Lxy$	
2		$\sim Laa$	
3	$\checkmark$	$\exists y Lay$	$1 \forall$
4		Lab	$3 \exists$
5	$\checkmark$	$\exists y Lby$	$1 \forall$
6		Lbc	$5 \exists$
7	$\checkmark$	$\exists y Lcy$	$1 \forall$
8		Lcd	$7 \exists$

$$\vdots$$

The tree never comes to an end. It is infinitely long. We apply the universal quantifier rule to 1 and thus generate a new existentially quantified formula at 3. Applying $\exists$ to this new formula at 4 introduces a new name letter, 'b', into the path. But since the universal formula at 1 is not checked, we must apply $\forall$ again at 5 for 'b'. This produces a new existential formula, which in turn introduces another new name letter, 'c', at line 6. But that requires us to apply $\forall$ to line 1 for 'c'—and so on. The form is in fact invalid, as one can see intuitively by interpreting 'L' as "is the father of"; from the assumption that everything has a father, it clearly does not follow that a is his own father. But the tree never finishes and hence never returns an answer.

This peculiar result illustrates the undecidability of predicate logic, which was mentioned at the beginning of this section. It is in fact impossible to produce a finite set of rules which returns the answer 'valid' or 'invalid' correctly for every case. *Notice, however, that this tree does not give the wrong answer; it does not give any. The answers which the refutation tree test does give are always correct.*

6.6 IDENTITY

Special symbols may be added to the language of predicate logic for special purposes. One of the most useful is the identity predicate, '$=$', which means "is identical to" or "is the same thing as." This predicate is used extensively in mathematics, primarily in application to numbers. In logic its application is broader; the name letters 'a' and 'b' in the formula '$a = b$' may denote any sort of object. We might interpret 'a', for example, as "Mark Twain" and 'b' as "Samuel Clemens," so that '$a = b$' means "Mark Twain is identical to (i.e., is the same individual as) Samuel Clemens." Or we might interpret 'a' as "the War of Secession" and 'b' as "the Civil War," so that '$a = b$' means "The War of Secession is identical to the Civil War."

The identity predicate is special because, like logical symbols and unlike ordinary predicate letters, its interpretation is fixed. It *always* means "is identical to." It is also syntactically peculiar in that, unlike other two-place predicates, it is written between the name letters to which it applies, rather than before them.[6] These peculiarities necessitate an additional formation rule:

The result of writing '$=$' between any pair of name letters is an atomic wff.

Together with the other four formation rules, this provides a rich variety of new expressions.

---

[6]Some authors surround identity formulas with brackets, especially when they are negated, thus further increasing the syntactic dissimilarity between '$=$' and other two-place predicates.

SOLVED PROBLEM

6.40   Formalize the following English sentences in the notation of predicate logic with identity, using the interpretation given below:

Symbol	Interpretation
Names	
c	Samuel Clemens
h	*Huckleberry Finn* (the book)
t	Mark Twain
One-Place Predicates	
A	is an American author
Two-Place Predicates	
B	is better (as an author) than
W	wrote

(*a*)   Mark Twain is not Samuel Clemens.

(*b*)   Mark Twain exists.

(*c*)   If Mark Twain is Samuel Clemens, then Samuel Clemens wrote *Huckleberry Finn*.

(*d*)   Only Mark Twain wrote *Huckleberry Finn*.

(*e*)   No American author is better than Mark Twain.

(*f*)   Mark Twain is the best American author.

(*g*)   Something exists.

(*h*)   At least two things exist.

(*i*)   At most one thing exists.

(*j*)   Exactly one thing exists.

(*k*)   Exactly two things exist.

(*l*)   There is exactly one writer of *Huckleberry Finn*.

(*m*)   There are at least two writers of *Huckleberry Finn*.

(*n*)   There are at most two writers of *Huckleberry Finn*.

(*o*)   There are exactly two writers of *Huckleberry Finn*.

Solution

(*a*) $\sim t = c$. This is often written '$t \neq c$'. Note that in accordance with the formation rules, there are no brackets around the atomic wff '$t = c$'; yet the negation sign applies to this entire wff, not just to 't'.

(*b*) $\exists x \, x = t$.

(*c*) $t = c \rightarrow Wch$. Outer brackets have been dropped from this formula.

(*d*) $\forall x (Wxh \rightarrow x = t)$.

(*e*) $\forall x (Ax \rightarrow \sim Bxt)$ (or equivalently $\sim \exists x (Ax \,\&\, Bxt)$).

(*f*) $At \,\&\, \forall x ((Ax \,\&\, \sim x = t) \rightarrow Btx)$. To say that Twain is the best American author is to say that he is an American author who is better than all *other* American authors. That is what this wff asserts. Note that if we had written the second conjunct without the qualification '$\sim x = t$', i.e., as '$\forall x (Ax \rightarrow Btx)$', it would assert that Twain is better than himself; this qualification is therefore crucial.

(*g*) $\exists x \, x = x$. This somewhat artificial rendering literally means "There exists something identical to itself."

(*h*) $\exists x \exists y \sim x = y$. Recall that in formulas like '$\exists x \exists y Lxy$', the variables 'x' and 'y' do not necessarily denote different objects. Thus, reading 'L' as "likes," this formula is true if even a single person likes him- or herself. Hence, to assert that at least two *different* objects exist, we need the qualification '$\sim x = y$'.

(i) $\forall x \forall y \, x = y$. This says, "For any objects x and y, x is identical to y," i.e., "All objects are identical," which is just to say that the universe contains at most one thing.

(j) $\exists x \forall y \, y = x$. This literally says, "There exists an object to which all objects are identical."

(k) $\exists x \exists y (\sim x = y \, \& \, \forall z (z = x \lor z = y))$. The clause '$\sim x = y$' asserts that x and y are two different things. The clause '$\forall z(z = x \lor z = y)$' says that for any object z, z is identical either to x or to y, i.e., that there are no objects other than x and y.

(l) $\exists x(Wxh \, \& \, \forall y(Wyh \rightarrow y = x))$.

(m) $\exists x \exists y((Wxh \, \& \, Wyh) \, \& \, \sim x = y)$.

(n) $\forall x \forall y \forall z(((Wxh \, \& \, Wyh) \, \& \, Wzh) \rightarrow ((x = y \lor x = z) \lor y = z))$.

(o) $\exists x \exists y(((Wxh \, \& \, Wyh) \, \& \, \sim x = y) \, \& \, \forall z(Wzh \rightarrow (z = x \lor z = y)))$.

Semantically, '$=$' is interpreted in every model as the identity relation. Thus, every object in the domain bears that relation to itself and to nothing else. The refutation tree technique can be extended to deal with this intended interpretation accordingly. This requires two new rules:

Identity (=):   If a wff of the form $\alpha = \beta$ appears on an open path, then if another wff ϕ containing either α or β appears unchecked on that path, write at the bottom of the path any wff not already on the path which is the result of replacing one or more occurrences of either of these name letters by the other in ϕ. Do not check either $\alpha = \beta$ or ϕ.

Negated Identity ($\sim$=):   Close any open path on which a wff of the form $\sim \alpha = \alpha$ occurs.

The idea that every object is self-identical is expressed by the negated identity rule. The identity rule expresses the idea that if an identity holds between the objects designated by two names α and β, then those names are interchangeable in every wff ϕ. In other words, every statement that involves α is equivalent to a statement where β is used in place of α (one or more times), and vice versa.

SOLVED PROBLEMS

6.41 Construct a refutation tree to decide whether the following form is valid:

$$a = b \vdash Fab \rightarrow Fba$$

Solution

1		$a = b$	
2		$\sim(Fab \rightarrow Fba)$	
3	$\checkmark$	$\sim(Faa \rightarrow Faa)$	$1, 2 =$
4		Faa	$3 \sim\!\rightarrow$
5		$\sim Faa$	$3 \sim\!\rightarrow$
6		X	$4, 5 \sim$

At step 3 we replace both occurrences of 'b' in '$\sim(Fab \rightarrow Fba)$' by '$a$', using the identity rule. The tree closes at line 5; hence the form is valid.

6.42 Construct a refutation tree to decide whether the following form is valid:

$$Fa, Fb \vdash a = b$$

Solution

1	Fa
2	Fb
3	$\sim a = b$

No rules apply here. In particular, neither identity rule operates on formulas of the form $\sim\!\alpha = \beta$ where α and β are distinct. The tree is therefore finished and the form is invalid. The open path represents a universe containing two distinct objects, both of which are F.

6.43 Construct a refutation tree to decide whether the following form is valid:

$$\sim\! a = b, \sim\! b = c \vdash \sim\! a = c$$

Solution

1	$\sim\! a = b$	
2	$\sim\! b = c$	
3 $\checkmark$	$\sim\!\sim\! a = c$	
4	$a = c$	$3 \sim\sim$
5	$\sim\! c = b$	$1, 4 =$
6	$\sim\! b = a$	$2, 4 =$

We apply the identity rule with line 4 to each formula to which it can be applied, but still the tree does not close. The form is therefore invalid. The tree shows us that the premises may be true and the conclusion false in a universe in which there are objects a, b, and c such that a and c are identical but distinct from b.[7]

6.44 Construct a refutation tree to decide whether the following form is valid:

$$\vdash \forall x \forall x (x = y \rightarrow y = x)$$

Solution

1 $\checkmark$	$\sim\!\forall x \forall x(x = y \rightarrow y = x)$	
2 $\checkmark$	$\exists x \sim\!\forall x(x = y \rightarrow y = x)$	$1 \sim\!\forall$
3 $\checkmark$	$\sim\!\forall x(a = y \rightarrow y = a)$	$2 \exists$
4 $\checkmark$	$\exists x \sim\!(a = y \rightarrow y = a)$	$3 \sim\!\forall$
5 $\checkmark$	$\sim\!(a = b \rightarrow b = a)$	$4 \exists$
6	$a = b$	$5 \sim\!\rightarrow$
7	$\sim\! b = a$	$5 \sim\!\rightarrow$
8	$\sim\! a = a$	$6, 7 =$
9	X	$8 \sim\! =$

The form is valid. At line 8 we replace the occurrence of 'b' in '$\sim\! b = a$' by 'a' to obtain '$\sim\! a = a$', which closes the tree by the negated identity rule. This form is known as the *law of symmetry of identity*.

6.45 Construct a refutation tree to decide whether the following form is valid:

$$\vdash \forall x \forall y \forall z((x = y \;\&\; y = z) \rightarrow x = z)$$

Solution

1 $\checkmark$	$\sim\!\forall x \forall y \forall z((x = y \;\&\; y = z) \rightarrow x = z)$	
2 $\checkmark$	$\exists x \sim\!\forall y \forall z((x = y \;\&\; y = z) \rightarrow x = z)$	$1 \sim\!\forall$
3 $\checkmark$	$\sim\!\forall y \forall z((a = y \;\&\; y = z) \rightarrow a = z)$	$2 \exists$
4 $\checkmark$	$\exists y \sim\!\forall z((a = y \;\&\; y = z) \rightarrow a = z)$	$3 \sim\!\forall$
5 $\checkmark$	$\sim\!\forall z((a = b \;\&\; b = z) \rightarrow a = z)$	$4 \exists$
6 $\checkmark$	$\exists z \sim\!((a = b \;\&\; b = z) \rightarrow a = z)$	$5 \sim\!\forall$
7 $\checkmark$	$\sim\!((a = b \;\&\; b = c) \rightarrow a = c)$	$6 \exists$

---

[7]Some authors allow the identity rule to be applied to an identity and itself, so that line 4 generates three more formulas, '$a = a$', '$c = c$', and '$c = a$'. Under this looser identity rule, the tree above would not be finished until these three formulas were added. But the extra formulas generated by this looser version of the rule are never needed to close a tree. Our statement of the rule stipulates that ϕ is a formula distinct from $\alpha = \beta$, which prevents the production of these superfluous formulas.

8	√	$a = b \;\&\; b = c$	$7 \sim\rightarrow$
9		$\sim a = c$	$7 \sim\rightarrow$
10		$a = b$	$8 \;\&$
11		$b = c$	$8 \;\&$
12		$a = c$	$10, 11 =$
13		X	$9, 12 \sim$

The form is valid. It is known as the *law of transitivity of identity*.

Table 6-1 summarizes the six refutation tree rules for predicate logic with identity.

Table 6-1   Refutation Tree Rules for Predicate Logic with Identity*

> Universal Quantification (∀):   If a wff of the form ∀βφ appears on an open path, then if α is a name letter that occurs in some wff on that path, write φ%ᵦ (the result of replacing all occurrences of β in φ by α) at the bottom of the path. If no wff containing a name letter appears on the path, then choose some name letter α and write φ%ᵦ at the bottom of the path. In either case, do not check ∀βφ.
>
> Negated Existential Quantification (~∃):   If an unchecked wff of the form ~∃βφ appears on an open path, check it and write ∀β~φ at the bottom of every open path that contains the newly checked wff.
>
> Negated Universal Quantification (~∀):   If an unchecked wff of the form ~∀βφ appears on an open path, check it and write ∃β~φ at the bottom of every open path that contains the newly checked wff.
>
> Existential Quantifier (∃):   If an unchecked wff of the form ∃βφ appears on an open path, check it. Then choose a name letter α that does not yet appear anywhere on that path and write φ%ᵦ (the result of replacing every occurrence of β in φ by α) at the bottom of every open path that contains the newly checked wff.
>
> Identity (=):   If a wff of the form α = β appears on an open path, then if another wff φ containing either α or β appears unchecked on that path, write at the bottom of the path any wff not already on the path which is the result of replacing one or more occurrences of either of these name letters by the other in φ. Do not check either α = β or φ.
>
> Negated Identity (~=):   Close any open path on which a wff of the form ~α = α occurs.

*Only rules specific to predicate logic are listed here. For a listing of propositional logic rules, see Table 3-1.

Supplementary Problems

I   Formalize the following English sentences, using the interpretation given below. (The last five require the identity predicate.)

Symbol	Interpretation
Names	
a	Al
b	Beth
f	fame
m	money

One-Place Predicates
    F   is famous
    G   is greedy
    H   is a human being

Two-Place Predicate
    L   likes

Three-Place Predicate
    P   prefers ... to

  (1)   Both Al and Beth like money.

  (2)   Neither Al nor Beth is famous.

  (3)   Al likes both fame and money.

  (4)   Beth prefers fame to money.

  (5)   Al prefers Beth to money and to fame.

  (6)   Beth prefers something to Al.

  (7)   Al prefers nothing to Beth.

  (8)   Some humans are both greedy and famous.

  (9)   Anything which is greedy likes money.

(10)   Not everything which likes money is greedy.

(11)   If Al and Beth are greedy, then a greedy human being exists.

(12)   Beth likes every human being.

(13)   Something does not like itself.

(14)   Something does not like something.

(15)   Nothing likes everything.

(16)   No human being likes everything.

(17)   Al likes every human being who likes him.

(18)   Some of the famous like themselves.

(19)   The famous do not all like themselves.

(20)   Not everyone likes everything that is famous.

(21)   If Al is greedy and Beth is not, then Al and Beth are not identical.

(22)   Beth likes everything except Al.

(23)   Al is the only human being who is not greedy.

(24)   Al prefers money to everything else.

(25)   Any human being who prefers money to everything else is greedy.

II   Determine which of the following formulas are wffs and which are not. Explain your answer.

   (1) (Fa)

   (2) Fab

   (3) $Fab \rightarrow Ga$

   (4) $\sim Fxy$

   (5) $\forall x \sim Fxy$

   (6) $\exists x \exists y \sim Fxy$

   (7) $(\exists x \exists y \sim Fxy)$

   (8) $\forall x Fx \rightarrow Lax$

   (9) $\forall x(Fx \rightarrow a = x)$

(10) $\forall x \forall y(Fx \rightarrow \sim y = x)$

III   Evaluate the following wffs in the model given below:

       Universe: the states of the U.S.

       c : California
       n : Nevada
       u : Utah
       E : the class of states on the East Coast
       A : the relation that holds between two adjacent states.
       C : the relation that holds between three states x, y, and z just in case the capital of x is closer to the capital of y than to the capital of z.

  (1) Ec

  (2) $\sim Acn$

  (3) $En \rightarrow Anc$

  (4) $Cucn \rightarrow Eu$

  (5) $Cucn \leftrightarrow Ccun$

  (6) $\exists x \sim Ex$

  (7) $\exists x (Ex \ \& \ Axx)$

  (8) $\exists x (Ex \rightarrow Axx)$

  (9) $\forall x \sim (Ex \rightarrow Axx)$

 (10) $\exists x \forall y Axy$

 (11) $\forall x \exists y Axy$

 (12) $\forall x \exists y \exists z Cxyz$

 (13) $\exists x \forall y (Axy \leftrightarrow \sim Ey)$

 (14) $\exists x \exists y ((Ex \ \& \ Ey) \ \& \ Axy)$

 (15) $\forall x (\sim Ex \rightarrow \exists y (Ey \ \& \ Axy))$

 (16) $\forall x (Ex \rightarrow \exists y (Ey \ \& \sim Axy))$

 (17) $\exists x (Ex \ \& \ \forall y (Ey \rightarrow Axy))$

 (18) $\forall x \forall y (Axy \leftrightarrow Ayx)$

 (19) $\forall x \forall y \forall z ((Axy \ \& \ Ayz) \rightarrow Axz)$

 (20) $\exists x (Ex \ \& \ \exists y \exists z ((Ey \ \& \ Ez) \ \& \ (Axy \ \& \ Axz)))$

IV   Determine which of the following argument forms is valid by examining whether there exists a model in which the premises are true and the conclusion false.

  (1) $Fa \vdash Fa$

  (2) $\exists x Fx \vdash \forall x Fx$

  (3) $\forall x Fx \vdash \exists x Fx$

  (4) $Fa \vdash \exists x Fx$

  (5) $\exists x Fx \vdash Fa$

  (6) $\forall x \sim Fx \vdash \sim \exists x Fx$

  (7) $\exists x \sim Fx \vdash \sim \forall x Fx$

  (8) $\sim \exists x Fx \vdash \forall x \sim Fx$

  (9) $\sim \forall x Fx \vdash \exists x \sim Fx$

 (10) $\exists x \exists y (Mx \ \& \ Ny) \vdash \exists x (Mx \ \& \ \exists y Ny)$

V   Construct a refutation tree for each of the following forms to decide whether it is valid.

  (1) $\exists x Fx \vdash Fa$

  (2) $\forall x Fx \vdash Fa$

(3) $Fa \vdash \exists x Fx$

(4) $Fa \vdash \forall x Fx$

(5) $\forall x Fx \vdash \sim\exists x \sim Fx$

(6) $\sim\exists x \sim Fx \vdash \forall x Fx$

(7) $\forall x \sim Fx \vdash \sim\forall x Fx$

(8) $\sim\forall x Fx \vdash \forall x \sim Fx$

(9) $\forall x Fx \lor \forall x Gx \vdash \forall x (Fx \lor Gx)$

(10) $\forall x (Fx \lor Gx) \vdash \forall x Fx \lor \forall x Gx$

(11) $\vdash \forall x (Fx \lor \sim Fx)$

(12) $\vdash \forall x \sim (Fx \rightarrow \sim Fx)$

(13) $\vdash \exists x Fx \leftrightarrow \sim\forall x \sim Fx$

(14) $\exists x (Fx \& \sim Fx) \vdash P$

(15) $\exists x Fx \& \exists x \sim Fx \vdash P$

(16) $\sim\exists x Fx \vdash \forall x (Fx \rightarrow P)$

(17) $\forall x \forall y (Lxy \rightarrow Lyx), \exists x Lax \vdash \exists x Lxa$

(18) $\exists x \exists y Lxy \vdash \exists x Lxx$

(19) $\forall x (Fx \rightarrow \forall y Gy) \vdash \forall x Gx$

(20) $\forall x (Fx \rightarrow \exists y Gy) \vdash Ga$

(21) $\vdash \sim\forall x \forall y \sim x = y$

(22) $\vdash \forall x \forall y (\sim x = y \leftrightarrow \sim y = x)$

(23) $\forall x (Fx \rightarrow Gx), \sim Fa \vdash \exists x \sim x = a$

(24) $\exists x \exists y Lxy \vdash \exists x \exists y (Lxy \& \sim x = y)$

(25) $\vdash \forall x \forall y ((Fxy \& x = y) \rightarrow Fyx)$

VI Formalize each of the following arguments, using the interpretation given below. Then construct a refutation tree to determine whether the form of the argument is valid or invalid.

Symbol	Interpretation
Names	
p	propositional logic
r	predicate logic
i	predicate logic with identity
One-Place Predicates	
R	is a set of rules
S	is a formal system
Two-Place Predicates	
F	is a formula of
P	is a part of
W	is a wff of

(1) Propositional logic is a part of predicate logic. Therefore, predicate logic is not a part of propositional logic.

(2) Propositional logic is a part of itself. Therefore, something is a part of propositional logic.

(3) Propositional logic is a part of predicate logic, but predicate logic is not a part of propositional logic. Consequently, predicate logic is not a part of itself.

(4) Everything is a part of itself. Therefore, propositional logic is a part of predicate logic.

(5) Every formal system is a set of rules. Therefore, every set of rules is a formal system.

(6) Since every formal system is a set of rules, anything which is not a set of rules is not a formal system.

(7) It is not true that formal systems do not exist, for predicate logic is a formal system.

(8) Predicate logic is not a formal system. Therefore, formal systems do not exist.

(9) Every formal system is a part of itself. Therefore, something is a part of itself.

(10) Formulas of predicate logic exist. Therefore, wffs of predicate logic exist, since all wffs of predicate logic are formulas of predicate logic.

(11) Predicate logic is a formal system. Therefore, all formal systems are formal systems.

(12) Not every formula of predicate logic is a wff of predicate logic. Therefore, some formulas of predicate logic are not wffs of predicate logic.

(13) Every wff of a formal system is a formula of that system. Therefore, there exists a formal system not all of whose wffs are formulas of that system.

(14) If one formal system is part of a second formal system, then every wff of the first is a wff of the second. Predicate logic is a part of predicate logic with identity, and both are formal systems. Therefore, every wff of predicate logic is also a wff of predicate logic with identity.

(15) If one thing is a part of another and that second thing is part of a third, then the first is a part of the third. Predicate logic is a part of predicate logic with identity. Therefore, if propositional logic is part of predicate logic, then propositional logic is a part of predicate logic with identity.

(16) Everything is a part of itself. Therefore, if one thing is not a part of another, the two are not identical.

(17) Propositional logic is a part of predicate logic. Predicate logic is not a part of propositional logic. Therefore, predicate logic is not identical to propositional logic.

(18) For any objects x and y, if x is a part of y and y is a part of x, then x is identical to y. Predicate logic is not identical to propositional logic. Therefore, predicate logic is not a part of propositional logic.

(19) Everything is a part of itself. Therefore, if predicate logic is identical to propositional logic, then predicate logic is a part of propositional logic and propositional logic is a part of predicate logic.

(20) Predicate logic and propositional logic are formal systems. Propositional logic is a part of predicate logic, but predicate logic is not a part of propositional logic. Therefore, there are at least two different formal systems.

Answers to Selected Supplementary Problems

I   (5) *Pabm & Pabf*

    (10) $\sim\forall x(Lxm \rightarrow Gx)$

    (15) $\sim\exists x\forall yLxy$, or, equivalently, $\forall x\sim\forall yLxy$

    (20) $\sim\forall x(Hx \rightarrow \forall y(Fy \rightarrow Lxy))$

    (25) $\forall x((Hx \& \forall y(\sim y = m \rightarrow Pxmy)) \rightarrow Gx)$

II (5) Not a wff. The variable 'y' occurs without a quantifier.

    (10) 'Fa' is a wff, by rule 1; and since '$b = a$' is a wff by the rule for identity, '$\sim b = a$' is a wff, by rule 2. Hence '$(Fa \rightarrow \sim b = a)$' is a wff, by rule 3, whence two applications of rule 4 it follows that '$\forall x\forall y(Fx \rightarrow \sim y = x)$' is a wff.

III (5) True

    (10) False

    (15) False

    (20) True

IV (5) This argument is invalid. Any model in which 'F' designates a nonempty class whose members do not include the interpretation of 'a' makes the premise '$\exists xFx$' true and the conclusion 'Fa' false.

(10)   If M is a model in which the premise is true, then there exists at least one *a*-variant M′ of M in which '∃y(Ma & Ny)' is true (by condition (5)). Hence there exists at least one *b*-variant M″ of M′ in which 'Ma & Nb' is true (by another application of condition (5)). Since 'Ma & Nb' is true in M″, so are 'Ma' and 'Na'. This means that 'Ma' is true in M′ too, since M′ agrees with M″ on everything except possibly for the interpretation of '*b*'. Moreover, '∃yNy' must be true in M′ by condition (5). Thus 'Ma & ∃yNy' is true in M′. And since M′ is an *a*-variant of *M*, this implies that '∃x(Mx & ∃yNy)' is true in M (condition (5)). Because M was arbitrarily chosen, we may now conclude that *every* model that makes '∃y(Ma & Ny)' true also makes '∃x(Mx & ∃yNy)' true. The argument is therefore valid.

V   (5)   1   ∀xFx
              2   √   ~~∃x~Fx
              3   √   ∃x~Fx   2 ~~
              4   ~Fa   3 ∃
              5   Fa   1 ∀
              6   X   4, 5 ~

The form is valid.

(10)   1   ∀x(Fx ∨ Gx)
       2   √   ~(∀xFx ∨ ∀xGx)
       3   √   ~∀xFx   2 ~∨
       4   √   ~∀xGx   2 ~∨
       5   √   ∃x~Fx   3 ~∀
       6   √   ∃x~Gx   4 ~∀
       7   ~Fa   5 ∃
       8   ~Gb   6 ∃
       9   √   Fa ∨ Ga   1 ∀
       10   √   Fb ∨ Gb   1 ∀

       11   Fa   9 ∨   Ga   9 ∨
       12   X   7, 11 ~
       13   Fb   10 ∨   Gb   10 ∨
       14   X   8, 13 ~

The form is invalid.

(15)   1   √   ∃xFx & ∃x~Fx
       2   ~P
       3   √   ∃xFx   1 &
       4   √   ∃x~Fx   1 &
       5   Fa   3 ∃
       6   ~Fb   4 ∃

The form is invalid.

(20)   1   ∀x(Fx → ∃yGy)
       2   ~Ga
       3   √   Fa → ∃yGy   1 ∀

       4   ~Fa 3 →   √ ∃yGy   3 →
       5   Gb   4 ∃
       6   √ Fb → ∃yGy   1 ∀

       7   ~Fb 6 →   √ ∃yGy   6 →
       8   Gc   7 ∃
       9   √ Fc → ∃yGy 1 ∀

       10   ~Fc 9 →   √ ∃yGy   9 →

The tree is infinite; because line 1 is never checked, the formula '∃yGy' appears over and over again (lines 4, 7, 10, etc.), and in each of its new appearances it generates a new name letter. We can nevertheless see that the form is invalid, since all paths except the leftmost one eventually end without closing, and any tree with even one complete open path represents an invalid form.

(25)
1	✓	$\sim\forall x\forall y((Fxy \& x = y) \to Fyx)$	
2	✓	$\exists x\sim\forall y((Fxy \& x = y) \to Fyx)$	$1 \sim\forall$
3	✓	$\sim\forall y((Fay \& a = y) \to Fya)$	$2\exists$
4	✓	$\exists y\sim((Fay \& a = y) \to Fya)$	$3\sim\forall$
5	✓	$\sim((Fab \& a = b) \to Fba)$	$4\exists$
6	✓	$Fab \& a = b$	$5\sim\to$
7		$\sim Fba$	$5\sim\to$
8		Fab	$6\&$
9		$a = b$	$6\&$
10		Fbb	$8, 9 =$
11		Fba	$9, 10 =$
12		X	$7, 11 \sim$

The form is valid.

VI (5)
1		$\forall x(Sx \to Rx)$	
2	✓	$\sim\forall x(Rx \to Sx)$	
3	✓	$\exists x\sim(Rx \to Sx)$	$2\sim\forall$
4	✓	$\sim(Ra \to Sa)$	$3\exists$
5		Ra	$4\sim\to$
6		$\sim Sa$	$4\sim\to$
7	✓	$Sa \to Ra$	$1\forall$

8   $\sim Sa$ 7→ Ra 7→

The form is invalid.

(10)
1	✓	$\exists xFxr$	
2		$\forall x(Wxr \to Fxr)$	
3	✓	$\sim\exists xWxr$	
4		Far	$1\exists$
5		$\forall x\sim Wxr$	$3\sim\exists$
6		$\sim War$	$5\forall$
7		$\sim Wrr$	$5\forall$
8	✓	$War \to Far$	$2\forall$
9	✓	$Wrr \to Frr$	$2\forall$

10            $\sim War$ 8→ Far 8→

11 $\sim Wrr$ 9→ Frr 9→ $\sim Wrr$ 9→ Frr 9→

The form is invalid. Notice that under the given interpretation some of the paths represent universes involving the arcane possibility that predicate logic is a formula of itself. Although this is not a possibility that we would normally regard as counting against the validity of the argument, from a purely formal point of view it must not be ignored. The other open paths, however, exhibit more mundane ways in which the premises may be true and the conclusion false.

(15)
1		$\forall x\forall y\forall z((Pxy \& Pyz) \to Pxz)$	
2		Pri	
3	✓	$\sim(Ppr \to Ppi)$	
4		Ppr	$3\sim\to$
5		$\sim Ppi$	$3\sim\to$
6		$\forall y\forall x((Ppy \& Pyz) \to Ppz)$	1
7		$\forall x((Ppr \& Prz) \to Ppz)$	6

8		√	$(Ppr \& Pri) \rightarrow Ppi$	7

9	√	$\sim(Ppr \& Pri)$	8 $\rightarrow$	Ppi	8 $\rightarrow$
10				X	5, 9 $\sim$
11	$\sim Ppr$	9 $\sim\&$	$\sim Pri$	9 $\sim\&$
12	X	4, 11 $\sim$	X	2, 10 $\sim$

The form is valid.

(20)

1		√	$Sr \& Sp$	
2		√	$Ppr \& \sim Prp$	
3		√	$\sim\exists x\exists y((Sx \& Sy) \& \sim x = y)$	
4			Sr	1 &
5			Sp	1 &
6			Ppr	2 &
7			$\sim Prp$	2 &
8			$\forall x \sim \exists y((Sx \& Sy) \& \sim x = y)$	3 $\sim\exists$
9		√	$\sim\exists y((Sr \& Sy) \& \sim r = y)$	8 $\forall$
10			$\forall y \sim((Sr \& Sy) \& \sim r = y)$	9 $\sim\exists$
11		√	$\sim((Sr \& Sp) \& \sim r = p)$	10 $\forall$

12		√	$\sim(Sr \& Sp)$	11 $\sim\&$		√	$\sim\sim r = p$	11 $\sim\&$
13							$r = p$	12 $\sim\sim$
14	$\sim Sr$	12 $\sim\&$	$\sim Sp$	12 $\sim\&$			$\sim Ppp$	7, 13 =
15	X	4, 14 $\sim$	X	5, 14 $\sim$			$\sim Ppr$	13, 14 =
16							X	6, 15 $\sim$

The form is valid.

The Predicate Calculus

7.1 REASONING IN PREDICATE LOGIC

In this chapter we present a system of inference rules that extends the propositional calculus of Chapter 4 to the full language of predicate logic with identity. The motivation for developing such as system—called the *predicate calculus*—is similar to that of Chapter 4: the characterization of validity in terms of truth values or refutation trees is in principle adequate, but does not capture the inferential process involved in a typical argument. Typically, an arguer tries to establish a conclusion by *deducing* it *from* the premises, not by issuing a challenge to describe a possible situation in which the premises are true and the conclusion false.

Since the language of predicate logic includes the truth-functional connectives, the predicate calculus will include all the basic introduction and elimination rules of the propositional calculus (and hence all the derived rules; for summaries of both the basic and derived rules, see the end of Chapter 4). In addition, the predicate calculus has new introduction and elimination rules for the quantifiers and for the identity predicate. It can be shown that this set of rules is both *valid* (in the sense that the rules generate only argument forms that are valid in virtue of the semantics of the quantifiers, the truth-functional connectives, and the identity predicate) and *complete* (in the sense that they generate all valid argument forms). As with the propositional calculus, however, a rigorous proof of this fact is beyond the scope of this book.

Validity and completeness are obviously very important properties, for they guarantee that the inferential techniques of the predicate calculus identify the same class of valid argument forms as the semantic techniques of Chapter 6. However, it also follows from this that the predicate calculus is undecidable, since the property of being a valid predicate logic argument form is undecidable. (See the remarks at the beginning of Section 6.5.) Thus, the inference rules of the predicate calculus do not always provide an answer to the question 'Is this form valid?'. If we are unable to construct a proof for a given form, that could mean either of two things: the form is in fact invalid, or it is valid but we are not skillful enough to produce a proof.

7.2 INFERENCE RULES FOR THE UNIVERSAL QUANTIFIER

Before introducing the quantifier rules, we first consider a proof in the predicate calculus which uses only rules of propositional logic.

SOLVED PROBLEM

7.1 Prove:

$$\sim Fa \lor \exists xFx, \ \exists xFx \to P \vdash Fa \to P$$

Solution

1	$\sim Fa \lor \exists xFx$	A
2	$\exists xFx \to P$	A
3	Fa	H (for $\to$I)
4	$\sim \sim Fa$	3 DN
5	$\exists xFx$	1, 4 DS
6	P	2, 5 $\to$E
7	$Fa \to P$	3–6 $\to$I

Notice that despite their greater internal complexity, formulas of predicate logic are treated just as formulas of propositional logic are by the propositional calculus rules.

We now introduce the first of our new rules, the universal quantifier elimination rule, ∀E. Universal elimination is a formal expression of the fact that whatever is true of everything must be true of a particular individual, and is somewhat analogous to the universal quantification rule of the refutation tree technique (Section 6.5).

> Universal Elimination (∀E):   From a universally quantified wff $\forall\beta\phi$ we may infer any wff of the form $\phi\%_\beta$ which results from replacing each occurrence of the variable β in ϕ by some name letter α.

This rule is sometimes called *universal instantiation*. It is used, for example, in proving the validity of:

> All men are mortal.
> Socrates is a man.
> ∴ Socrates is mortal.

Using 'M_1' for 'is a man', 'M_2' for 'is mortal', and 's' for 'Socrates', we may formalize this argument as:

$$\forall x(M_1x \rightarrow M_2x), M_1S \vdash M_2s$$

SOLVED PROBLEMS

7.2   Prove the validity of this form.

Solution

1	$\forall x(M_1x \rightarrow M_2x)$	A
2	M_1s	A
3	$M_1s \rightarrow M_2s$	1 ∀E
4	M_2s	2, 3 →E

In the application of ∀E at line 3, we derive the form of the proposition 'If Socrates is a man, then Socrates is mortal' from the form of 'All men are mortal'. With respect to the rule ∀E, β is 'x', α is 's', ϕ is '$(M_1x \rightarrow M_2x)$', and $\phi\%_\beta$ is '$(M_1s \rightarrow M_2s)$', from which we drop the outer brackets according to our convention.

7.3   Prove:

$$\forall x(Fx \rightarrow Gx), \forall xFx \vdash Ga$$

Solution

1	$\forall x(Fx \rightarrow Gx)$	A
2	$\forall xFx$	A
3	$Fa \rightarrow Ga$	1 ∀E
4	Fa	2 ∀E
5	Ga	3, 4 →E

Since the conclusion contains the name letter 'a', this is obviously the letter to use in performing the two steps of ∀E.

7.4   Prove:

$$\sim Fa \vdash \sim\forall xFx$$

Solution

1	~Fa	A
2	∀xFx	H (for ~I)
3	Fa	2 ∀E
4	Fa & ~Fa	1, 3 &I
5	~∀xFx	2–4 ~I

The conclusion is negated, so the obvious strategy is *reductio ad absurdum*. We hypothesize the conclusion without its negation sign at line 2 and show that this leads to a contradiction at line 4. That enables us to discharge the hypothesis and infer the conclusion at line 5.

7.5   Prove:

$\forall x \forall y Fxy \vdash Faa$

Solution

1	∀x∀yFxy	A
2	∀yFay	1 ∀E
3	Faa	2 ∀E

Note that two separate steps of ∀E are required to remove the two quantifiers. This derivation illustrates remark 1 near the end of Section 6.2.

The introduction rule ∀I for the universal quantifier is a bit more complicated than ∀E. The fundamental idea is this: If we can prove something about an individual *a* without making any assumptions that distinguish *a* from any other individual, then what we have proved of *a* could have been proved of anything at all. We may therefore conclude that it holds for everything.

Universal introduction is used in proving universally quantified conclusions. Consider, for example, the valid (but unsound) argument:

    All fish are guppies.
    All guppies are handsome.
∴ All fish are handsome.

This argument is a categorical syllogism, and so we could show its validity by a Venn diagram as in Chapter 5. But here we will use the more powerful techniques of the predicate calculus. The argument may be formalized as:

$\forall x(Fx \to Gx), \forall x(Gx \to Hx) \vdash \forall x(Fx \to Hx)$

SOLVED PROBLEM

7.6   Construct a proof for this form.

Solution

1	∀x(Fx → Gx)	A
2	∀x(Gx → Hx)	A
3	Fa → Ga	1 ∀E
4	Ga → Ha	2 ∀E
5	Fa → Ha	3, 4 HS
6	∀x(Fx → Hx)	5 ∀I

The name letter 'a', which is introduced by ∀E at steps 3 and 4, designates an individual about which we have made no assumptions ('a' does not occur in assumptions 1 and 2).[1] This enables the individual a to function in the proof as a representative of all individuals. For the purposes of the proof it does not matter who or what a is. All that matters is that we have made no assumptions about a, so that whatever we ultimately prove about a could have been proved as well about any individual. At step 5 we prove that if a is a fish, then a is handsome. Because we have made no assumptions about a, our proof is perfectly general; we could replace each occurrence of the name letter 'a' in the proof with any other name letter without disrupting the validity of the proof down through line 5. Suppose, for example, that we replace 'a' with 'b'. Then at line 5 we would have proved that if b is a fish then b is handsome. We could prove the same thing for the individual c, the individual d—and in general for any individual. Thus, because we have made no special assumptions about a, lines 1 to 5 function in effect as a proof that for any object x, if x is a fish, then x is handsome. That is just what ∀I enables us to conclude at line 6.

Formally, the rule ∀I may be stated as follows.

> <u>Universal Introduction (∀I):</u>   From a wff ϕ containing a name letter α not occurring in any assumption or in any hypothesis in effect at the line on which ϕ occurs, we may infer any wff of the form $\forall\beta\phi^\beta{}_\alpha$, where $\phi^\beta{}_\alpha$ is the result of replacing all occurrences of α in ϕ by some variable β not already in ϕ.

In the application of ∀I at line 6 of Problem 7.6, the wff ϕ is '$Fa \to Ha$', the name letter α is 'a', the variable β is 'x', and the formula $\phi^\beta{}_\alpha$ is '$(Fx \to Hx)$'.

∀I is sometimes called *universal generalization*. The requirement that α not occur in any assumption or in any hypothesis in effect at the line on which ϕ occurs ensures that we assume nothing that distinguishes the individual designated by α from any other individual. The requirement that the variable β not occur in ϕ ensures that $\forall\beta\phi^\beta{}_\alpha$ will be well formed (see formation rule 4).

∀I must be applied strictly as stated. In particular, the following qualifications are crucial:

(1) *The name letter α may not appear in any assumption.* The following derivation, for example, ignores this qualification and in consequence is invalid:

$$
\begin{array}{lll}
1 & Fa & A \\
2 & \forall xFx & 1 \ \forall\text{I (incorrect)}
\end{array}
$$

From the assumption, say, that Albert is a fish, it certainly does not follow that everything is a fish.

(2) *The name letter α may not appear in any hypothesis in effect at the line on which ϕ occurs.* (Recall that a hypothesis is *in effect* at a line if it is introduced before that line and is not yet discharged at that line; in other words, it is in effect if the vertical line beginning with the hypothesis extends down through that line.) This prevents the sort of invalidity exhibited above from occurring in the hypothetical parts of a proof. The following derivation is invalid because it violates this clause:

$$
\begin{array}{lll}
1 & \forall x(Fx \to Gx) & A \\
2 & Fa \to Ga & 1 \ \forall\text{E} \\
3 & \quad Fa & H \ (\text{for} \to\text{I}) \\
4 & \quad Ga & 2, 3 \to\text{E} \\
5 & \quad \forall xGx & 4 \ \forall\text{I (incorrect)} \\
6 & Fa \to \forall xGx & 3\text{--}5 \to\text{I}
\end{array}
$$

---

[1] Some versions of the predicate calculus employ special names called "arbitrary names" for this purpose. Others use unquantified variables.

(Here φ is '*Ga*' and α is '*a*'.) Clearly, from the assumption that all fish are guppies, it does not follow that if Albert is a fish, then everything is a guppy. The mistake is at step 5, where ∀I is applied to '*Ga*' even though '*a*' occurs in the hypothesis '*Fa*', which is still in effect.

(3)   φ^β⁄$_α$ *is the result of replacing all occurrences of* α *in* φ *by some variable* β. The emphasis here is on the word 'all'. This clause prevents the following sort of mistake:

1	∀x*Lxx*	A
2	*Laa*	1 ∀E
3	∀z*Lza*	2 ∀I (incorrect)

(Here φ is '*Laa*', β is '*z*', and α is '*a*'.) From the assertion that everything loves itself, it does not follow that everything loves Albert. φ^β⁄$_α$ should be '*Lzz*', not '*Lza*'. Thus ∀I would be correctly used if instead of '∀z*Lza*' we concluded '∀z*Lzz*'.

The following proofs illustrate correct uses of the universal quantifier rules.

SOLVED PROBLEMS

7.7   Prove:

$$\forall x(Fx \,\&\, Gx) \vdash \forall xFx \,\&\, \forall xGx$$

Solution

1	∀x(*Fx* & *Gx*)	A
2	*Fa* & *Ga*	1 ∀E
3	*Fa*	2 &E
4	*Ga*	2 &E
5	∀x*Fx*	3 ∀I
6	∀x*Gx*	4 ∀I
7	∀x*Fx* & ∀x*Gx*	5, 6 &I

To utilize the assumption '∀x(*Fx* & *Gx*)', we instantiate it at line 2, using '*a*' to designate a representative individual. Since we make no assumptions or hypotheses about *a*, the applications of ∀I at steps 5 and 6 are legitimate.

7.8   Prove:

$$\forall x(Fx \rightarrow (Gx \lor Hx)), \forall x{\sim}Gx \vdash \forall xFx \rightarrow \forall xHx$$

Solution

1	∀x(*Fx* → (*Gx* ∨ *Hx*))	A
2	∀x~*Gx*	A
3	*Fa* → (*Ga* ∨ *Ha*)	1 ∀E
4	~*Ga*	2 ∀E
5	⎜ ∀x*Fx*	H (for →I)
6	⎜ *Fa*	5 ∀E
7	⎜ *Ga* ∨ *Ha*	3, 6 →E
8	⎜ *Ha*	4, 7 DS
9	⎜ ∀x*Hx*	8 ∀I
10	∀x*Fx* → ∀x*Hx*	5–9 →I

Again, the name letter '*a*' is introduced to designate a representative individual. Since the conclusion is a conditional, we hypothesize its antecedent at 5 in order to derive its consequent at 9 and then apply →I at 10. The use of ∀I at step 9 obeys the ∀I rule, since none of our assumptions or hypotheses contains '*a*'. Notice that a different strategy would be required if instead of the conditional conclusion '∀x*Fx* → ∀x*Hx*' we wished to prove the universally quantified conclusion '∀x(*Fx* → *Hx*)'; this is illustrated in the next problem.

7.9 Prove:

$$\forall x(Fx \rightarrow (Gx \vee Hx)), \forall x\sim Gx \vdash \forall x(Fx \rightarrow Hx)$$

Solution

1	$\forall x(Fx \rightarrow (Gx \vee Hx))$	A
2	$\forall x\sim Gx$	A
3	$Fa \rightarrow (Ga \vee Ha)$	1 $\forall$E
4	$\sim Ga$	2 $\forall$E
5	$\quad Fa$	H (for $\rightarrow$I)
6	$\quad Ga \vee Ha$	3, 5 $\rightarrow$E
7	$\quad Ha$	4, 6 DS
8	$Fa \rightarrow Ha$	5–7 $\rightarrow$I
9	$\forall x(Fx \rightarrow Hx)$	8 $\forall$I

Here we need to obtain the conditional '$Fa \rightarrow Ha$' in order to get '$\forall x(Fx \rightarrow Hx)$' by $\forall$I at step 8, hence we hypothesize its antecedent 'Fa' instead of '$\forall xFx$' for $\rightarrow$I at step 5.

7.10 Prove:

$$\forall xFax, \forall x\forall y(Fxy \rightarrow Gyx) \vdash \forall xGxa$$

Solution

1	$\forall xFax$	A
2	$\forall x\forall y(Fxy \rightarrow Gyx)$	A
3	Fab	1 $\forall$E
4	$\forall y(Fay \rightarrow Gya)$	2 $\forall$E
5	$Fab \rightarrow Gba$	4 $\forall$E
6	Gba	3, 5 $\rightarrow$E
7	$\forall xGxa$	6 $\forall$I

In step 3 we introduce 'b' to stand for a representative individual. Any other name letter would do as well, except for 'a'. We could, of course, legitimately infer 'Faa' by $\forall$E from line 1 at line 3, but that would block any later attempt to use $\forall$I with 'a', since 'a' occurs in the assumption at 1. The use of $\forall$I with 'b' at step 7, however, is legitimate, since we have made no assumptions or hypotheses containing 'b'. Note that two separate steps of $\forall$E (lines 4 and 5) are required to remove the quantifiers from '$\forall x\forall y(Fxy \rightarrow Gyx)$'.

7.11 Prove:

$$\forall xFx \rightarrow \forall xGx, \sim Ga \vdash \sim\forall xFx$$

Solution

1	$\forall xFx \rightarrow \forall xGx$	A
2	$\sim Ga$	A
3	$\quad \forall xFx$	H (for $\sim$I)
4	$\quad \forall xGx$	1, 3 $\rightarrow$E
5	$\quad Ga$	4 $\forall$E
6	$\quad Ga \,\&\sim Ga$	2, 5 &I
7	$\sim\forall xFx$	3–6 $\sim$I

Since the conclusion is a negated formula, a *reductio* strategy is in order. Thus we hypothesize '$\forall xFx$' at line 3. Notice that $\forall$E cannot be performed directly on '$\forall xFx \rightarrow \forall xGx$', since it is a conditional, not a universally quantified wff. (This is a bit easier to see if we recall that its true form is '$(\forall xFx \rightarrow \forall xGx)$', restoring the outer brackets dropped by convention.)

7.3 INFERENCE RULES FOR THE EXISTENTIAL QUANTIFIER

Like the universal quantifier, the existential quantifier has both an elimination and an introduction rule. The existential introduction rule, ∃I, is straightforward: from the premise that an individual has a certain (simple or complex) property, it follows that *something* (the individual in question, if nothing else) has that property. In formal terms:

> Existential Introduction (∃I):   Given a wff φ containing some name letter α, we may infer any wff of the form ∃βφβ⁄α, where φβ⁄α is the result of replacing one or more occurrences of α in φ by some variable β not already in φ.

For example, given the assumption '*Fa & Ga*' (the assumption that the individual *a* has the complex property of being both *F* and *G*), we may infer '∃x(Fx & Gx)'. This is a simple two-step proof:

```
1  Fa & Ga          A
2  ∃x(Fx & Gx)      1 ∃I
```

With respect to the formal statement of the rule, φ in this case is '*Fa & Ga*', α is '*a*', β is '*x*', and φβ⁄α is '(*Fx & Gx*)'. There are several things to note about this rule:

(1) *Unlike ∀I, ∃I places no restrictions on previous occurrences of the name letter* α; α may legitimately appear in an undischarged hypothesis or in an assumption, as in step 1 above.

(2) *Again in contrast to the case for ∀I, the variable β need not replace all occurrences of α in φ; it need replace only one or more.* Thus the following proofs are both correct:

```
1  Fa & Ga          A
2  ∃x(Fx & Ga)      1 ∃I

1  Fa & Ga          A
2  ∃x(Fa & Gx)      1 ∃I
```

From the assumption that Alf is a frog and Alf is green, it certainly follows that *something* is such that *it* is a frog and Alf is green, and also that *something* is such that Alf is a frog and *it* is green.

(3) *Like ∀I, ∃I allows us to introduce only one quantifier at a time, and only at the leftmost position in a formula.* The following inference, for example, is incorrect:

```
1  Fa → Ga          A
2  ∃xFx → Ga        1 ∃I (incorrect)
```

To see the error clearly, note that '∃xFx → Ga' is not officially a wff; it is a conventional abbreviation for the wff '(∃xFx → Ga)', from which it is obtained by dropping the outer brackets. This makes it clear that the quantifier does not occupy the leftmost position, which it must have if ∃I is applied correctly; the leftmost symbol is a bracket. It is also apparent to intuition that this inference is invalid; from the assumption that if Alf is a frog then Alf is green it clearly does not follow that if *something* is a frog then Alf is green.

The ∃I rule brings with it two important presuppositions both of which are implicit in the semantics of the language of predicate logic:

(1) All proper names refer to existing individuals.
(2) At least one individual exists.

The first presupposition is a result of the fact that ∃I may be applied to any name letter. Since the existential quantifier asserts existence, we must always interpret name letters as referring to existing individuals in order to guarantee that all applications of ∃I are valid. To see how invalidity may result if we ignore this presupposition, suppose we interpret '*a*' as the name of a nonexistent individual, say the Greek god Apollo, and 'M' as the predicate 'is mythological'. Then by $\exists$I we can obtain the following *in*valid derivation:

```
1   Ma        A
2   ∃xMx      1 ∃I
```

The premise is intuitively true (Apollo is a mythological being); but the conclusion is false, since mythological beings do not actually exist. We have followed the $\exists$I rule strictly here; our mistake consists not in misapplication of the rule but in using the name of a nonexistent thing. The standard predicate calculus is simply not equipped to deal with such names, and it is important for users of the predicate calculus to be aware of this limitation. Various modifications of the calculus (known as *free logics*) have been developed to deal with non-designating names or names that designate nonexistent things, but we shall not discuss them here.

The second presupposition is a consequence of the first. Since name letters are always interpreted as referring to existing things, use of the calculus on any permissible interpretation presupposes the existence of at least one thing. But mere use of the calculus does not presuppose the existence of more than one (nonlinguistic) thing, even though (through the device of subscripts) it possesses a potential infinity of name letters. For two or more or even infinitely many names may all refer to the same individual. It is *not* presumed that different names must refer to different things. (See Section 6.4.)

The following proofs further illustrate the use of $\exists$I.

SOLVED PROBLEMS

7.12　Prove:

$$\forall x(Fx \lor Gx) \vdash \exists x(Fx \lor Gx)$$

Solution

```
1   ∀x(Fx ∨ Gx)        A
2   Fa ∨ Ga            1 ∀E
3   ∃x(Fx ∨ Gx)        2 ∃I
```

We introduce 'a' in step 2 to designate a representative individual. (We may presume that such an individual exists because the predicate calculus presupposes the existence of at least one thing.) Since we've assumed that everything is either F or G, it follows that a is either F or G and hence that *something* is either F or G.

7.13　Prove:

$$\forall x(Fx \lor Gx) \vdash \exists xFx \lor \exists xGx$$

Solution

```
 1   ∀x(Fx ∨ Gx)              A
 2   Fa ∨ Ga                  1 ∀E
 3  │ Fa                      H (for →I)
 4  │ ∃xFx                    3 ∃I
 5  │ ∃xFx ∨ ∃xGx             4 ∨I
 6   Fa → (∃xFx ∨ ∃xGx)       3–5 →I
 7  │ Ga                      H (for →I)
 8  │ ∃xGx                    7 ∃I
 9  │ ∃xFx ∨ ∃xGx             8 ∨I
10   Ga → (∃xFx ∨ ∃xGx)       7–9 →I
11   ∃xFx ∨ ∃xGx              2, 6, 10 ∨E
```

The conclusion of Problem 7.12 was an existentially quantified disjunction. Here the conclusion is a disjunction of two existentially quantified statements; thus a different proof strategy is required. Having instantiated '$\forall x(Fx \lor Gx)$' at step 2, we then proceed by two hypothetical proofs (lines 3–6 and 7–10) to deduce the conditionals needed to prove the conclusion by ∨E.

7.14   Prove

$$\sim\exists xFx \vdash \forall x\sim Fx$$

Solution

1	$\sim\exists xFx$	A
2	Fa	H (for $\sim$I)
3	$\exists xFx$	2 ∃I
4	$\exists xFx \;\&\; \sim\exists xFx$	1, 3 &I
5	$\sim Fa$	2–4 $\sim$I
6	$\forall x\sim Fx$	5 ∀I

To obtain the universally quantified conclusion '$\forall x\sim Fx$', we need to get '$\sim Fa$' and then apply ∀I. Since '$\sim Fa$' is negated, a $\sim$I strategy is indicated. We therefore hypothesize 'Fa' at step 2 in order to derive a contradiction at step 4, which allows us to obtain '$\sim Fa$' at 5. Our hypothesis contains the name letter 'a', but since it is no longer in effect at line 5 and 'a' does not occur in the assumption, the application of ∀I at 6 is correct.

7.15   Prove:

$$\sim\exists x(Fx \;\&\; \sim Gx) \vdash \forall x(Fx \rightarrow Gx)$$

Solution

1	$\sim\exists x(Fx \;\&\; \sim Gx)$	A
2	Fa	H (for $\rightarrow$I)
3	$\sim Ga$	H (for $\sim$I)
4	$Fa \;\&\; \sim Ga$	2, 3 &I
5	$\exists x(Fx \;\&\; \sim Gx)$	4 ∃I
6	$\exists x(Fx \;\&\; \sim Gx) \;\&\; \sim\exists x(Fx \;\&\; \sim Gx)$	1, 5 &I
7	$\sim\sim Ga$	3–6 $\sim$I
8	Ga	7 $\sim$E
9	$Fa \rightarrow Ga$	2–8 $\rightarrow$I
10	$\forall x(Fx \rightarrow Gx)$	9 ∀I

As in Problem 7.14, the conclusion is universally quantified, so we follow a similar overall strategy. The strategy is to obtain '$Fa \rightarrow Ga$' first and then apply ∀I. Since '$Fa \rightarrow Ga$' is a conditional, we proceed by $\rightarrow$I, hypothesizing its antecedent 'Fa' at line 2. There is no straightforward way to derive 'Ga' from 'Fa', and so we try negation introduction, hypothesizing '$\sim Ga$' in hopes of obtaining a contradiction. We get the contradiction at line 6, which allows us to obtain 'Ga' at line 8 and hence to apply $\rightarrow$I at line 9. Both of our hypotheses contain 'a', but since neither is in effect at line 9 and our assumption does not contain 'a', the use of ∀I at line 10 is legitimate.

To reason from existential premises, we need a second existential quantifier rule, existential elimination (∃E).[2] Like $\rightarrow$I and $\sim$I, ∃E uses hypothetical reasoning. An existential premise asserts that at least one thing has a (simple or complex) property. To reason from such a premise, we choose an individual to represent one of the things having this property and hypothesize that this individual in fact has this property. Then, without making any additional assumptions about this individual, we derive from our hypothesis whatever conclusion we are trying to prove. Since we have assumed nothing about

---

[2]Some versions of the predicate calculus have a rule called *existential instantiation*, which serves roughly the same purpose as ∃E but operates differently. The two should not be confused.

our representative individual except that it has the property which the existential premise ascribes to *something*, the fact that we have derived the desired conclusion from our hypothesis shows that no matter which individual has this property, our conclusion must be true. We are therefore entitled to discharge the hypothesis and assert our conclusion on the strength of the existential premise alone. Existential elimination is what officially enables us to do this. (The rationale for this rule is similar to the rationale for the existential quantification rule of the refutation tree technique; see Section 6.5.)

For example, to prove the conclusion '$\exists xFx$' from the existential premise '$\exists x(Fx \& Gx)$', we proceed as follows:

1	$\exists x(Fx \& Gx)$	A
2	$Fa \& Ga$	H (for $\exists$E)
3	Fa	2 &E
4	$\exists xFx$	3 $\exists$I
5	$\exists xFx$	1, 2–4 $\exists$E

The existential premise '$\exists x(Fx \& Gx)$' asserts that at least one thing has the complex property of being both F and G, but it does not tell us which thing or things do. We choose the individual a to represent one of these things and hypothesize at step 2 that a has this property. (This is merely a hypothesis; the individual a, whatever it is, need not in fact have this property.) Line 2 says in effect, "Let's suppose that a is one of the things that are both F and G and see what follows." In the hypothetical derivation (lines 2 to 4), we show that the desired conclusion '$\exists xFx$' follows. Because we make no assumptions about a except for the hypothesis that it is both F and G, the hypothetical derivation has a special sort of generality; we could have used any other name letter in the hypothesis and still have derived the same conclusion. This generality is the key to the proof, for it guarantees that no matter what it is that is both F and G (so long as *something* is), the conclusion '$\exists xFx$' must be true. Thus it guarantees that '$\exists xFx$' validly follows from '$\exists x(Fx \& Gx)$'. The $\exists$E rule allows us to affirm this by discharging the hypothesis and reasserting the conclusion on the strength of '$\exists x(Fx \& Gx)$' alone. As justification, we cite the existential premise, line 1, and the lines containing the hypothetical derivation, lines 2 to 4.

We may now state the $\exists$E rule precisely:

> Existential Elimination ($\exists$E):   Given an existentially quantified wff $\exists\beta\phi$ and a derivation of some conclusion ψ from a hypothesis of the form ϕ^α_β (the result of replacing each occurrence of the variable β in ϕ by some name letter α not already in ϕ), we may discharge ϕ^α_β and reassert ψ. *Restriction:* The name letter α may not occur in ψ, nor in any assumption, nor in any hypothesis that is in effect at the line at which $\exists$E is applied.

In the derivation above, $\exists\beta\phi$ is '$\exists x(Fx \& Gx)$', β is 'x', α is 'a', ϕ^α_β is '$Fa \& Ga$', and ψ is '$\exists xFx$'. We shall call ϕ^α_β a *representative instance of* $\exists\beta\phi$, since in $\exists$E, α is taken to represent one of the things that have the property indicated by ϕ.

Various aspects of this rule require elaboration:

(1)   *The name letter α must not already occur in ϕ.* This is to prevent mistakes such as the following:

1	$\forall x\exists yFyx$	A
2	$\exists yFya$	1 $\forall$E
3	Faa	H (for $\exists$E)
4	$\exists xFxx$	3 $\exists$I
5	$\exists xFxx$	2, 3–4 $\exists$E (incorrect)

Here $\exists\beta\phi$ is '$\exists yFya$', β is 'y', α is 'a', ϕ^α_β is 'Faa', and ψ is '$\exists xFxx$'. This derivation is obviously invalid; from the assumption that everything has a father ($\forall x\exists yFyx$), it does not follow that something is its own father ($\exists xFxx$). The mistake is in the application of $\exists$E at step 5. Because 'a' already occurs in '$\exists yFya$', introducing 'a' at line 3 to represent the individual or individuals designated by the variable 'y' in this formula deprives the hypothetical derivation (lines 3 and 4) of generality and thus invalidates the application of $\exists$E at 5. For if 'y' were replaced by any other name letter in forming the hypothesis at line 3, we could not derive '∃xFxx' at 4. Thus the hypothetical derivation does not show that '∃xFxx' (someone is his own father) must be true no matter which individual we suppose to be *a*'s father. (Indeed, this can be deduced only under the special hypothesis '*Faa*'.) So we have not shown that the mere fact that *a* has a father (∃yFya) implies that someone is his own father (∃xFxx), which is why our use of ∃E is incorrect.

(2)   *The name letter* α *must not occur in* ψ *(the conclusion of the hypothetical derivation)*. If this restriction is violated, the following sort of mistake may result:

    1   ∃xHxx   A
    2 | Haa   H (for ∃E)
    3 | ∃xHax   2 ∃I
    4   ∃xHax   1, 2–3 ∃E (incorrect)

This is invalid; from the assumption that something hit itself (∃xHxx), it does not follow that, say, Alice hit something (∃xHax). The mistake, once again, is in the application of ∃E. The formula ψ (in this case, '∃xHax') contains the name letter α (in this case, '*a*'), which deprives the hypothetical derivation of generality. For if we had introduced some other name letter, say '*b*', to designate our representative individual at step 2, we could not have derived the conclusion that Alice hit something (∃xHax); at step 3 we could derive only '∃xHbx'. Thus the hypothetical derivation does not show that no matter what individual hit itself, Alice must have hit something, and so the use of ∃E is incorrect.

(3)   α *must not occur in any assumption*. Here is an example of what happens if we violate this provision:

    1   ∃xGx   A
    2   Fa   A
    3 | Ga   H (for ∃E)
    4 | Fa & Ga   2, 3 &I
    5 | ∃x(Fx & Gx)   4 ∃I
    6   ∃x(Fx & Gx)   1, 3–5 ∃E (incorrect)

Once again, the derivation is invalid. Given the assumptions that something is a giraffe (∃xGx) and Amos is a frog (*Fa*), we should not be able to deduce that something is both a frog and a giraffe (∃x(Fx & Gx)). The mistake is that the name letter α (in this case '*a*') occurs in the assumption '*Fa*'. This destroys the generality of the hypothetical derivation and thus invalidates the use of ∃E at line 6. For if we had used any name letter other than '*a*' to designate our representative individual at step 2, we would not have been able to obtain '∃x(Fx & Gx)' at line 5. Thus the hypothetical derivation does not show that no matter which individual is a giraffe, something must be both a frog and a giraffe.

(4)   α *must not occur in any hypothesis in effect at the line at which* ∃E *is applied*. (Recall that a hypothesis is in effect at a line if it is introduced before that line but not yet discharged at that line.) This is essentially the same provision as condition 3, but applied to hypotheses rather than assumptions. To see the similarity with condition 3, consider the following derivation:

    1   ∃xGx   A
    2 | Fa   H (for →I)
    3 | | Ga   H (for ∃E)
    4 | | Fa & Ga   2, 3, &I
    5 | | ∃x(Fx & Gx)   4 ∃I
    6 | ∃x(Fx & Gx)   1, 3–5 ∃E (incorrect)
    7   Fa → ∃x(Fx & Gx)   2–6 →I

Like the previous example, this is invalid; given the assumption that something is a giraffe (∃xGx), it should not follow that if Amos is a frog, then something is both a frog and a giraffe (Fa → ∃x(Fx & Gx)). This mistake lies in the fact that the name letter α (in this case '*a*') occurs in the hypothesis at line 2, which is still in effect when ∃E is applied at line 6. This destroys the generality of the hypothetical derivation in the same way that the assumption of '*Fa*' did in the previous example—and thus makes the use of ∃E at line 6 incorrect.

We now consider some proofs which illustrate correct applications of ∃E.

SOLVED PROBLEMS

7.16 Prove:

$$\forall x(Fx \rightarrow Gx), \exists xFx \vdash \exists xGx$$

Solution

1	$\forall x(Fx \rightarrow Gx)$	*A*
2	$\exists xFx$	*A*
3	Fa	*H* (for ∃E)
4	$Fa \rightarrow Ga$	1 ∀E
5	Ga	3, 4 →E
6	$\exists xGx$	5 ∃I
7	$\exists xGx$	2, 3–6 ∃E

The assumption at line 2 asserts that something has the property *F*. At step 3 we hypothesize that *a* is one such thing. We derive our conclusion from this hypothesis at step 6. This hypothetical derivation obeys all the strictures on the ∃E rule and in consequence is perfectly general. We could have derived the same conclusion from the hypothesis that any other individual has the property *F*. Hence the final step of ∃E is legitimate. Notice that we must perform ∃I at step 6 *before* applying ∃E. If we reversed the order of these two steps, the conclusion of our hypothetical derivation would be '*Ga*', which contains '*a*' and hence would prevent application of ∃E (see condition 2, above).

7.17 Prove:

$$\exists x(Fx \vee Gx) \vdash \exists xFx \vee \exists xGx$$

Solution

1	$\exists x(Fx \vee Gx)$	*A*
2	$Fa \vee Ga$	*H* (for ∃E)
3	Fa	*H* (for →I)
4	$\exists xFx$	3 ∃I
5	$\exists xFx \vee \exists xGx$	4 ∨I
6	$Fa \rightarrow (\exists xFx \vee \exists xGx)$	3–5 →I
7	Ga	*H* (for →I)
8	$\exists xGx$	7 ∃I
9	$\exists xFx \vee \exists xGx$	8 ∨I
10	$Ga \rightarrow (\exists xFx \vee \exists xGx)$	7–9 →I
11	$\exists xFx \vee \exists xGx$	2, 6, 10 ∨E
12	$\exists xFx \vee \exists xGx$	1, 2–11 ∃E

The assumption is an existentially quantified disjunction. To use it, we hypothesize a representative instance for ∃E at step 2. This representative instance is the disjunction '*Fa* ∨ *Ga*', and to reason from a disjunction the typical strategy is ∨E. But this requires the two conditionals listed on lines 6 and 10, so we obtain these conditionals by →I. This gives us the conclusion by ∨E at line 11. We complete the proof by discharging the hypothesis and reasserting the conclusion by ∃E at line 12. The reader should verify that step 12 obeys the ∃E rule.

7.18   Prove:

$$\exists xFx \lor \exists xGx \vdash \exists x(Fx \lor Gx)$$

Solution

1	$\exists xFx \lor \exists xGx$	A
2	$\exists xFx$	H (for →I)
3	Fa	H (for ∃E)
4	$Fa \lor Ga$	3 ∨I
5	$\exists x(Fx \lor Gx)$	4 ∃I
6	$\exists x(Fx \lor Gx)$	2, 3–5 ∃E
7	$\exists xFx \to \exists x(Fx \lor Gx)$	2–6 →I
8	$\exists xGx$	H (for →I)
9	Ga	H (for ∃E)
10	$Fa \lor Ga$	9 ∨I
11	$\exists x(Fx \lor Gx)$	10 ∃I
12	$\exists x(Fx \lor Gx)$	8, 9–11 ∃E
13	$\exists xGx \to \exists x(Fx \lor Gx)$	8–12 →I
14	$\exists x(Fx \lor Gx)$	1, 7, 13 ∨E

This is just the converse of Problem 7.17. Here we are proving an existential conclusion from a disjunction, so the dominant strategy is ∨E, rather than ∃E. Again, this requires us to obtain two conditionals (lines 7 and 13) by →I. Each →I proof embodies an ∃E strategy. Both uses of ∃E are legitimate, even though the name letter 'a' occurs in hypotheses at lines 3 and 9, since these hypotheses are not in effect at the lines at which ∃E is applied (see condition 4, above).

7.19   Prove:

$$\exists x\forall yLxy \vdash \forall x\exists yLyx$$

Solution

1	$\exists x\forall yLxy$	A
2	$\forall yLay$	H (for ∃E)
3	Lab	2 ∀E
4	$\exists yLyb$	3 ∃I
5	$\forall x\exists yLyx$	4 ∀I
6	$\forall x\exists yLyx$	1, 2–5 ∃E

Since the assumption is existentially quantified, we hypothesize a representative instance of it at step 2. The use of ∀I at 5 is correct, since the name letter 'b' does not occur in any assumption or hypothesis. Likewise (as the reader should confirm), the step of ∃E at line 6 meets all the requirements of the ∃E rule. The last two steps of the proof could have been interchanged, like this:

5	$\exists yLyb$	1, 2–4 ∃E
6	$\forall x\exists yLyx$	5 ∀I

The proof would still be correct. It is useful to compare this solution with Problem 6.37, to see the analogy between inference rules and refutation rules.

7.20   Prove:

$$\forall x(Fx \to \exists yLxy), \exists x(Fx \mathbin{\&} Gx) \vdash \exists x\exists y(Gx \mathbin{\&} Lxy)$$

Solution

1	$\forall x(Fx \rightarrow \exists yLxy)$	A
2	$\exists x(Fx \mathrel{\&} Gx)$	A
3	$\quad Fa \mathrel{\&} Ga$	H (for $\exists$E)
4	$\quad Fa \rightarrow \exists yLay$	1 $\forall$E
5	$\quad Fa$	3 &E
6	$\quad \exists yLay$	4, 5 $\rightarrow$E
7	$\quad\quad Lab$	H (for $\exists$E)
8	$\quad\quad Ga$	3 &E
9	$\quad\quad Ga \mathrel{\&} Lab$	7, 8 &I
10	$\quad\quad \exists y(Ga \mathrel{\&} Lay)$	9 $\exists$I
11	$\quad\quad \exists x\exists y(Gx \mathrel{\&} Lxy)$	10 $\exists$I
12	$\quad \exists x\exists y(Gx \mathrel{\&} Lxy)$	6, 7–11 $\exists$E
13	$\exists x\exists y(Gx \mathrel{\&} Lxy)$	2, 3–12 $\exists$E

This proof requires two uses of $\exists$E, one for each existential quantifier in the assumptions. We begin at step 3 by hypothesizing a representative instance of '$\exists x(Fx \mathrel{\&} Gx)$'. Instantiating '$\forall x(Fx \rightarrow \exists yLxy)$' with '$a$' at step 4 enables us to derive '$\exists yLay$' at line 6. To use this existential formula, we hypothesize a representative instance of it at 7, introducing the new name letter 'b'. (We could not use 'a' again and write 'Laa' as a representative instance of the formula at line 6, since that would block correct application of $\exists$E later on.) From the hypothesis in step 7 we derive the conclusion at line 11 and then discharge the two hypotheses by two steps of $\exists$E at lines 12 and 13. Even though the name letter 'b' occurs in a hypothesis at 7, the step of $\exists$E at 12 is legitimate because this hypothesis is no longer in effect at 12. Likewise, even though 'a' occurs in hypotheses at 3 and 7, the step of $\exists$E at 13 is legitimate, since these hypotheses are no longer in effect at 13. We need not consider occurrences of 'a' in checking step 12, since the relevant name letter α at this step is 'b', not 'a'; likewise, we need not consider occurrences of 'b' in checking step 13, since α is 'a' (see the formal statement of the $\exists$E rule).

7.21 Prove:

$$\forall x(Fx \rightarrow \sim Gx) \vdash \sim\exists x(Fx \mathrel{\&} Gx)$$

Solution

1	$\forall x(Fx \rightarrow \sim Gx)$	A
2	$\quad \exists x(Fx \mathrel{\&} Gx)$	H (for $\sim$I)
3	$\quad\quad Fa \mathrel{\&} Ga$	H (for $\exists$E)
4	$\quad\quad Fa \rightarrow \sim Ga$	1 $\forall$E
5	$\quad\quad Fa$	3 &E
6	$\quad\quad \sim Ga$	4, 5 $\rightarrow$E
7	$\quad\quad Ga$	3 &E
8	$\quad\quad P \mathrel{\&} \sim P$	6, 7 CON
9	$\quad P \mathrel{\&} \sim P$	2, 3–8 $\exists$E
10	$\sim\exists x(Fx \mathrel{\&} Gx)$	2–9 $\sim$I

Since the conclusion is negated, we adopt a *reductio* strategy, hypothesizing '$\exists x(Fx \mathrel{\&} Gx)$' at 2 and working toward a contradiction. Since '$\exists x(Fx \mathrel{\&} Gx)$' is existentially quantified, to use it we must hypothesize a representative instance of it at 3. The contradiction '$Ga \mathrel{\&} \sim Ga$' could have been obtained at step 8, but that would have prevented the application of $\exists$E at line 9, since '$Ga \mathrel{\&} \sim Ga$' contains the name 'a' (see condition 2 for the $\exists$E rule). And since we cannot perform $\sim$I until we have first discharged the hypothesis of line 3 by $\exists$E, we would have reached an impasse. So at this point some ingenuity is required. We recall that the derived rule CON allows us to infer any wff from a wff together with its negation. We use this rule at step 8 to infer a new contradiction, '$P \mathrel{\&} \sim P$', which does not contain 'a'. (Any other contradiction would have done as well, as long as it did not contain 'a'.) This opens the way to a correct use of $\exists$E at step 9, which allows us to complete our *reductio* strategy at step 10.

We conclude this section with some important points that should be kept in mind when constructing proofs:

(1) *All four quantifier rules operate only at the leftmost position of a formula, i.e., only on a quantifier whose scope is the whole formula.* The rules ∃I and ∀I enable us to introduce a quantifier as the leftmost symbol. Similarly, ∀E allows us to remove a quantifier only if it is the leftmost symbol, and an ∃E hypothesis may be constructed only by removing the leftmost quantifier from an existentially quantified wff (in each case, of course, the variable governed by the quantifier is replaced by a name letter). Special care must be taken when '&', '∨', '→', or '↔' is the main operator of a formula, since a formula "officially" has outer brackets, though these are dropped by convention. Its leftmost symbol is thus officially the left outer bracket, so that when a quantifier is introduced it must go to the left of this bracket, and the bracket and its mate on the right must now appear explicitly, since they are no longer outer brackets (compare condition 3 for the ∃I rule).

(2) *To prove an existentially or universally quantified conclusion, the typical strategy is first to prove a formula from which this conclusion can be obtained by ∃I or ∀I.* For example,

To Prove:	First Prove:
∃xFx	Fa
∀x(Fx → Gx)	Fa → Ga
∀x∼Fx	∼Fa
∀x∃yFxy	∃yFay
∃yFay	Fab
∃xFxx	Faa

Then adopt a substrategy based on the form of the conclusion needed for ∃I or ∀I. For example, in proving '∀x(Fx → Gx)', we first prove 'Fa → Ga', and to prove this we use →I as a substrategy (see Problems 7.6 and 7.9). Similarly, in proving '∀x∼Fx', we first prove '∼Fa'—and since this is a negated formula, we would generally begin by hypothesizing 'Fa' for ∼I.

(3) *If the conclusion is in the form of a negation, conjunction, disjunction, conditional, or biconditional, then it is usually best to employ the propositional calculus strategy for proving conclusions of that form (see Table 4-1, Section 4.3).* For example, to prove the negated formula '∼∀xFx', use a ∼I strategy: hypothesize '∀xFx' and try to derive a contradiction (see Problems 7.4 and 7.11). Note that it will not do to derive '∼Fa' and then apply ∀I, since even if this is possible under the assumptions given for the problem, the result will be '∀x∼Fx', which is not the intended conclusion. Similarly, to prove the conditional '∀xFx → ∀xGx', hypothesize its antecedent and use →I (see Problem 7.8). It will not do to derive 'Fa → Ga' and then apply ∀I, since even if this is possible, it will yield only '∀x(Fx → Gx)'.

7.4   THEOREMS AND QUANTIFIER EQUIVALENCE RULES

As in the propositional calculus, it is possible in the predicate calculus to prove some wffs without making assumptions. These wffs are the *theorems* of the predicate calculus. It can be shown by metalogical reasoning that these theorems are exactly the *logical truths* of predicate logic, i.e., those wffs which are true in every model. Logical truths are formulas whose truth is necessary regardless of the meanings assigned to their nonlogical symbols. Because the predicate calculus incorporates the propositional calculus, all theorems of the propositional calculus are also theorems of the predicate calculus. But the predicate calculus has additional theorems that belong to it alone.

SOLVED PROBLEMS

7.22   Prove the theorem:

$$\vdash \forall x(Fx \to Fx)$$

Solution

1	Fa	H (for $\to$I)
2	$Fa \to Fa$	1–1 $\to$I
3	$\forall x(Fx \to Fx)$	2 $\forall$I

Here, at step 2, we apply $\to$I to a single line which is both the antecedent and consequent of the conditional '$Fa \to Fa$', which we are trying to prove. The final step of $\forall$I is legitimate, since although 'a' occurs in the hypothesis 'Fa', that hypothesis is discharged at line 2. This theorem can be proved even more efficiently using the derived rule TI (see Section 4.5):

1	$Fa \to Fa$	TI Chap. 4, Supp. Prob. IV(1)
2	$\forall x(Fx \to Fx)$	1 $\forall$I

The theorem used here is '$P \to P$', of which '$Fa \to Fa$' is a substitution instance. Note that line 1 is neither a hypothesis nor an assumption, so that even though it contains 'a', the application of $\forall$I at line 2 is legitimate. This becomes clear when we note that '$Fb \to Fb$', '$Fc \to Fc$', and so on, are all substitution instances of this theorem; indeed, no matter what name we put in place of 'a', we get a substitution instance. Thus the theorem clearly implies that for anything x, $Fx \to Fx$.

7.23   Prove the theorem:

$$\vdash \forall xFx \to Fa$$

Solution

1	$\forall xFx$	H (for $\to$I)
2	Fa	1 $\forall$E
3	$\forall xFx \to Fa$	1, 2 $\to$I

The theorem is a conditional statement, so we use a conditional proof strategy.

7.24   Prove the theorem:

$$\vdash \sim(\forall xFx \,\&\, \exists x \sim Fx)$$

Solution

1	$\forall xFx \,\&\, \exists x{\sim}Fx$	H (for $\sim$I)
2	$\forall xFx$	1 &E
3	$\exists x{\sim}Fx$	1 &E
4	$\sim Fa$	H (for $\exists$E)
5	Fa	2 $\forall$E
6	$P \,\&\, {\sim}P$	4, 5 CON
7	$P \,\&\, {\sim}P$	3, 4–6 $\exists$E
8	$\sim(\forall xFx \,\&\, \exists x{\sim}Fx)$	1–7 $\sim$I

The theorem is a negation, so we hypothesize '$\forall xFx \,\&\, \exists x{\sim}Fx$' for $\sim$I. This is broken into its conjuncts as lines 2 and 3. To use '$\exists x{\sim}Fx$' we must proceed by $\exists$E; we hypothesize a typical instance of '$\exists x{\sim}Fx$' at line 4. The contradiction '$Fa \,\&\, {\sim}Fa$' could be obtained at 6, but we could not then discharge the hypothesis of step 4, since '$Fa \,\&\, {\sim}Fa$' contains 'a'. To overcome this difficulty, we instead obtain a new contradiction not containing 'a' by CON (compare Problem 7.21). This allows us to discharge the $\exists$E hypothesis of step 4 at line 7 and apply $\sim$I at 8.

7.25   Prove the theorem:

$$\vdash \forall xFx \lor \exists x \sim Fx$$

Solution

1	$\sim\forall xFx$	H (for $\rightarrow$I)
2	$\sim\exists x\sim Fx$	H (for $\sim$I)
3	$\sim Fa$	H (for $\sim$I)
4	$\exists x\sim Fx$	3 $\exists$I
5	$\exists x\sim Fx\ \&\ \sim\exists x\sim Fx$	2, 4 &I
6	$\sim\sim Fa$	3–5 $\sim$I
7	Fa	6 $\sim$E
8	$\forall xFx$	7 $\forall$I
9	$\forall xFx\ \&\ \sim\forall xFx$	1, 8 &I
10	$\sim\sim\exists x\sim Fx$	2–9 $\sim$I
11	$\exists x\sim Fx$	10 $\sim$E
12	$\sim\forall xFx\rightarrow\exists x\sim Fx$	1–11 $\rightarrow$I
13	$\sim\sim\forall xFx\vee\exists x\sim Fx$	12 MI
14	$\forall xFx\vee\exists x\sim Fx$	13 DN

The strategy here is indirect. The theorem is a disjunction, and there is no obvious way to prove a disjunction. (See Table 4.1.) However, the equivalences MI and DN imply that the theorem is equivalent to the conditional '$\sim\forall xFx\rightarrow\exists x\sim Fx$'. Thus we can prove the theorem by proving this conditional (step 12) and then applying MI and DN. We begin our proof by hypothesizing the antecedent in the conditional, '$\sim\forall xFx$', for $\rightarrow$I. To prove its consequent, '$\exists x\sim Fx$', we proceed by $\sim$I, hypothesizing the negation of the consequent at step 2 and introducing yet another $\sim$I hypothesis at step 3 to obtain the necessary contradictions. The use of $\forall$I at 8 is permissible, since although 'a' appears in the hypothesis of step 3, this hypothesis is no longer in effect at line 8.

A number of important equivalences are provable in the predicate calculus. The first of these shows the logical equivalence of the formulas '$\forall x(Fx\rightarrow\sim Gx)$' and '$\sim\exists x(Fx\ \&\ Gx)$', which are both ways of expressing the E-form proposition 'No F are G'.

SOLVING PROBLEM

7.26   Prove the equivalence:

$$\vdash\forall x(Fx\rightarrow\sim Gx)\leftrightarrow\sim\exists x(Fx\ \&\ Gx)$$

Solution

1	$\forall x(Fx\rightarrow\sim Gx)$	H (for $\rightarrow$I)
2	$\sim\exists x(Fx\ \&\ Gx)$	1 Prob. 7.21
3	$\forall x(Fx\rightarrow\sim Gx)\rightarrow\sim\exists x(Fx\ \&\ Gx)$	1–2 $\rightarrow$I
4	$\sim\exists x(Fx\ \&\ Gx)$	H (for $\rightarrow$I)
5	Fa	H (for $\rightarrow$I)
6	Ga	H (for $\sim$I)
7	$Fa\ \&\ Ga$	5, 6 &I
8	$\exists x(Fx\ \&\ Gx)$	7 $\exists$I
9	$\exists x(Fx\ \&\ Gx)\ \&\ \sim\exists x(Fx\ \&\ Gx)$	4, 8 &I
10	$\sim Ga$	6–9 $\sim$I
11	$Fa\rightarrow\sim Ga$	5–10 $\rightarrow$I
12	$\forall x(Fx\rightarrow\sim Gx)$	11 $\forall$I
13	$\sim\exists x(Fx\ \&\ Gx)\rightarrow\forall x(Fx\rightarrow\sim Gx)$	4–12 $\rightarrow$I
14	$\forall x(Fx\rightarrow\sim Gx)\leftrightarrow\sim\exists x(Fx\ \&\ Gx)$	3, 13 $\leftrightarrow$I

Table 7-1 lists four equivalences which express important relationships between the universal and existential quantifiers. The first two of these equivalences are proved below. The third and fourth are left as exercises for the reader. It is also useful to compare these equivalences with the refutation rules for negated quantifiers (see Section 6.5).

Table 7-1   Quantifier Equivalences

$$\vdash \sim\forall x\sim Fx \leftrightarrow \exists xFx$$
$$\vdash \sim\forall xFx \leftrightarrow \exists x\sim Fx$$
$$\vdash \forall x\sim Fx \leftrightarrow \sim\exists xFx$$
$$\vdash \forall xFx \leftrightarrow \sim\exists x\sim Fx$$

SOLVED PROBLEMS

7.27   Prove the equivalence:

$$\vdash \sim\forall x\sim Fx \leftrightarrow \exists xFx$$

Solution

1	$\sim\forall x\sim Fx$	H (for $\rightarrow$I)
2	$\sim\exists xFx$	H (for $\sim$I)
3	Fa	H (for $\sim$I)
4	$\exists xFx$	3 $\exists$I
5	$\exists xFx \ \& \sim\exists xFx$	2, 4 &I
6	$\sim Fa$	3–5 $\sim$I
7	$\forall x\sim Fx$	6 $\forall$I
8	$\forall x\sim Fx \ \& \sim\forall x\sim Fx$	1, 7 &I
9	$\sim\sim\exists xFx$	2–8 $\sim$I
10	$\exists xFx$	9 $\sim$E
11	$\sim\forall x\sim Fx \rightarrow \exists xFx$	1–10 $\rightarrow$I
12	$\exists xFx$	H (for $\rightarrow$I)
13	Fa	H (for $\exists$E)
14	$\forall x\sim Fx$	H (for $\sim$I)
15	$\sim Fa$	14 $\forall$E
16	$Fa \ \& \sim Fa$	13, 15 &I
17	$\sim\forall x\sim Fx$	14–16 $\sim$I
18	$\sim\forall x\sim Fx$	12, 13–17 $\exists$E
19	$\exists xFx \rightarrow \sim\forall x\sim Fx$	12–18 $\rightarrow$I
20	$\sim\forall x\sim Fx \leftrightarrow \exists xFx$	11, 19 $\leftrightarrow$I

7.28   Prove the equivalence:

$$\vdash \sim\forall xFx \leftrightarrow \exists x\sim Fx$$

Solution

1	$\sim\forall xFx$	H (for $\rightarrow$I)
2	$\sim\exists x\sim Fx$	H (for $\sim$I)
3	$\sim Fa$	H (for $\sim$I)
4	$\exists x\sim Fx$	3 $\exists$I
5	$\exists x\sim Fx \ \& \sim\exists x\sim Fx$	2, 4 &I
6	$\sim\sim Fa$	3–5 $\sim$I
7	Fa	6 $\sim$E
8	$\forall xFx$	7 $\forall$I
9	$\forall xFx \ \& \sim\forall xFx$	1, 8 &I
10	$\sim\sim\exists x\sim Fx$	2–9 $\sim$I
11	$\exists x\sim Fx$	10 $\sim$E

12	$\sim\forall x Fx \rightarrow \exists x \sim Fx$	1–11 $\rightarrow$I
13	$\exists x \sim Fx$	H (for $\rightarrow$I)
14	$\sim Fa$	H (for $\exists$E)
15	$\forall x Fx$	H (for $\sim$I)
16	Fa	15 $\forall$E
17	$Fa \, \& \sim Fa$	14, 16 &I
18	$\sim\forall x Fx$	15–17 $\sim$I
19	$\sim\forall x Fx$	13, 14–18 $\exists$E
20	$\exists x \sim Fx \rightarrow \sim\forall x Fx$	13–19 $\rightarrow$I
21	$\sim\forall x Fx \leftrightarrow \exists x \sim Fx$	12, 20 $\leftrightarrow$I

The four quantifier equivalences are the basis for the most important derived rule of the predicate calculus. To state this rule clearly, we need a bit more terminology. A formula which results from a wff when we remove an initial quantifier-plus-variable is called an *open formula on the variable*. For example, if we remove '$\exists x$' from '$\exists x(Fx \, \& \, Gx)$', the result, '$(Fx \, \& \, Gx)$', is an open formula on 'x'. Or, again, if we remove '$\forall z$' from '$\forall z\forall x(Fxz \rightarrow Gzx)$', the result, '$\forall x(Fxz \rightarrow Gzx)$', is an open formula on 'z'. An open formula is not itself a wff, since our formation rules do not permit unquantified variables. (See Section 6.3.)

Now notice that in the proofs of Problems 7.27 and 7.28, all occurrences of the formula 'Fx' could have been replaced by occurrences of any other open formula on 'x', and the result would still be a valid proof. Thus, for example, by the reasoning used in Problem 7.27, we would have proved not only '$\vdash \forall x Fx \leftrightarrow \sim\exists x \sim Fx$', but also '$\vdash \forall x(Fx \, \& \, Gx) \leftrightarrow \sim\exists x \sim(Fx \, \& \, Gx)$', '$\vdash \forall x \exists y Lxy \leftrightarrow \sim\exists x \sim \exists y Lxy$', and so on. That is, we could have proved that any wff consisting of an open formula on 'x' prefixed by '$\forall x$' is logically equivalent to the wff consisting of that same open formula prefixed by '$\sim\exists x\sim$'. Since 'F' can be interpreted as standing for any property, it can be interpreted as standing for any property expressible by an open formula on 'x'. Hence Problem 7.27 is in fact a proof of all these equivalences.

It is, moreover, clear that the variable 'x' could have been uniformly replaced in Problems 7.27 and 7.28 by any other variable without affecting the validity of the proof. Hence, for example, we may also regard Problem 7.27 as a proof of the equivalences '$\vdash \forall z Fz \leftrightarrow \sim\exists z \sim Fz$', '$\vdash \forall y(Fy \, \& \, Gy) \leftrightarrow \sim\exists y \sim(Fy \, \& \, Gy)$', and so on.

In general, then, Problem 7.27 shows that any wff of the form $\forall\beta\phi$, where β is a variable and ϕ an open formula on that variable, is logically equivalent to $\sim\exists\beta\sim\phi$. The other equivalences listed in Table 7-1 have the same generality. Thus, since logically equivalent formulas may validly be substituted for one another in any context, these four equivalences give rise to the following derived rule:

Quantifier Exchange (QE):   Let ϕ be an open formula on a variable β. Then if one member of the following pairs is a subwff of some wff ψ, we may validly infer from ψ the result of replacing one or more occurrences of this member by the other member of the pair:

$\sim\forall\beta\sim\phi$, $\exists\beta\phi$

$\sim\forall\beta\phi$, $\exists\beta\sim\phi$

$\forall\beta\sim\phi$, $\sim\exists\beta\phi$

$\forall\beta\phi$, $\sim\exists\beta\sim\phi$

Like the derived rules of the propositional calculus (Sections 4.4 to 4.6), QE enables us to shorten and simplify proofs, but it allows us to prove nothing new. Anything provable by QE is already provable by the four quantifier introduction and elimination rules, together with the ten basic rules of the propositional calculus. To illustrate how QE can shorten and simplify proofs, we re-prove below the form of Problem 7.21:

SOLVED PROBLEM

7.29 Prove:

$$\forall x(Fx \rightarrow \sim Gx) \vdash \sim \exists x(Fx \& Gx)$$

Solution

1	$\forall x(Fx \rightarrow \sim Gx)$	A
2	$Fa \rightarrow \sim Ga$	1 $\forall$E
3	$\sim(Fa \& \sim\sim Ga)$	2 Prob. 4.46
4	$\sim(Fa \& Ga)$	3 DN
5	$\forall x\sim(Fx \& Gx)$	4 $\forall$I
6	$\sim\exists x(Fx \& Gx)$	5 QE

Compare this with the proof of Problem 7.21.

Some additional problems using QE follow. In each case, we shall also rely on some derived rules of the propositional calculus.

SOLVED PROBLEMS

7.30 Prove the theorem:

$$\vdash \forall xFx \lor \exists x\sim Fx$$

Solution

1	$\forall xFx \lor \sim\forall xFx$	TI 4.44
2	$\forall xFx \lor \exists x\sim Fx$	1 QE

7.31 Prove:

$$\forall x\sim Fx \rightarrow \forall x\sim Gx \vdash \exists xGx \rightarrow \exists xFx$$

Solution

1	$\forall x\sim Fx \rightarrow \forall x\sim Gx$	A
2	$\sim\exists xFx \rightarrow \forall x\sim Gx$	1 QE
3	$\sim\exists xFx \rightarrow \sim\exists xGx$	2 QE
4	$\exists xGx \rightarrow \exists xFx$	3 TRANS

7.32 Prove:

$$\sim\exists xFx, \sim\exists xGx \vdash \sim\exists x(Fx \lor Gx)$$

Solution

1	$\sim\exists xFx$	A
2	$\sim\exists xGx$	A
3	$\forall x\sim Fx$	1 QE
4	$\forall x\sim Gx$	2 QE
5	$\sim Fa$	3 $\forall$E
6	$\sim Ga$	4 $\forall$E
7	$\sim Fa \& \sim Ga$	5, 6 &I
8	$\sim(Fa \lor Ga)$	7 DM
9	$\forall x\sim(Fx \& Gx)$	8 $\forall$I
10	$\sim\exists x(Fx \& Gx)$	9 QE

It should be noted that our QE rules are somewhat more restricted than need be. They allow quantifier exchange only when a quantifier-plus-variable prefixes an open formula on the variable. They do not, for example, license the inference

1   $\forall x(Fx \rightarrow \exists y \sim Lxy)$ A
2   $\forall x(Fx \rightarrow \sim \forall y Lxy)$ 1 QE

since 'Lxy' contains 'y' in addition to 'x' and so is not an open formula just on 'x'. This inference, however, is valid, as is any inference in which '$\exists y \sim$' is replaced by '$\sim \forall y$' anywhere in a wff. (The same is true for the other QE rules.) Some authors use wholly unrestricted QE rules, which allow inferences like the one above. Others restrict QE even more, allowing it to be used only on initial quantifiers. Still others do not permit QE at all. None of these variations affect what can be proved in the logic, since anything provable by any version of QE is also provable by the four introduction and elimination rules.

7.5   INFERENCE RULES FOR THE IDENTITY PREDICATE

Our proof techniques may also be applied to forms containing the identity predicate. This requires two new introduction and elimination rules, both of which are related in an obvious way to the identity rules of the refutation tree method (Section 6.6). The identity introduction rule is like the derived rule TI (Section 4.5) in that it introduces formulas into proofs without deriving them from previous lines.

Identity Introduction (=I):   For any name letter α, we may assert $\alpha = \alpha$ at any line of a proof.

Identity introduction is used in the following proofs.

SOLVED PROBLEMS

7.33   Prove the theorem:

$\vdash \forall x\, x = x$

Solution

1   $a = a$ =I
2   $\forall x\, x = x$ 1 $\forall$I

Note that the use of $\forall$I at step 2 is legitimate, since the proof uses no assumptions or hypotheses containing 'a'. This theorem is known as the *law of reflexivity of identity*.

7.34   Prove the theorem:

$\vdash \exists x\, a = x$

Solution

1   $a = a$ =I
2   $\exists x\, a = x$ 1 $\exists$I

This theorem asserts the existence of the object denoted by 'a'. The same reasoning could be repeated, of course, using any name letter other than 'a'. It therefore reflects the presupposition, noted in Section 7.3, that all names denote existing things.

The identity elimination rule (the rule that allows us to reason from identities as premises) is simply the familiar idea that if $a = b$, the names 'a' and 'b' are interchangeable. This rule is also called *substitutivity of identity*.

Identity Elimination (=E):   If ϕ is a wff containing a name letter α, then from ϕ and either $\alpha = \beta$ or $\beta = \alpha$ we may infer $\phi \beta / \alpha$, the result of replacing one or more occurrences of α in ϕ by β.

SOLVED PROBLEMS

7.35 Prove:

$$Fa, a = b \vdash Fb$$

Solution

1	*Fa*	*A*
2	*a = b*	*A*
3	*Fb*	1, 2 =E

7.36 Prove:

$$Fa, \sim Fb \vdash \sim a = b$$

Solution

1	*Fa*	*A*
2	$\sim Fb$	*A*
3	*a = b*	*H* (for ~I)
4	*Fb*	1, 3 =E
5	*Fb &* $\sim Fb$	2, 4 &I
6	$\sim a = b$	3–5 ~I

7.37 Prove the theorem:

$$\vdash \forall x \forall y \, (x = y \rightarrow y = x)$$

Solution

1	*a = b*	*H* (for →I)
2	*a = a*	=I
3	*b = a*	1, 2 =E
4	$a = b \rightarrow b = a$	1–3 →I
5	$\forall y(a = y \rightarrow y = a)$	4 ∀I
6	$\forall x \forall y(x = y \rightarrow y = x)$	5 ∀I

At step 3 we use the identity '*a = b*' (line 1) to replace the first occurrence of '*a*' in '*a = a*' (line 2) with '*b*'. That is, with respect to the statement of =E, '*a = a*' is ϕ and '*b = a*' is $\phi^b/_a$. This theorem expresses the *law of symmetry of identity* whose semantic validity has been proved in Problem 6.44.

7.38 Prove the theorem:

$$\vdash \forall x \forall y \forall z((x = y \, \& \, y = z) \rightarrow x = z)$$

Solution

1	*a = b & b = c*	*H* (for →I)
2	*a = b*	1 &E
3	*b = c*	1 &E
4	*a = c*	2, 3 =E
5	$(a = b \, \& \, b = c) \rightarrow a = c$	1–4 →I
6	$\forall x((a = b \, \& \, b = z) \rightarrow a = z)$	5 ∀I
7	$\forall y \forall z((a = y \, \& \, y = z) \rightarrow a = z)$	6 ∀I
8	$\forall x \forall y \forall z((x = y \, \& \, y = z) \rightarrow x = z)$	7 ∀I

This theorem expresses the *law of transitivity of identity*. (Compare Problem 6.45.)

Tables 7-2 and 7-3 summarize the inference rules (basic and derived) for predicate logic with identity.

Table 7-2   Basic Rules of Inference for Predicate Logic with Identity*

Universal Elimination (∀E):   From a universally quantified wff ∀βφ infer any wff of the form φ%ᵦ which results from replacing each occurrence of the variable β in φ by some name letter α.

Universal Introduction (∀I):   From a wff φ containing a name letter α not occurring in any assumption or in any hypothesis in effect at the line on which φ occurs, infer any wff of the form ∀βφᵝ%, where φᵝ% is the result of replacing all occurrences of α in φ by some variable β not already in φ.

Existential Introduction (∃I):   Given a wff φ containing some name letter α, infer any wff of the form ∃βφᵝ%, where φᵝ% is the result of replacing one or more occurrences of α in φ by some variable β not already in φ.

Existential Elimination (∃E):   Given an existentially quantified wff ∃βφ and a derivation of some conclusion ψ from a hypothesis of the form φ%ᵦ (the result of replacing each occurrence of the variable β in φ by some name letter α not already in φ), discharge φ%ᵦ and reassert ψ. *Restriction:* The name letter α may not occur in ψ, nor in any assumption, nor in any hypothesis that is in effect at the line at which ∃E is applied.

Identity Introduction (=I):   For any name letter α, we may assert α = α at any line of a proof.

Identity Elimination (=E):   If φ is a wff containing a name letter α, then from φ and either α = β or β = α we may infer φᵝ%, the result of replacing one or more occurrences of α in φ by β.

*Only rules specific to the predicate calculus are listed here. For a listing of propositional calculus rules, both basic and derived, see Tables 4-3 and 4-4 at the end of Chapter 4.

Table 7-3   Derived Rules of Predicate Logic

Quantifier Exchange (QE):   Let φ be an open formula on variable β. Then if one member of any of the following pairs is a subwff of some wff ψ, infer from ψ the result of replacing one or more occurrences of this member by the other member of the pair:

> ~∀β~φ, ∃βφ
> ~∀βφ, ∃β~φ
> ∀β~φ, ~∃βφ
> ∀βφ, ~∃β~φ

Supplementary Problems

Construct a proof for each of the following forms, using both basic and derived rules.

(1)   ∀xFx ⊢ Fa & (Fb & (Fc & Fd))

(2)   ∀x(Fx ∨ Gx), ~Fa ⊢ Ga

(3)   ~Fa ⊢ ~∀x(Fx & Gx)

(4)   ∀x(Fx ↔ R), R ⊢ Fa

(5)   ∀x(~Fx ∨ ~Gx) ⊢ ~(Fa & Ga)

(6)   ∀x(Fx → Gx) ⊢ ∀x(~Gx → ~Fx)

(7)   ∀x(Fx → Gx) ⊢ ∀x~Gx → ∀x~Fx

(8)   ∀x∀yFxy ⊢ ∀xFxx

(9) $\forall xFx \vdash \forall xGx \rightarrow \forall x(Fx \,\&\, Gx)$

(10) $\forall x\forall y(Fxy \rightarrow \sim Fyx) \vdash \forall x \sim Fxx$

(11) $\forall xFx \vdash \exists xFx$

(12) $\sim\exists xFx \vdash \sim Fa$

(13) $\exists x \sim Fx \vdash \sim\forall xFx$

(14) $\forall x(Fx \rightarrow Gx) \vdash \exists xFx \rightarrow \exists xGx$

(15) $\sim\exists x(Fx \,\&\, Gx) \vdash \forall x(\sim Fx \lor \sim Gx)$

(16) $\sim\forall x(Fx \,\&\, Gx) \vdash \exists x(\sim Fx \lor \sim Gx)$

(17) $\sim\exists x\exists yLxy \vdash \forall x \sim Lxx$

(18) $\exists xFx \vdash \exists x\exists y(Fx \,\&\, Fy)$

(19) $\forall x \sim Fx \vdash \forall x(Fx \rightarrow Gx)$

(20) $\forall x \sim Fx \vdash \forall x(Fx \rightarrow \sim Gx)$

(21) $\forall x\forall y\forall z((Lxy \,\&\, Lyz) \rightarrow \sim Lxz) \vdash \forall x \sim Lxx$

(22) $\vdash \sim\exists x(Fx \,\&\, \sim Fx)$

(23) $\vdash \exists xFx \lor \exists x \sim Fx$

(24) $\vdash \exists xFx \lor \forall x \sim Fx$

(25) $\vdash \sim\exists x\forall y(Lxy \leftrightarrow \sim Lyy)$

(26) $\vdash \forall x\exists y\, x = y$

(27) $\vdash \forall x\forall y(x = y \leftrightarrow y = x)$

(28) $Fb,\ \sim Fa \vdash \sim\forall x\forall y\, x = y$

(29) $\forall x(x = a \lor x = b),\ \exists xFx,\ \sim Fa \vdash Fb$

(30) $\vdash \forall x\forall y(x = y \rightarrow (Fx \leftrightarrow Fy))$

Answers to Selected Supplementary Problems

(5)
1	$\forall x(\sim Fx \lor \sim Gx)$	A
2	$\sim Fa \lor \sim Ga$	1 $\forall$E
3	$\sim(Fa \,\&\, Ga)$	2 DM

(10)
1	$\forall x\forall y(Fxy \rightarrow \sim Fyx)$	A
2	Faa	H (for $\sim$I)
3	$\forall y(Fay \rightarrow \sim Fya)$	1 $\forall$E
4	$Faa \rightarrow \sim Faa$	3 $\forall$E
5	$\sim Faa$	2, 4 $\rightarrow$E
6	$Faa \,\&\, \sim Faa$	3, 5 &I
7	$\sim Faa$	2–6 $\sim$I
8	$\forall x \sim Fxx$	7 $\forall$I

(15)
1	$\sim\exists x(Fx \,\&\, Gx)$	A
2	$\forall x \sim(Fx \,\&\, Gx)$	1 QE
3	$\sim(Fa \,\&\, Ga)$	2 $\forall$E
4	$\sim Fa \lor \sim Ga$	3 DM
5	$\forall x(\sim Fx \lor \sim Gx)$	4 $\forall$I

(20)
1	$\forall x \sim Fx$	A
2	Fa	H (for $\rightarrow$I)
3	$\sim Fa$	1 $\forall$E
4	Ga	2, 3 CON
5	$Fa \rightarrow Ga$	2–4 $\rightarrow$I
6	$\forall x(Fx \rightarrow Gx)$	5 $\forall$I

(25)

1	$\exists x \forall y(Lxy \leftrightarrow \sim Lyy)$	H (for $\sim$I)
2	$\forall y(Lay \leftrightarrow \sim Lyy)$	H (for $\exists$E)
3	$Laa \leftrightarrow \sim Laa$	2 $\forall$E
4	Laa	H (for $\sim$I)
5	$Laa \rightarrow \sim Laa$	3 $\leftrightarrow$E
6	$\sim Laa$	4, 5 $\rightarrow$E
7	$Laa \ \& \sim Laa$	4, 6 &I
8	$\sim Laa$	4–7 $\sim$I
9	$\sim Laa \rightarrow Laa$	3 $\leftrightarrow$E
10	Laa	8, 9 $\rightarrow$E
11	$P \ \& \sim P$	8, 10 CON
12	$P \ \& \sim P$	1, 2–11 $\exists$E
13	$\sim\exists x \forall y(Lxy \leftrightarrow \sim Lyy)$	1–12 $\sim$I

This surprising theorem has a number of interesting instances. It can be interpreted as stating, for example, that there does not exist someone x who for all y likes y if and only if y does not like him- or herself. This, as the derivation shows, is necessarily true! ('Lxy' can also be interpreted as "y is predicable of x," thinking of x and y as properties, or as "y is a member of x," thinking of x and y as sets, so that this argument formalizes the reasoning of the Russell antinomy; see Section 11.2.) To understand the derivation intuitively, it is perhaps easiest to interpret 'L' as "likes." Then, once again, what is to be proved is that no one likes all and only those who don't like themselves. To prove this, we begin with the hypothesis that someone does like all and only those who don't like themselves and then hypothesize further that this individual is a (lines 1 and 2). It follows by $\forall$E (step 3) that a likes himself if and only if he doesn't like himself. Does a, then, like himself or not? Suppose that he does (line 4). Then a contradiction quickly follows; so by $\sim$I he does not (step 8). But this too leads to a contradiction (steps 8 and 10); hence our original hypothesis leads unavoidably to contradiction. The contradiction exhibited on lines 8 and 10, however, cannot be used for $\exists$E to complete the proof, since it contains the name letter 'a', which appears in the $\exists$E hypothesis at line 2. So we can use CON to convert this contradiction into one that can be used for $\exists$E (we arbitrarily choose '$P \ \& \ P$'). This enables us to perform the $\exists$E step at line 12 and hence to obtain our conclusion at line 13 by $\sim$I (see Problems 7.21 and 7.24).

(30)

1	$a = b$	H (for $\rightarrow$I)
2	Fa	H (for $\rightarrow$I)
3	Fb	1, 2 =E
4	$Fa \rightarrow Fb$	2–3 $\rightarrow$I
5	Fb	H (for $\rightarrow$I)
6	Fa	1, 5 =E
7	$Fb \rightarrow Fa$	5–6 $\rightarrow$I
8	$Fa \leftrightarrow Fb$	4, 7 $\leftrightarrow$I
9	$a = b \rightarrow (Fa \leftrightarrow Fb)$	1–8 $\rightarrow$I
10	$\forall y(a = y \rightarrow (Fa \leftrightarrow Fy)$	9 $\forall$I
11	$\forall x \forall y(x = y \rightarrow (Fx \leftrightarrow Fy))$	10 $\forall$I

Chapter 8

Fallacies

8.1 CLASSIFICATION OF FALLACIES

In Chapters 3 to 7 we have focused on the task of testing the deductive validity of a wide class of argument forms. However, deductive validity was only one of the criteria for argument evaluation that were discussed in Chapter 2. The methods of Chapters 3 to 7 are of no help with respect to the other criteria: the actual truth or falsity of premises, the degree of relevance, and the effect of suppressing evidence that bears upon the conclusion. Moreover, truth tables, refutation trees, Venn diagrams, and inference rules establish validity or invalidity only for argument forms, not for specific arguments. The informal evaluation techniques of Chapter 2 do apply to specific arguments, but they rely heavily on intuition. In this and the following two chapters, we shall consider various techniques for evaluating nondeductive arguments with greater clarity and precision. We begin in this chapter with the study of fallacies. This is also informal, but it augments intuition by providing general accounts of some of the more common mistakes of ordinary reasoning. The next two chapters will then focus on inductive validity and the assessment of probabilistic reasoning.

Fallacies (in the broadest sense) are simply mistakes that occur in arguments and affect their cogency. In Latin, the verb *fallere* means 'to deceive'. Fallacious arguments may be deceptive, because they often superficially appear to be good arguments. But deception is not a necessary condition of a fallacy, as we use that term here. Whenever we reason invalidly or irrelevantly, accept premises we should not, or fail to make appropriate use of relevant facts at our disposal, we commit a fallacy.

There is no universally accepted definition of 'fallacy'. Many authors use definitions narrower than the one just given, but not infrequently an author's actual use of the term is at odds with his or her definition. Likewise, there is no universally accepted classification of fallacies. We shall divide fallacies into six classes: fallacies of relevance, circular reasoning, semantic fallacies, inductive fallacies, formal fallacies, and fallacies of false premises. Other authors use different classificatory schemes.

Fallacies of relevance occur when the premises of an argument have no bearing upon its conclusion. In addition, such fallacies often involve a distractive element which diverts attention away from this very problem.

Circular reasoning is the fallacy of assuming what we are trying to prove.

Semantic fallacies result when the language employed to construct arguments has multiple meanings or is excessively vague in a way that interferes with assessment of the argument.

Inductive fallacies occur when the probability of an argument's conclusion, given its premises—i.e., its inductive probability—is low, or at least less than the arguer supposes.

Formal fallacies occur when we misapply a valid rule of inference or else follow a rule which is demonstrably invalid.

And finally, there is a class of mistakes traditionally classified as fallacies which consist in an argument's having *false premises*.

The next six sections correspond to each of these classifications. Our survey of fallacies is representative, not exhaustive. Since people are ingenious, they can come up with new and unsuspected ways of committing logical mistakes. But the recurrence of similar kinds of mistakes over the two millennia since Aristotle drew up the first catalog of fallacies is evidence that bad reasoning habits obey definite patterns.

Many logic texts still employ Latin expressions to label certain fallacies. We furnish Latin terms (and their English equivalents) wherever such usage is customary.

8.2   FALLACIES OF RELEVANCE

Fallacies of relevance occur when the premises of an argument have no bearing upon its conclusion. Such arguments are often called *non sequiturs* (from the Latin phrase '*non sequitur*', meaning "it does not follow"). We will distinguish a number of different fallacies of relevance, though the generic error is the same in each case. (For a general discussion of the role of relevance in argument evaluation, see Section 2.4.)

Lack of relevance is not the only thing wrong with the arguments we will discuss in this section. Most fallacies of relevance occur in arguments that are invalid and have low inductive probabilities as well. In this section, however, we focus exclusively on the problem of irrelevance.

Ad hominem arguments try to discredit a claim or proposal by attacking its proponents instead of providing a reasoned examination of the proposal itself. '*Ad hominem*' means "against the person." *Ad hominem* arguments come in at least five varieties.

(1)   *Ad hominem abusive* arguments attack a person's age, character, family, gender, ethnicity, social or economic status, personality, appearance, dress, behavior, or professional, political, or religious affiliations. The implication is that there is no reason to take the person's views seriously.

SOLVED PROBLEM

8.1   What is wrong with the following argument?

> Jones advocates fluoridation of the city water supply.
> Jones is a convicted thief.
> ∴ We should not fluoridate the city water supply.

Solution

Even if Jones is a convicted thief, this has no bearing on whether the water supply should be fluoridated. To dismiss Jones' view simply because Jones is reprehensible is to commit the *ad hominem* abusive fallacy.

In practice, the conclusions of such arguments are often left unstated or conveyed by innuendo. Some arguments against the person are based on libel, slander, or character assassination. In such cases the premises are false as well as irrelevant.

SOLVED PROBLEMS

8.2   Does the argument below commit an abusive *ad hominem*?

> Jones says he saw my client commit a crime.
> Jones is a habitual drunkard.
> ∴ Jones' testimony is worthless.

Solution

This is a borderline case. If Jones is a habitual drunkard, that surely is somewhat relevant to the reliability of his testimony, but the relevance in this case is not extremely strong or direct. Jones' observation might well have taken place when he was sober.

8.3   Is this argument any better?

> Jones says he saw my client commit a crime on the night of April 21.
> Jones was so drunk on the night of April 21 that he was incapable of observing events accurately.
> ∴ Jones' testimony about this event is unreliable.

Solution

Yes; if its premises are true, this argument presents good reasons for discarding the witness's testimony. There is no *ad hominem* fallacy here.

(2) The fallacy of *guilt by association* is the attempt to repudiate a claim by attacking not the claim's proponent, but the company he or she keeps, or by questioning the reputations of those with whom he or she agrees. This is also known as *poisoning the well*, which suggests a fitting scenario:

SOLVED PROBLEM

8.4 What is wrong with the following argument?

Jones advocates fluoridation of the city water supply.
Jones spends much of his free time hanging around with known criminals, drug addicts, and deviants.
∴ We should not fluoridate the city water supply.

Solution

The premises are irrelevant to the conclusion; even if Jones has detestable friends, what he advocates may well be true. Notice that it would not be a fitting reply merely to contest the second premise ("Jones actually spends most of his free time helping the elderly and performing volunteer work in hospitals"). The central logical issue here is not whether Jones is the victim of a smear, but the failure to produce any premises germane to the conclusion.

(3) *Tu quoque* ("you too") arguments attempt to refute a claim by attacking its proponent on the grounds that he or she is a hypocrite, upholds a double standard of conduct, or is selective and therefore inconsistent in enforcing a principle. The implication is that the arguer is unqualified to make the claim, and hence that there is no reason to take the claim seriously.

SOLVED PROBLEM

8.5 Criticize the following argument.

Jones believes we should abstain from liquor.
Jones is a habitual drunkard.
∴ We should not abstain from liquor.

Solution

Jones' actions have no bearing on the truth or falsity of his belief, even if he holds the belief hypocritically; the argument commits a *tu quoque* fallacy.

It requires self-discipline to distinguish between what someone says and what that person does; to stifle one's natural resentment against insincere, opportunistic, or morally weak individuals; and to avoid becoming defensive when someone else recommends a course of action. But a rational person is one who can do this.

(4) *Vested interest* arguments attempt to refute a claim by arguing that its proponents are motivated by the desire to gain something (or avoid losing something). The implication is that were it not for this vested interest, the claim's proponents would hold a different view, and hence that we should discount their argument.

SOLVED PROBLEM

8.6   What is wrong with the following argument?

> Jones supports the fluoridation bill pending in Congress.
> He does so because he owns a major fluoridation firm, which will reap huge dividends if the bill passes.
> ∴ We should not support this bill.

Solution

Once again, the premises are irrelevant to the conclusion. Fluoridation may well be justified independently of Jones' allegedly selfish motives. The argument commits the vested interest fallacy. Whether Jones stands to gain or lose from fluoridation is immaterial. What counts is whether fluoridation is hygienically desirable, whether it is cost-effective, and so forth.

(5)   *Circumstantial ad hominem* fallacies are sometimes grouped in a single category with vested interest fallacies, but there is a distinction between them. The circumstantial version of the *ad hominem* fallacy is the attempt to refute a claim by arguing that its proponents endorse two or more conflicting propositions. The implication is that we may therefore safely disregard one or all of those propositions.

SOLVED PROBLEM

8.7   Criticize the following argument:

> Jones says he abhors all forms of superstition.
> Jones also says that breaking a mirror brings bad luck.
> ∴ There probably is something to superstition after all.

Solution

This is an *ad hominem* circumstantial fallacy. Jones' claims, whether consistent or not, have by themselves no bearing on the truth of the conclusion.

All five kinds of *ad hominem* arguments attempt to refute a claim by attacking its proponents. *Straw man* arguments, by contrast, attempt to refute a claim by confusing it with a less plausible claim (the straw man) and then attacking that less plausible claim instead of addressing the original issue. The term comes from medieval fencing, where participants warmed up by practicing against dummies (straw men) before facing flesh-and-blood adversaries.

A straw man argument may in fact provide good reasons against the less plausible claim that it confuses with the real issue, but these reasons will be irrelevant to the real issue itself.

SOLVED PROBLEM

8.8   Criticize the following argument:

> There can be no truth if everything is relative.
> ∴ Einstein's theory of relativity cannot be true.

Solution

The premise is completely irrelevant to the conclusion, because Einstein's theory does not assert that everything is relative (whatever that means). The claim that everything is relative is a straw man, and the argument implicitly attacks this straw man rather than examining Einstein's actual theory. Thus even if the premise is true (which is certainly doubtful; it is difficult even to see what this premise could mean), it offers no support for the conclusion.

Notice that to correctly diagnose a straw man argument we generally need to know something about the claim in question. In this case, we need to know that Einstein's theory does not assert that everything is relative.

Ad baculum arguments (also called *appeals to force, appeals to the stick*) are attempts to establish a conclusion by threat or intimidation.

SOLVED PROBLEM

8.9   Criticize the following argument:

> If you don't vote for me, Ms. Jones, I'll tell everybody you are a liar.
> ∴ You ought to vote for me.

Solution

Once again, the premise is irrelevant to the justification of the conclusion. Coercion, threats, and intimidation may be persuasive in some cases, but they have no place in any scheme of rational appraisal. Note that it makes no difference how Jones responds to this threat; even if she capitulates, that does not change the fact that this sort of "reasoning" is logically unacceptable.

Typically an *ad baculum* occurs as part of a conversation and may often be stated incompletely. In the case of Problem 8.9, for example, the conclusion might be left implicit. Thus *ad baculum* fallacies are difficult to distinguish from mere threats. The "offer you can't refuse" is immoral, but is it illogical? Consider this example:

SOLVED PROBLEM

8.10   Is the following a fallacious *ad baculum* argument?

> Parent: Clean up your room OR ELSE . . .
> Child: Or else what?
> Parent: Or else you'll get spanked.

Solution

No. There is no argument here, merely an ultimatum. The parent is not trying to justify or support a conclusion. Where there is no argument, there can be no fallacy (at least in the sense of 'fallacy' which we are using here).

Ad verecundiam arguments (*appeals to authority*) occur when we accept (or reject) a claim merely because of the prestige, status, or respect we accord its proponents (or opponents).

SOLVED PROBLEMS

8.11   Why is this bare appeal to authority inadequate?

> My teacher says that I should be proud to be an American.
> ∴ I should be proud to be an American.

Solution

Without some further evidence that the teacher's statement is correct or justified, the premise is simply irrelevant to the conclusion. Saying something doesn't make it so, no matter how eminent the speaker.

8.12   Is this argument any better?

>   Her teacher says that I should be ashamed to be an American.
>   ∴ I should be ashamed to be an American.

Solution

No. It's another flat-footed appeal to authority. The fact that it reaches a diametrically opposite conclusion doesn't help at all.

Our two examples show that the inadequacy of bare appeals to authority is especially apparent when authorities' opinions are in conflict. It is a simple strategy, however, to add a premise to each of these examples, thereby forging a link of relevance and making each of the arguments valid.

SOLVED PROBLEM

8.13   Do these arguments commit the fallacy of appeal to authority?

>   My teacher says that I should be proud to be an American.
>   Everything my teacher says is true.
>   ∴ I should be proud to be an American.
>
>   Her teacher says that I should be ashamed to be an American.
>   Everything her teacher says is true.
>   ∴ I should be ashamed to be an American.

Solution

No. The premises are now fully relevant to the conclusions; and moreover, each argument is now deductively valid. However, the second premise in each case can hardly be true. These arguments are still fallacious; we have merely shifted the location of the fallacy. The fundamental problem with these two arguments as well as the arguments in Problems 8.11 and 8.12 is that they ask us to take someone else's word for the conclusion, rather than considering it on its own merits.

There are, however, many occasions when an appeal to authority is either justified or else unavoidable. Few of us know enough physics to verify the equation $E = mc^2$ for ourselves. Without the required mathematical and experimental background to confirm it, it is certainly reasonable to take Einstein's word (and, indeed, the word of the entire community of contemporary physicists) for it. In a complex society, where labor is divided and expertise segregated into specialties, it is difficult if not impossible for any individual to know enough to make unaided decisions on such matters. Even Einstein didn't master physics by himself; he was trained at a university, and learned the techniques of his predecessors.

Thus much of our knowledge is based unavoidably on appeals to authority. These appeals are not fallacious, provided that we have good evidence that the authorities themselves have adequate justification for their views. Appeals to authority are nonetheless fallacious if they demand uncritical acceptance of the authority's pronouncements without providing evidence of the authority's reliability (as in Problems 8.11 and 8.12) or if they overestimate the authority's reliability (as in Problem 8.13). (In the latter case, the fallacy is not the fallacy of appeal to authority per se, but simply a false premise.)

In contemporary North America, perhaps the most prevalent form of appeal to authority is the *testimonial*, exemplified by celebrities who appear in advertisements and commercials endorsing products, services, or brands of consumer goods. The image of the "star" actor, actress, or athlete surrounded by the brandished commodity is all that is necessary. Overt claims need not be made. This presents a problem, since without claims there can be no argument, and hence no potential fallacy to expose. Moreover, we don't want to twist ads around, or put words into the mouths of commercial spokespersons. Provided we heed these warnings, however, we can represent the gist of testimonials as arguments without creating straw men.

SOLVED PROBLEM

8.14 As we know, the opinions of "authorities" don't always coincide. But the fallacious appeal they convey is constant. Consider these two cases:

> Leslie Nielsen urges us to buy a new Chrysler.
> ∴ We should buy a new Chrysler.

> Kevin Costner urges us to buy a new Ford.
> ∴ We should buy a new Ford.

What is objectionable in each case?

Solution

Both arguments are testimonial versions of the fallacy of appeal to authority. In neither case does the premise have any relevance to the conclusion.

Notice that the fallacy would remain unchanged if the actor merely urged us to take a test drive. The fallacy is committed whenever we reason merely by the form:

> X says that P.
> ∴ P.

It doesn't matter whether the results are beneficial, harmful, or neither.

Testimonials usually exploit fame or notoriety rather than special knowledge. But this doesn't have to be the case. Suppose Nielsen or Costner urged aspiring thespians to enroll at the Jack Webb actor's school. Since Nielsen and Costner are both successful actors, this opinion (like the physics community's endorsing $E = mc^2$) would thereby gain some relevance. As a general rule, an appeal to authority is relevant (and hence reasonable) in proportion to the reliability of the authority in the field in which the appeal is made.

Ad populum arguments (*appeals to the people*) occur when we infer a conclusion merely on the grounds that most people accept it. This fallacy has the form:

> X says that P.
> ∴ P.

which makes it analogous to appeal to authority. However, the 'X' in this case stands for majority opinion, not the views of a specialized (or glamorous) person or minority.

SOLVED PROBLEM

8.15 Why is this appeal to popular belief illegitimate?

> Everybody believes that premarital sex is wrong.
> ∴ Premarital sex is wrong.

Solution

The issue is not whether "everybody" in fact believes this (though that premise is in fact false), but what we may infer from such a premise. The argument fails to establish any relevant connection between premises and conclusion and hence it is fallacious.

As with authority, however, there are times when an appeal to "common sense" or "common knowledge" is in order, provided that we have some reason to believe that this "common knowledge" is itself well grounded or reliable.

Because most of us want peer approval and because pressures to conform to social dictates are always strong, appeals to the people often encourage the *bandwagon effect*, which asks us to join forces with others.

SOLVED PROBLEM

8.16   What is wrong with the following argument?

> More people drive Chevrolets than any other car.
>
> ∴ Shouldn't you? (A rhetorical question indicating that you should.)

Solution

This is a bandwagon version of the *ad populum* fallacy. The premise is simply irrelevant. It provides no evidence that Chevrolets are good cars or the right car for you to drive, or even that you should be driving at all. (Perhaps it would be better to ride a bike!) The popularity of Chevrolets *may* result from their quality, but the argument gives us no reason to believe that it does. It could as well be a function of the effectiveness of the advertising used to promote them or of other factors unrelated to their suitability for your needs.

Besides climbing on the bandwagon, advertisers often aim their commercial pitch at *cognoscenti*, those alleged to be "in the know" on a particular topic. If you want to be in the know, too, you will do as the ad recommends.

SOLVED PROBLEM

8.17   Explain the fallacy of relevance in this argument:

> Discriminating palates prefer wine *x*.
>
> ∴ You should drink wine *x*.

Solution

No reason is given why you should drink what "discriminating palates" prefer, so the conclusion simply does not follow. This is an *ad populum* fallacy which appeals to the cognoscenti. The premise of this argument is also questionable; exactly who (or what?) counts as a "discriminating palate," and to what things do these "discriminating palates" prefer this wine? (We'll have more to say about this premise in Problem 8.27.)

Ad misericordiam arguments (*appeals to pity*) ask us to excuse or forgive an action on the grounds of extenuating circumstances. They seek clemency for breaches of duty, or sympathy for someone whose poor conduct or noncompliance with a rule is already established. An appeal to pity may be either legitimate or fallacious, depending on whether or not the allegedly extenuating circumstances are genuinely relevant to the case.

SOLVED PROBLEMS

8.18   Explain the fallacy in the following argument:

> Oh, officer, you see my baby here was crying for some candy and I took her to the candy store before I came back to my car.
>
> ∴ You shouldn't give me a parking ticket.

Solution

This is an *ad misericordiam* fallacy. The arguer appeals to the officer's pity, but the appeal is clearly irrelevant. This is no reason not to get a parking ticket.

8.19   Other cases are not so clear-cut: Is the professor vindicated by the following argument?

> Although I received low teaching evaluations, most of my students had no prior knowledge of the subject and had trouble comprehending the reading.
>
> During the semester, I was burdened with other pressing duties, including a heavy schedule of committee work and participation in several major conferences.
>
> I had no teaching assistant or grader, and I had to grade a very large number of papers every week without assistance.
>
> ∴ These evaluations should not determine or play a significant role in my reappointment.

Solution

We don't have enough information to tell. The premises are more relevant in this case than in Problem 8.18, but the degree to which they excuse the bad teaching evaluations is unclear. To decide whether this is an *ad misericordiam* fallacy, we need to know exactly how bad the evaluations were, how extraordinary the mitigating circumstances were, and by what standards these factors are to be weighed.

Ad ignorantiam arguments (*appeals to ignorance*) have one of the following two forms:

It has not been proved that P.   It has not been proved that $\sim P$.
∴ $\sim P$.   ∴ P.

Here are two classic examples:

SOLVED PROBLEM

8.20   What is wrong with these arguments?

> No one has ever proved that God exists.
> ∴ God does not exist.

> No one has ever proved that God does not exist.
> ∴ God exists.

Solution

Both are fallacious appeals to ignorance. Nothing about the existence of God follows from our inability to prove God's existence or nonexistence (i.e., from our ignorance about the matter).

These arguments suggest a false dichotomy: either our evidence for a claim is conclusive or the claim itself is false. Quite obviously, however, a claim may be true even if our evidence for it is inconclusive. In the absence of proof, the rational approach is to weigh the available evidence, and, if the preponderance of the evidence favors one conclusion, to adopt that conclusion *tentatively*. Sometimes, however, the available evidence is not sufficient even to favor a tentative conclusion. In that case, it is often best simply to suspend judgment.

Ignoratio elenchi (*missing the point*) occurs when the premises of an argument warrant a different conclusion from the one the arguer draws. This can be very embarrassing, especially if the conclusion which does follow contradicts or undermines the one actually drawn. The expression "missing the point" is also used as a general catchphrase to describe the fallacies of relevance. However, we shall reserve it for the precise mistake just mentioned.

SOLVED PROBLEM

8.21 What point does the arguer miss?

> Any amount of inflation is bad for the economy.
> Last month inflation was running at an annual rate of 10 percent.
> This month the inflation rate is only 7 percent.
> ∴ The economy is on the upswing.

Solution

Given the premises, what follows is only that the *rate* of inflation is slowing down. Inflation is still occurring (i.e., things are still getting worse), but more slowly than before. This is very different from maintaining that the economy is improving; indeed, it suggests the opposite conclusion.

Missing the point can be devastating, because it implies that the arguer doesn't understand what he or she is saying. Not all cases of missing the point have a conclusion that is in some sense opposite to what the premises license. But all do involve drawing the wrong conclusion, one which the premises fail to authorize.

A *red herring* is an extraneous or tangential matter used purely to divert attention away from the issue posed by an argument. (The phrase stems from a method used to train hunting dogs to track down a scent.) Because it is irrelevant, a red herring contributes nothing to an argument, though it may mislead its audience into thinking otherwise. Red herrings are rhetorical devices. They enable those who use them to mask other defects hampering an argument, and thus to evade the real issue.

SOLVED PROBLEM

8.22 What is wrong with this argument?

> Some members of the police force may be corrupt, but there are corrupt politicians, corrupt plumbers, corrupt salespeople, and even corrupt preachers.
> There are also lots of honest cops on the job.
> ∴ Let's put police corruption in perspective (the implication being, of course, that police corruption is not as bad as it may seem).

Solution

Rhetorical grandiloquence is here employed to throw us "off the scent" of the real issue, which is what to do about police who accept bribes. The argument attempts to lull its audience into complacency about this issue. Notice that the first premise contains a whiff of *tu quoque*. (Why single out police for punishment? Corruption exists in all walks of life.) The second premise invites us to feel sympathy for honest police officers who are beleaguered by the bad reputation of their brethren (another diversionary tactic). Furthermore, the real conclusion (that police corruption is not as bad as it may seem) is not explicitly stated, so that we are less inclined to notice that the premises are irrelevant to it.

8.3 CIRCULAR REASONING

Circular reasoning (also called *petitio principii* and *begging the question*) occurs when an argument assumes its own conclusion. Such an argument is always valid (since if the assumptions are all true, the conclusion must surely be true!). And it is not lacking in relevance (for what could be more relevant to a statement than that statement itself?). Furthermore, if all the assumptions happen to be true, the argument is actually sound. But it is useless as a means of *proving* its conclusion. To see this, we must consider the context of the argument. Either the argument is offered in a context in which the conclusion is already known to be true (in which case there is no point in trying to prove it), or it is offered in a context in which the conclusion is doubtful. But if the conclusion is doubtful, then so (to precisely the same degree) is the assumption which is identical to it. An argument which employs doubtful assumptions (even if they are, unbeknownst to us, true) provides no more credibility for its conclusion than its premises already have (see Section 2.2). Hence, in neither context does a question-begging argument advance the credibility of its conclusion.

Some authors use the term 'begging the question' in a very broad sense, to designate any reasoning based on disputed premises. We shall insist, however, that the disputed premise be the conclusion itself—or at least a statement very closely related to it (see Problem 8.24, below). In practice, question-begging arguments are usually disguised, either by restating one of the premises in different words as the conclusion or by keeping one of the two identical statements implicit.

SOLVED PROBLEMS

8.23   Analyze the mistake in the following argument:

> Public nudity is immoral because it is just plain wrong.

Solution

The premise is 'Public nudity is just plain wrong' and the conclusion is 'Public nudity is immoral'. These two sentences say basically the same thing. The argument therefore begs the question. We can see the structure of the argument clearly by diagraming it as in Chapter 1. Since the conclusion and the premise are really the same proposition, we give them the same number (1), so that the diagram looks like this:

1
↓
1

or, perhaps more illuminatingly, like this:

which provides a graphic illustration of the argument's circularity.

8.24   Analyze the mistake in this argument:

> Capital punishment is justified. For our country is full of criminals who commit barbarous acts of murder and kidnapping. And it is perfectly legitimate to punish such inhuman people by putting them to death.

Solution

This argument argues for the conclusion that capital punishment is justified by assuming that it is legitimate to put certain criminals to death. Obviously these two statements say the same thing, even though the alleged truth of the latter is psychologically strengthened by the statement of the other premise. So the argument begs the question. This can be made explicit by bracketing and numbering the statements of the argument as in Chapter 1:

① [Capital punishment is justified.] For ② [our country is full of criminals who commit barbarous acts of murder and kidnapping.] And ① [it is perfectly legitimate to punish such inhuman people by putting them to death.]

We could then diagram the argument in either of the ways indicated below:

$$\begin{array}{cc} \dfrac{2+1}{\downarrow} & \dfrac{2+1}{} \\ 1 & \circlearrowleft \end{array}$$

It is also possible to regard the statement that it is legitimate to put criminals to death as a restatement of the conclusion, rather than a premise. In that case the argument is not question-begging, but premise 2 fails to provide logical support for the conclusion.

Question-begging epithets are phrases that prejudice discussion and thus in a sense assume the very point at issue. Often they suggest an *ad hominem* abusive attack: 'godless communist', 'knee-jerk liberal', 'bleeding heart', 'Neanderthal conservative', 'reactionary', 'bra burner', 'redneck'. They need not, however, be part of a question-begging argument. They often amount to nothing more than name-calling.

Complex questions are a rhetorical trick somewhat akin to question-begging arguments. The question 'Have you stopped beating your spouse?', for example, presupposes an answer to the logically prior question 'Did you ever beat your spouse?'. If this implicit question has not been answered affirmatively, then the original question is illegitimate, a verbal trap to lure the unwary. It is illegitimate because it presupposes what has not been established, namely, that the person being questioned has beaten his or her spouse. A complex question, of course, is not even a statement, let alone an argument. But its resemblance to question-begging arguments invites brief comment.

8.4 SEMANTIC FALLACIES

Semantic fallacies occur when the language employed to express an argument has multiple meanings or is excessively vague in ways that interfere with assessment of the argument's cogency.

Ambiguity (or *equivocation*) is multiplicity of meaning. Most words and many phrases have more than one meaning. This alone does not inhibit understanding, since context usually makes the intended meaning clear. We expect 'ball' to mean something different at an outdoor stadium from what it means inside a dance pavilion. When we call someone a dummy, we don't mean that he or she is made of wood. 'Puppet on a string' has figurative as well as literal connotations.

Despite contextual cues, ambiguity can sometimes cause trouble. Relatively abstract terms such as 'right' and 'law' are prime candidates for troublesome equivocation, because we tend to run their different meanings together. A political right is not the same thing as a legal or moral right—or as *the* political right (i.e., conservatism). These senses of 'right', in turn, are distinct from 'right' as a synonym for truth. A law of physics is not a law in the same sense as a socially instituted law.

Ambiguity generates fallacies when the meaning of an expression shifts during the course of an argument, causing a misleading appearance of validity.

SOLVED PROBLEM

8.25 Evaluate the following argument:

It is silly to fight over mere words.
Discrimination is just a word.
∴ It is silly to fight over discrimination.

Solution

In this context 'discrimination' can have one of two meanings: (1) action or policy based on prejudice or partiality or (2) the word 'discrimination' itself. (In the latter case, the word should in fact be enclosed in quotation marks, as explained in Section 1.6; but unfortunately, many people ignore that convention.) Now if we take 'discrimination' to mean the word itself in both its occurrences, there is no fallacy, and the argument is valid and perhaps even sound. (It seems a bit odd, however, to speak of fighting over this word.) On the other hand, if we take 'discrimination' in both its occurrences to mean discriminating action or policy, then although the argument is again valid, the second premise is certainly false; hence the argument fails. What seems most likely, however, is that the arguer intended 'discrimination' to mean the word in the premise and the action or policy in the conclusion. In that case, the premises may both be true, but they are irrelevant to the conclusion and the argument is invalid. However, confusion with the other two interpretations can give rise to a deceptive appearance of both relevance and validity. Under this third interpretation, the arguer commits a fallacy of ambiguity.

Our analysis of this example made reference to several different interpretations and judged the argument differently under each. Ambiguous reasoning frequently requires such treatment. No single interpretation can lay claim to being the argument that is *really* expressed unless we have conclusive evidence that the arguer intended it and not the others. In the absence of such evidence, careful analysis requires attention to all plausible interpretations. In fact, there is yet a fourth possible interpretation of the argument in Problem 8.25 that we did not consider, the interpretation under which 'discrimination' means the action or policy in the second premise and the word in the conclusion. This interpretation is so implausible, however (since it makes the second premise both false and irrelevant and invalidates the reasoning), that it can safely be ignored.

Amphiboly is ambiguity at the level of sentence structure, i.e., ambiguity not traceable to any particular word in a sentence but to the way the words are assembled. One logically troublesome source of amphiboly is the occurrence of both universal and existential quantifiers in the same English sentence. As noted in Section 6.2, this generates ambiguity that can be represented in predicate logic by changing the relative scopes of the quantifiers. For example, the sentence 'Some number is greater than any number' has two meanings:

(1) For any number x, there is some number or other (not necessarily the same in each case) which is greater than x.

(2) There is some number y which is greater than all numbers.[1]

Notice that this duality of meaning is not traceable to any single ambiguous word in the original statement, but is instead a product of that statement's structure. Under interpretation 1, the statement is true. For every number x, there *is* a greater number: $x + 1$, for example. But under interpretation 2 it is false. For interpretation 2 implies that y is greater than itself (since y is greater than all numbers and is itself a number), which of course is impossible.

SOLVED PROBLEM

8.26   Evaluate the following argument:

> Some number is greater than any number.
> ∴ Some number is greater than itself.

---

[1]Using 'N' for 'is a number' and 'G' for 'is greater than', these can be represented respectively in predicate logic as follows:

(1) $\forall x(Nx \rightarrow \exists y(Ny \mathbin{\&} Gyx))$

(2) $\exists y(Ny \mathbin{\&} \forall x(Nx \rightarrow Gyx))$

Notice the shift in relative positions of the quantifiers.

Solution

The premise is amphibolous in the way indicated above. Under interpretation 1, the premise is true, but the inference is invalid (obviously, since the conclusion is false). And under interpretation 2, the inference is valid but the premise is false. Hence under no interpretation is the argument sound.

Vagueness is indistinctness of meaning, as opposed to multiplicity of meaning. Consider this variation on Problem 8.17:

SOLVED PROBLEM

8.27   What is wrong with this argument?

Discriminating palates prefer wine x.

I have a discriminating palate.

∴   I should drink wine x.

Solution

This version (in contrast to the version of Problem 8.17) is valid, and the premises are relevant to the conclusion. However, the meanings of the premises are so indistinct that their truth is in question. Exactly what is a "discriminating palate," and who (besides the arguer) has one? Anyone who likes wine x? That would be a criterion, but it would be question-begging. In advertising, the meanings of words are often left indistinct so that consumers may interpret them in whatever way they find most flattering. But that is not the end of a vagueness here. Even if we could decide who has a "discriminating palate," to what things are such people supposed to prefer wine x? To all other wines? To the cheapest wines? To rotten meat? Unless these questions are answered, nothing definite has been asserted. We cannot tell whether or not the premises are true, and so we should not accept the argument.

An extreme version of vagueness is *doublethink*, in which every sentence cancels out its predecessor and contradicts its successor. George Orwell's novel *1984* describes a fictional society (Oceania) whose rulers invent a language (Newspeak) which operates on this self-destructive principle. Newspeak is also a systematized form of lying in which every utterance is unverifiable. There is no access to facts except through distorted records, all composed in Newspeak. The aim of Newspeak is to make thought, and therefore dissent, impossible. The fraud underlying Newspeak cannot be grasped, even by the officials who introduce and maintain it. Orwell's concepts have inspired considerable research into the extent to which everyday conversation, scientific jargon, and military, political, corporate, and bureaucratic discourse in contemporary society incorporate elements of Orwellian doublethink.

Accent refers to emphases that generate multiple (and often misleading) interpretations. Newspaper headlines, contractual fine print, commercial "giveaways," and deceptive contest entry forms are frequent sources of fallacies of accent.

SOLVED PROBLEM

8.28   Why is this newspaper headline deceptive?

STUDENTS DEMONSTRATE
New Laser Beam Techniques
Used to Retrieve Coins from Vending Machines

Solution

The first line encourages the reader to believe that (college?) students are engaged in political protest. The second makes us expect a major scientific breakthrough. The third again forces us to reinterpret the entire message. Reliance on the first line or first and second lines alone leads to erroneous conclusions. To draw such conclusions is to commit the fallacy of accent.

8.5 INDUCTIVE FALLACIES

Inductive fallacies occur when the inductive probability of an argument (i.e., the probability of its conclusion given its premises) is low, or at least lower than the arguer thinks it is. (For a general discussion of inductive probability, see Section 2.3.)

Hasty generalization means fallaciously inferring a conclusion about an entire class of things from inadequate knowledge of some of its members. Hasty generalizations are usually fallacious statistical or inductive generalizations (two common inductive forms discussed in Sections 9.3 and 9.4, respectively). Occasionally, the term is used in a broader sense, to describe *any* fallacious extrapolation from observed to unobserved data.

SOLVED PROBLEM

8.29   Evaluate the following argument:

Last Monday I wrecked my car.
The Monday before that my furnace broke.
∴ Bad things always happen to me on Mondays.

Solution

The inductive probability of this argument is extremely low. Two bad events on two Mondays are hardly sufficient to warrant a sweeping conclusion about all Mondays. This is a fallacy of hasty generalization.

Hasty statistical generalizations often stem from biased, unrepresentative, or inadequate sampling techniques. This is a problem for social scientists, pollsters, and survey takers. Insufficient observation is another source of hasty generalization. What we regard as true of the world may reflect only our limited experience.

Faulty analogy is an inductive fallacy associated with analogical reasoning. In analogical reasoning we assert that object x has certain similarities with object(s) y, and that y has a further property P. We then conclude that x has P. This form of reasoning sometimes has a fairly high inductive probability. For example, we may reason that since rats have many relevant physiological similarities to human beings, and since a certain chemical has been shown to cause cancer in rats, that same chemical is likely to cause cancer in humans. The inductive probability of analogical reasoning depends quite sensitively, however, on the degree and relevance of the similarity.[2] If the similarity is slight or not particularly relevant, then a fallacy is likely to result.

SOLVED PROBLEM

8.30   Is this analogical argument a good one?

The American colonies justly fought for their independence in 1776.
Today the American Football Alliance is fighting for its independence.
∴ The Alliance's cause is just too.

---

[2]It depends as well on other factors, such as the relative strengths of the premises and conclusion; these additional factors are discussed more thoroughly in Chapter 9.

Solution

No. The disputes surrounding the Revolutionary War included religious liberty, repressive taxation, quartering of troops, and national sovereignty, whereas the (imagined) football alliance presumably seeks the right to compete for players, customers, and lucrative television contracts. The analogy is extremely weak at best. The two causes really have little in common, except for the word 'independence' and the aura it promotes. As a result, the argument has quite a low inductive probability (which is not, of course, to say that its conclusion is false; the alliance's causes may well be just, even though this argument for it is weak). Anyone who offers such an argument thinking that it lends much support to its conclusion thereby commits the fallacy of faulty analogy.

The *gambler's fallacy* is an argument of the form:

  x has not occurred recently.
  ∴ *x* is likely to happen soon.

This sort of reasoning is fallacious if '*x*' designates a type of event whose occurrences are more or less independent (i.e., one occurrence of which does not affect the likelihood of others). Separate occurrences of rain or of a favored outcome in a game of chance, for example, are independent events. The fallacy gets its name from the propensity of gamblers who have had a run of bad luck to think "my luck is bound to change soon."

SOLVED PROBLEMS

8.31 Explain the fallacy in the following argument:

  On each of the last ten throws, this coin has come up heads.
  ∴ If tossed again, it is almost certain to come up tails.

Solution

This is an instance of the gambler's fallacy. If the coin in question is fair, then the tosses are independent events, and the probability of tails on the next toss is .5, regardless of what has happened in the past. A long string of heads, however, may provide evidence that the coin or the method by which it is tossed is not fair, but biased in favor of heads. In that case, the probability of tails on the next toss is even less than .5. In neither case does a run of heads *increase* the probability of tails on a succeeding toss.

8.32 Why is this *not* an instance of the gambler's fallacy?

  This clock chimes every hour.
  It has not chimed for roughly 57 minutes.
  ∴ It is bound to chime soon.

Solution

The chimings are not independent events, but are dependent on one another in the way indicated by the first premise. This argument commits no fallacy.

'*False cause*' is a term covering a variety of logical sins. Most simply, it means confusing a cause with an effect. But it may also mean offering an immediate causal explanation for an event without considering alternatives. Another variant is *post hoc ergo propter hoc* (*after this, therefore because of this*—often abbreviated to *post hoc*), in which a causal relationship is inferred merely from the temporal proximity of two or more events. What is common to all false-cause fallacies is that their conclusions are causal claims which are inadequately supported by their premises. (For a discussion of how causal conclusions may be adequately supported see Section 9.6.)

SOLVED PROBLEMS

8.33   What is wrong with the following argument?

> Every prophet or messiah is a charismatic leader.
> ∴ The exercise of talent for leadership is a road to religious inspiration.

Solution

Even if the premise were true (which is doubtful), it would not make the conclusion probable. A correlation between leadership and inspiration does not mean that leadership causes inspiration. In this case, it is obviously far more likely to be the other way around: the religious inspiration motivates the leadership. But even that is not a foregone conclusion. There are yet other logical possibilities. Perhaps some independent third factor (genetic, social, or even supernatural) is responsible for each of these two qualities. Or perhaps the correlation is a coincidence. Nothing in the premise rules out any of these possibilities, and so the probability of the conclusion, given the premise, is quite low.

8.34   Evaluate this argument:

> Johnny Touchdown is a great athlete.
> Johnny Touchdown takes amphetamines regularly.
> ∴ Amphetamines help make Johnny a great athlete.

Solution

Again we have a correlation of two factors, but no evidence to indicate that this correlation is causal. The probability of the conclusion, given the premise, is low. In this case, the argument also suppresses important evidence (see Section 2.5), since it is common knowledge that although amphetamines may temporarily help to combat fatigue, they do not enhance playing performance; in the long run they actually impair it.

8.35   Evaluate this reasoning:

> The patient became violently ill immediately after eating lunch.
> There were no signs of illness prior to eating, and she was in good spirits during the meal.
> She is in good health overall, and her medical history shows no record of physical problems.
> ∴ She was the victim of food poisoning.

Solution

This is essentially a case of *post hoc* reasoning. Though the premises do tell us a bit more than just that the illness occurred right after lunch, the information they contain is hardly sufficient to render the conclusion very probable. We need to know more. The patient should be carefully examined. It would help to inspect or analyze the food she ate. We would also want to know whether anyone else who ate the same food (or used the same dishes, utensils, or serving bowls) has become ill. Even so, there is a variety of alternative explanations (cardiac arrest, stroke, asphyxiation, or sudden onset of a disease which has undergone a prior incubation period) available to the physician.

However, two important qualifications need to be made with respect to this example. First, to label the reasoning *post hoc* (a variety of false cause) is not to say that its conclusion is false. In this sense 'false cause' is a misnomer. It means merely that the evidence contained in the premises does not *by itself* make the conclusion very probable. Second, although the argument does not support its conclusion very well, it may nevertheless be prudent to consider it in deciding what action to take. The physician must make a fast decision; the patient's life may depend on it. Though not yet well confirmed, food poisoning is in this case a reasonable tentative hypothesis (though there are others as well). It might, for example, be a good idea to pump the patient's stomach. If this tentative hypothesis proves wrong, the patient will undergo acute but short-term discomfort. If it proves right, the patient's life may be saved. It would be a mistake, however, to act as if food poisoning were the only reasonable explanation. Heart attacks, for example, may follow a similar pattern. If the patient were having a heart attack, it would be a grave mistake to do nothing but pump her stomach.

In addition to the fallacies discussed above, the fallacy of *suppressed evidence* is often classified as an inductive fallacy. This is the fallacy of ignoring evidence bearing negatively on an inductively inferred conclusion. It is, however, unlike the other fallacies discussed in this section, in that it may occur even if the argument's inductive probability is quite high and is not overestimated by the arguer. The fallacy of suppressed evidence was discussed in Section 2.5 and will not be considered further here.

8.6 FORMAL FALLACIES

Formal fallacies occur when we misapply a valid rule of inference or else follow a rule which is demonstrably invalid. An invalid rule of inference is an argument form that has some invalid instances. Thus in identifying formal fallacies, it is not sufficient simply to show that an argument has an invalid form. If a formal fallacy is suspected, it is important to ascertain *both* that the rule on which the reasoning seems to be based is invalid (by using the methods of formal logic) *and* that the argument itself is invalid. Typically we show that the argument itself is invalid by finding a *counterexample*, an actual or logically possible case in which the premises of the argument are true and its conclusion is false.

SOLVED PROBLEM

8.36   Is a formal fallacy committed in the following argument?

> If it rains heavily tomorrow, the game will be postponed.
> It will not rain heavily tomorrow.
> ∴ The game will not be postponed.

Solution

Yes. The argument has an invalid form:

$$R \rightarrow P$$
$$\sim R$$
$$\therefore \sim P$$

and is itself invalid. The invalidity of the form can be verified by truth tables or refutation trees (Chapter 3). We can see that the argument itself is invalid by finding a counterexample. A counterexample for this argument is a situation in which the premises are both true (i.e., if it rains heavily tomorrow, the game will be postponed; but it's not going to rain heavily tomorrow) and yet the conclusion is false. There are a number of ways in which the conclusion could be false (i.e., the game *will* be postponed) under these conditions. Perhaps it will snow heavily tomorrow, so that the game will be postponed because of snow. Perhaps the visiting team will miss its incoming flight, so that the game has to be postponed for that reason. These possibilities are counterexamples, and there are many others as well. The arguer has simply made a mistake in reasoning, supposing that the conclusion validly follows from these premises when in fact it does not.

The invalid rule by which the arguer of Problem 8.36 reasoned is called *denying the antecedent*, and an invalid argument which has this form is said to commit the *fallacy of denying the antecedent*. The name comes from the second premise. In denying the antecedent, there is always a conditional premise and a second premise which is the denial (negation) of the antecedent of the conditional. The conclusion is the negation of the consequent of the conditional.

SOLVED PROBLEM

8.37   Is the following argument invalid?

> If anyone knows what happened, Richard knows.
> No one knows what happened.
> ∴ Richard does not know what happened.

Solution

No. Although it does have the form of denying the antecedent—

$$A \rightarrow R$$
$$\sim A$$
$$\therefore \ \sim R$$

it is nevertheless valid. If the premises are true, then (since the second premise asserts that no one knows) certainly the conclusion that Richard does not know must be true. There is no counterexample. One can also see that the argument is valid by noting that in addition to the invalid form above, it also has the valid predicate logic form:

$$\exists x Kx \rightarrow Kr$$
$$\sim \exists x Kx$$
$$\therefore \ \sim Kr$$

The validity of this form can be demonstrated by using refutation trees or by constructing a predicate calculus proof. (Indeed, the conclusion follows from the second premise alone.)

Any argument with at least one valid form is valid. But as Problem 8.37 illustrates, valid arguments can (and generally do) have invalid forms as well. (For further discussion of this point, see Section 3.6.) Thus, to repeat, having an invalid form does not mean that an argument commits a formal fallacy. A formal fallacy is clearly committed only if the argument is in fact invalid. If it happens that an arguer uses an invalid rule to produce a valid argument, and if we somehow know that the invalid rule was the one that the arguer was relying on (perhaps without realizing that it is invalid), then we may justly convict the arguer of committing a formal fallacy even though the argument is valid. However, cases of this sort are rare.

Another formal fallacy closely related to denying the antecedent is *affirming the consequent*, whose invalidity was proved in Problem 3.18. This fallacy occurs when we reason according to the following rule:

$$P \rightarrow Q$$
$$Q$$
$$\therefore \ P$$

Once again, the name comes from the second premise, which affirms the consequent of the first.

SOLVED PROBLEM

8.38   Evaluate this argument:

> If Smith inherited a fortune, then she is rich.
> She is rich.
> ∴ She inherited a fortune.

Solution

This is a fallacy of affirming the consequent. The argument has the form indicated above and is moreover clearly invalid. A typical counterexample for this argument is a scenario in which both the premises are true but in which it is false that Smith inherited her fortune; rather, she made it by creating a successful software corporation. (There are, of course, any number of other ways in which Smith could have gotten rich; all of these constitute counterexamples.)

Denying the antecedent and affirming the consequent are often confused with *modus tollens* and *modus ponens*, respectively. (For discussion of these valid rules, see Chapters 3 and 4.) Such confusion is often the source of these fallacies. More particularly, the confusion may arise from mistaking 'if' for 'only if'. If the first premise of Problem 8.36 read '*Only* if it rains heavily tomorrow will the game be postponed', then the argument would be valid. Similarly, in Problem 8.38, if the first premise were '*Only* if Smith inherited a fortune is she rich', this argument would be valid as well.

The fallacy of *composition* occurs when we invalidly impute characteristics of one or more parts of a thing to the whole of which they are parts.[3] This fallacy has the following form:

$p_1, \ldots, p_n$ are parts of w.

$p_1, \ldots, p_n$ have property F.

$\therefore$ w has property F.

(Here $p_1, \ldots, p_n$ may be any number of objects, possibly just a single object.)

SOLVED PROBLEMS

8.39   Evaluate the following argument:

Every sentence in this book is well written.

$\therefore$ This book is well written.

Solution

This is a typical fallacy of composition. It has the form indicated above (if we prefix it with the obvious assumption that sentences in the book are parts of the book) and is in fact invalid. To see its invalidity, consider the following counterexample. We can imagine or compose a book in which each sentence satisfies grammatical and aesthetic criteria while the sentences and paragraphs bear no relation to one another, resulting in incoherence. This would make the conclusion false while the premise is true.

8.40   What about this argument?

My liver is part of my body.

My liver contains at least a million cells.

$\therefore$ My body contains at least a million cells.

Solution

This argument has the form of compositional fallacies, but in spite of this it is valid. There is no counterexample and no fallacy.

As with other formal fallacies, fallacies of composition cannot be spotted by considering the argument form alone, since (as Problem 8.40 shows) the form of the fallacy of composition has some valid instances.

---

[3]This fallacy, like the fallacy of division discussed next, is often labeled as semantic fallacy. However, since it employs invalid rules of inference and need not involve any confusion about meaning, we prefer to call it formal fallacy.

The converse of the fallacy of composition is the fallacy of *division*. Whereas in composition we invalidly impute characteristics of the parts to the whole, in division we invalidly impute characteristics of the whole to the parts. This fallacy has the following form:

w has property F,

$p_1, \ldots, p_n$ are parts of w.

$\therefore$ $p_1, \ldots, p_n$ have property F.

SOLVED PROBLEMS

8.41   Evaluate this argument:

> This book is written in English.
>
> $\therefore$ Every sentence in this book is in English.

Solution

Fallacy of division. (The obvious assumption that the sentences of the book are parts of the book is suppressed.) A counterexample is that though the book is written in English, one or more sentences (e.g., quotations) may be in another language. Notice, however, that the converse argument (which has the form of fallacies of composition) is valid:

> The sentences of this book are parts of this book.
>
> Every sentence of this book is in English.
>
> $\therefore$ This book is in English.

8.42   Evaluate this argument:

> California is in the United States.
>
> Los Angeles and San Francisco are in California.
>
> $\therefore$ Los Angeles and San Francisco are in the United States.

Solution

This argument has the form of fallacies of division, but it is not fallacious. It is in fact sound.

8.7   FALLACIES OF FALSE PREMISES

Fallacies of false premises are mistakes of just the sort the name suggests. So arguments that commit this fallacy may be valid, but are never sound. One common instance is the *false dichotomy*. This fallacy is committed when we make the false assumption that only one of a number of alternatives holds.

SOLVED PROBLEM

8.43   What is wrong with the following argument?

> Either you're with us or you're against us.
>
> You're not with us.
>
> $\therefore$ You must be against us.

Solution

In many contexts, the first premise would be false. There is a third option—namely, the option of remaining neutral in the (unnamed) issue at hand. Notice that the reasoning is perfectly correct. The argument is an instance of disjunctive syllogism and is therefore valid. Moreover, the premises are quite relevant to the conclusion. The mistake, then, is merely the falsity of the first premise.

A *slippery slope* fallacy occurs when the conclusion of an argument rests upon an alleged chain reaction, suggesting that a single step in the wrong direction will result in a disastrous or otherwise undesirable outcome. We may represent this form of reasoning as follows:

$$A_1 \rightarrow A_2$$
$$A_2 \rightarrow A_3$$
$$\cdot$$
$$\cdot$$
$$\cdot$$
$$A_n \rightarrow A_{n+1}$$

It should not be the case that A_{n+1}.

∴ It should not be the case that A_1.

The phrase 'it should not be the case that' is to be understood in a moral sense (and hence is an operator of deontic logic; see Section 11.7). Although this form may seem valid, in fact its validity depends on the interpretation of the conditional operator '→' and the way it interacts with the deontic operator.[4] However, the fallacy usually labeled "slippery slope" is a problem not only with the validity of such reasoning, but also with the truth of its premises. That is why we include slippery slope in this section rather than the previous one. The examples below illustrate both problems.

SOLVED PROBLEMS

8.44   Criticize the following argument:

> If you step on a crack, you break your mother's back.
> If you break your mother's back, you commit a moral outrage.
> You should not commit a moral outrage.
> ∴ You should not step on a crack.

Solution

This is a transparent slippery-slope fallacy. The first premise is false. The second may be as well, since breaking your mother's back by accident would not be a moral outrage. Moreover, depending on the interpretations of the conditionals, the argument may be invalid in a rather subtle way. Imagine a scenario in which you are going to break your mother's back, but not as a result of stepping on a crack. That is, you are going to do it whether you step on a crack or not. Then, under certain interpretations of 'if . . . then' (the material interpretation, for one), the first premise is true. Suppose the other premises are true as well. Still, it may be false that you should not step on a crack, since stepping on a crack does not actually result in breaking your mother's back. Thus under certain interpretations of 'if . . . then', the argument has a counterexample and is therefore invalid.

---

[4]The conditional operator has several possible meanings. So far we have discussed only one, the material conditional (Section 3.1); others are mentioned in Section 11.7. However, a full discussion of the varieties of conditionals would involve technicalities that are beyond the scope of this book.

8.45   Consider this argument:

> If Paul comes back to Old Bricks, Mary will certainly go away.
>
> If Mary goes away, then Sue will certainly accept her job offer and move to Chicago.
>
> If Sue moves to Chicago, her best friends will go away too.
>
> If they go, then eventually every single person will abandon Old Bricks.
>
> We should not allow that to happen.
>
> ∴ Paul should not come back.

Solution

This is reminiscent of the so-called "domino theory", which was often used in an attempt to justify American involvement in the Vietnam war. (The arguer would hold that pulling out of Vietnam would result in the fall of Vietnam to communism, which would start a chain of events leading to the fall of all neighboring countries, like the fall of a series of domino pieces.) This is not a transparent slippery-slope fallacy, since it is by no means easy to assess the truth of the premises. No time limit is specified, and it is impossible to establish in advance whether the predictions are correct. It would be a mistake, however, to accept any of these premises uncritically.

Moreover, as in the previous problem, the argument may have invalid interpretations. To see this, suppose that every single person will eventually abandon Old Bricks for some reason or other, regardless of Paul's whereabouts. Then the conditional premises would all be true, at least according to the material interpretation. We may also agree on the last premise for the sake of the argument. Yet the conclusion could still be rejected. If Paul's decision is not going to halt the inevitable exodus from Old Bricks, then perhaps his coming back is for the best—*somebody* at least will live there. This is a logical point about the validity of the argument, not a factual or historical claim. In considering this sort of counterexample, we are saying not that things are fated to happen in this way, but that whether they are or are not must be carefully considered in evaluating the argument. As with the domino theory about Vietnam, closer consideration of the facts and circumstances is necessary to determine the truth of the matter. We should leave that to the historians.

Supplementary Problems

Rewrite the arguments below in standard form, and discuss the fallacy or fallacies (if any) committed in each. If there is no explicit argument but the passage seems to suggest one, write the suggested argument in standard form and evaluate it.

(1)   Marriages are like playing the horses. Some are sure things, while others are losers going out of the gate. The moral is, check out the contestants before you place your bets.

(2)   Every fish in this lake was biting when I baited my hooks with fresh worms two weeks ago. Fresh worms should enable me to catch many fish again today.

(3)   How dare you criticize my logic? You commit fallacies every time you write a paper or turn in an exam.

(4)   Sarah Socket recommends Louella Balsam shampoo. Sarah is a very popular actress. Therefore, I ought to try her shampoo.

(5)   Those who don't fight communism help it along, and Jones isn't fighting communism.

(6)   If the government doesn't stop terrorism now, it will eventually spread all over the world. If it spreads all over the world, there will be a nuclear conflagration. We can't have that. Therefore, the government must halt terrorism now.

(7)   Enormous State has the best graduate and undergraduate programs on the entire continent. Therefore, it is the best school on the continent.

(8)   There is no definite link between smoking and lung cancer, despite the surgeon general's report and years of scientific studies. Therefore smoking is not harmful to your lungs.

(9)   Dante, Leonardo da Vinci, and Fellini were all Italian. They were also great artists and writers. Clearly, Italians are more creative than the general population.

(10)   If we don't agree to the city's new rent proposal, the city will condemn the land and evict us. Therefore, we ought to sign the contract today.

(11)   I decided not to join BEST after I discovered that the founder of the organization is a former used-car salesman. That told me everything I needed to know about BEST. Don't become a member.

(12)   God has a providential plan for humankind. We know this because we know that God is real. And we know that God is real because providence is real.

(13)   Every time I eat at Joe's Diner I get sick. I'm going to stop eating there. Then I won't get sick anymore.

(14)   Nuclear weapons have not been used since 1945. Therefore, the danger of their use is currently very high.

(15)   Nuclear weapons have not been used since 1945. Therefore, it is unlikely that they will ever be used again.

(16)   When did you embezzle the company's pension funds?

(17)   Once you legalize abortion, inevitably you wind up with child pornography, child abuse, and abuse of the aged. Complete lack of respect for life breeds complete lack of respect for the living. Therefore, we must outlaw abortion.

(18)   I'm having so much fun on my vacation, I'm beginning to get nervous. Something is bound to go wrong before my trip is over.

(19)   Since World War II, most Americans have lived in cities. After 1946, most Americans began eating processed cereal for breakfast. Therefore, the move to the city was responsible for changes in American dietary behavior.

(20)   As long as I'm paying your college tuition, you won't be allowed to major in philosophy. You should major in business or suffer the consequences. Since you know what's good for you, I'm sure you'll follow my advice.

(21)   We have a tradition of satisfied customers. No one has ever complained about our service. If you're in the market, stop by our showroom. You won't regret making your next purchase here.

(22)   Eating this piece of pie now is good for my digestion. Therefore, eating an entire pie at once is good for my digestion.

(23)   Officer, I was only driving three miles over the speed limit. That's hardly breaking the law. Besides, the car in the next lane over was zooming much faster, yet you pulled me over. I don't deserve a speeding ticket.

(24)   Since we are purposeful creatures and we are parts of the universe, the universe must be purposeful as well.

(25)   Since the universe as a whole lacks purpose and we belong to the universe, our perception of being purposeful must be a mere illusion.

(26)   In *The Second Self*, Sherry Turkle claims that both children and professional "hackers" personify computers, treating machines as though they were human beings endowed with feelings and thoughts. This research shows that "hackers" have the same mentality as children.

(27)   If Darwin's theory of evolution was correct, then his own remote ancestors were apes. This proves how absurd Darwin's theory is.

(28)   I should never watch the *Yankees* on TV. Whenever I do, they lose.

(29)   If the murder weapon was a gun, police should have found powder marks in the vicinity of the body. The police found several powder marks on the carpet near where the victim fell. Therefore, the murder weapon was a gun.

(30) Limp-wristed, quiche-eating Californians are incapable of serious thought. To learn anything about politics or the real world, talk to a New Yorker.

(31) Logicians claim that the study of argument is an indispensable part of everyone's education. Since this helps to keep them employed, it's not surprising that they endorse such views. Let's not be influenced by thinking which does not represent the majority of students, parents, or teachers themselves.

(32) Timmy the Sheik thinks the Bears will repeat in the Super Bowl. So I've already asked my bookie to put $500 on the Bears next January. By the way, did you catch Timmy's one-minute spot for the American Cancer Association? He's right. The odds are against you if you smoke. I'm quitting cigarettes today.

(33) An exhaustive survey of the class of 2000 revealed that 81 percent believe in God, 91 percent believe in the importance of family life, and 95 percent respect both of their parents. This survey also showed that over three-fourths (77 percent) of entering students placed "having a professional career" high on the list of their objectives. Clearly, the return to traditional American moral values is responsible for the new vocational emphasis among our college-age youth.

(34) We cannot allow our school system to become bilingual. If we permit instruction in either of two languages, what's to prevent introducing a third or a fourth? Besides, how can we in good conscience tell representatives of one ethnic community that their language isn't good enough to be taught in the schools, when we've already decided to grant somebody else's children that very privilege? Consequently, we should remain monolingual.

(35) Television watching is directly to blame for the alarming decline in literacy. In the United States, the average person (adult or child) stares at the tube for 6.5 hours every day of the year. No other activity except sleep consumes so much time. Moreover, television is passive, requiring no thought or effort. It allows us to put our minds at rest. In that respect, TV is no different from sleep, with which it almost competes for recreational favor. During those 6.5 hours, the idiot box hypnotizes us and dulls our critical awareness of the world; it makes us incapable of thought and reason. Small wonder that Americans can no longer read and write; apart from everything else, TV drains all their leisure time.

(36) In the ruin of a family, its immemorial laws perish; and when the laws perish, the whole family is overcome by lawlessness. And when lawlessness prevails, . . . the women of the family are corrupted, . . . [and] a mixture of caste arises. . . . And this confusion brings the family itself to hell and those who have destroyed it; for their ancestors fall, deprived of their offerings of rice and water. By the sins of those who destroy a family and create a mixture of caste, the eternal laws of the caste and the family are destroyed. (*The Bhagavad-Gita*, I, 40–43)

(37) It is mistaken to suppose that two wrongs don't ever make a right. For when we engage in civil disobedience to protest against an unjust law, we nominally break the law as soon as we protest, and the protest is therefore a second "wrong." Yet surely civil disobedience is justified in such cases. Therefore, two wrongs sometimes do make a right. (Example courtesy of Leo Groarke)

(38) First offenders usually commit petty theft. Often they are teenagers, still attending high school. They may have done nothing more than go on a high-speed joyride with their companions. They are emotionally shaken by the arrest and fingerprinting process, and their parents are quite distraught when they learn the news and post bail. For these reasons, law enforcement officers are inclined to go easy on first offenders. We agree. Think about the mistakes you made as a teen. Then give a kid a break.

(39) While the nuclear power plant has been shut down since the last radiation leak, local residents are still getting their electricity from alternative sources. The good weather and pretty scenery in the area, plus ideal shopping and recreation, mean that citizens should not worry excessively about the on again, off again status of a single reactor.

(40) Professor X was fired for refusing to give student athletes passing grades, despite pressure from the head football coach. In the Jan Kemp case, decided in 1986 in the Georgia courts, a college administrator dismissed for failing student athletes was awarded back pay and damages. Kemp's job was to teach remedial English to members of the football team, whereas X is a nuclear physicist. Yet both were asked to compromise their principles. Both refused. Clearly the courts should reinstate professor X and compensate him adequately for being willing to lose his job than compromise his integrity.

Answers to Selected Supplementary Problems

(1)   Some horses are sure things, while others are losers going out of the gate.

            Some marriages are sure things, while others are losers going out of the gate.

            You should check out horses before you place your bets.

      ∴ You should check out the "participants" in marriage before you place your bets.

This argument commits fallacies of vagueness and false analogy. (We have rewritten it to fit the analogical format.) The language of the argument is extremely vague. What does it mean to "place your bets on" a marriage? Who does this? The couple themselves? Onlookers? Relatives? What sort of marriage is a "loser"? Moreover, the analogy is weak and not clearly relevant to the conclusion (mainly as a result of the conclusion's vagueness). The homely advice contained in the conclusion may in some sense be sound, but not for the reasons given by this argument.

(3)   You commit fallacies every time you write a paper or turn in an exam.

      ∴ Your criticism of my logic is inappropriate.

(We've replaced the rhetorical question with a statement that approximates its implication.) This is a *tu quoque* variant of the *ad hominem* fallacy.

(5)   Those who don't fight communism help it along.

            Jones is not fighting communism.

      ∴ Jones is helping communism.

False dichotomy.

(8)   There is no definite link between smoking and lung cancer, despite the surgeon general's report and years of scientific studies.

      ∴ Smoking cigarettes is not harmful to your lungs.

Fallacy of ignorance. Moreover, the premise is false.

(11)   The founder of BEST is a former used-car salesman.

      ∴ One should not become a member of BEST.

This is a fallacy of guilt by association, one of the varieties of *ad hominem*. The organization is attacked on the basis of a popular stereotype concerning used-car salesmen.

(14)   Nuclear weapons have not been used since 1945.

      ∴ The danger of their use is currently very high.

Gambler's fallacy. There is no evidence that uses of nuclear weapons are events dependent on one another in this way.

(16)   This is not an argument, but it is a complex question.

(17)   Once you legalize abortion, inevitably you wind up with child pornography, child abuse, and abuse of the aged.

            These things must not occur.

      ∴ We must outlaw abortion.

This is the nub of the argument. The remaining statement, 'Complete lack of respect for life breeds complete lack of respect for the living', is perhaps offered as justification for the first premise. If so, then the inference from this statement to the first premise is a simple *non sequitur*. The portion of the argument represented above is a slippery slope whose first premise is certainly questionable and whose validity is questionable on deontic grounds as well.

(21)   We have a tradition of satisfied customers.

            No one has ever complained about our service.

            You won't regret making your next purchase here.

      ∴ If you're in the market, you should stop by our showroom.

This argument is not deductively valid, but it is reasonably strong. None of the fallacies discussed in this chapter is committed here.

(23)   There are two arguments with the same conclusion here. The first commits the fallacy of special pleading:

> I was only driving three miles over the speed limit.
> That's hardly breaking the law.
> ∴ I don't deserve a speeding ticket.

The second commits a *tu quoque* fallacy:

> The car in the next lane over was zooming much faster, yet you pulled me over.
> ∴ I don't deserve a speeding ticket.

(27)   If Darwin's theory of evolution was correct, then his own remote ancestors were apes.

> A theory produced by a creature whose remote ancestors were apes is absurd.
> ∴ Darwin's theory is absurd.

With the addition of the implicit second premise, the argument is seen to be question-begging. This fallacy is combined with a quaint *ad hominem* abusive fallacy (as though it were shameful to be descended from apes!).

(30)   Limp-wristed, quiche-eating Californians are incapable of serious thought.

> ∴ To learn anything about politics or the real world, you should talk to a New Yorker.

Ad hominem abusive fallacy with a completely irrelevant conclusion. There is also a hint here of false dichotomy. Is the arguer assuming that everyone worth considering is either a Californian or a New Yorker?

(34)   This passage suggests two slippery-slope arguments, each with questionable premises:

> If we permit instruction in either of two languages, we may (will?) have to introduce a third or a fourth.
> We must not introduce a third or fourth language.
> ∴ We cannot allow our school system to become bilingual.
> If we do not remain monolingual, we will have to tell the representatives of one ethnic community that their language isn't good enough to be taught in the schools, when we've already decided to grant somebody else's children that very privilege.
> We cannot in good conscience tell representatives of one ethnic community that their language isn't good enough to be taught in the schools, when we've already decided to grant somebody else's children that very privilege.
> ∴ We should remain monolingual.

(37)   When we engage in civil disobedience to protest against an unjust law, we nominally break the law as soon as we protest, and the protest is therefore a second "wrong."

> Surely civil disobedience is justified in such cases.
> ∴ Two wrongs sometimes do make a right.

The argument equivocates betwen the legal and moral senses of the term 'wrong'. An unjust law is morally wrong; justified civil disobedience against an unjust law is legally wrong but morally right. A fallacy is committed if the conclusion is intended to mean (as seems likely) that two moral wrongs sometimes make a right, since if the term 'wrong' in the first premise is intended in the moral sense, then that premise is false.

(40)   Professor X was fired for refusing to give student athletes passing grades, despite pressure from the head football coach.

> In the Jan Kemp case, decided in 1986 in the Georgia courts, a college administrator dismissed for failing student athletes was awarded back pay and damages.
> Kemp's job was to teach remedial English to members of the football team, whereas X is a nuclear physicist.
> Yet both were asked to compromise their principles.
> Both refused.
> ∴ The courts should reinstate professor X and compensate him adequately for being willing to lose his job rather than compromise his integrity.

This is an analogical argument, and the analogy is quite strong and relevant. (There is a disanalogy in that X taught physics while Kemp taught English, but that is not a relevant disanalogy.) The argument tacitly assumes that the Georgia decision was correct. If that assumption is true and there is no suppressed evidence, then this is a good inductive argument.

Chapter 9

Induction

9.1 STATEMENT STRENGTH

This chapter concerns some common kinds of inductive reasoning. In an inductive argument, the conclusion need not follow from the premises as a matter of logical necessity. Rather, in inductive reasoning we are concerned with the probability of the conclusion, given the premises, i.e., the *inductive probability* of the argument (see Section 2.3). Since inductive probability depends in turn on the relative strengths of the premises and conclusion, we begin with a discussion of statement strength.

The strength of a statement is determined by what the statement says. The more it says, the *stronger* it is, regardless of whether what it says is true. A strong statement is true only under specific circumstances; the world must be just so in order for it to be true. A weak statement is true only under a wide variety of possible circumstances; it says nothing very specific, and it demands little of the world for its truth.

SOLVED PROBLEM

9.1 Some of the following statements are quite strong; the rest are quite weak. Distinguish the strong statements from the weak ones.

(*a*)   There are exactly 200 cities with populations over 100,000 in the United States.

(*b*)   Something is happening somewhere.

(*c*)   Something exists.

(*d*)   Hobbits are humanoid creatures, rarely over a meter tall, with ruddy faces and woolly toes, that inhabit burrows on hillsides in a land called The Shire.

(*e*)   If there are any crows, then some of them are male.

(*f*)   Jim's house is the third house on the left as you turn from Concord Street onto Main.

(*g*)   Every organism now alive has obtained its genetic material from once-living organisms.

(*h*)   Some people are sort of weird.

(*i*)   Every vertebrate has a heart.

(*j*)   It is not true that Knoxville, Tennessee, has exactly 181,379 inhabitants this very second.

Solution

(*a*)   Strong

(*b*)   Weak

(*c*)   Weak

(*d*)   Strong

(*e*)   Weak

(*f*)   Strong

(*g*)   Strong

(*h*)   Weak

(*i*)   Strong

(*j*)   Weak

It may seem surprising that statement (j) is weak, since this statement is very specific with respect to time, place, and number. But note that all it says is that there were *not* exactly this many people at this place and time; it therefore says very little, since it allows the population of Knoxville to be anything but 181,379 (and is true even if Knoxville doesn't exist). If we omit the phrase 'It is not true that', this statement becomes very strong, but when this phrase is included it is very weak.

Indeed, this holds in all cases; the negation of a weak statement is strong and the negation of a strong statement is weak. Consider, for example, the negation of the weak statement 'Something exists'. This is 'Nothing at all exists', which says a great deal, for it asserts that the universe is utterly devoid of anything. Even the slightest little flicker of existence would make it false. Thus it is quite informative and, fortunately, quite false.

The strength of a statement is approximately inversely related to what is called its *a priori* probability, that is, probability prior to or in the absence of evidence. (We say "approximately" because there are various conceptions of a priori probability which differ in detail, and there is no agreement as to which, if any, is the exact inverse of statement strength.) The stronger a statement is, the less inherently likely it is to be true; the weaker it is, the more inherently probable it is.

The strongest possible statements are those which say so much that they cannot be true, i.e., self-contradictions. The statement 'Waldo both is and is not a cat', for example, is maximally strong and thus has an a priori probability of zero. The strength of this statement is perhaps best appreciated by the observation that since it is self-contradictory it logically implies *every* statement.

The weakest possible statements are those which are logically necessary. Because logically necessary statements are true under all possible circumstances, in a sense they say nothing at all. The tautology 'If grass is green, then grass is green', for example, is logically necessary and therefore maximally weak. Its a priori probability is 1.

Comparisons of strength are not always possible. Among the five strong statements in Problem 9.1, for instance, it is impossible to say which is the strongest. All are stronger than any of the five weak statements, but no method is known for making precise comparisons among such diverse statements.

It is possible, however, to rank some sets of statements with respect to relative strength. This is achieved by the following rules:

Rule 1: If statement A deductively implies statement B but B does not deductively imply A, then A is stronger than B.

Rule 2: If statement A is logically equivalent to statement B (i.e., if A and B deductively imply one another), then A and B are equal in strength.

The justification of these two rules rests on the fact that if statement A logically implies statement B, then there are no possible circumstances in which A is true and B is false. Thus the possible circumstances in which A is true are a subset of the possible circumstances in which B is true. Now if (as in rule 1) B does not imply A, then there are possible circumstances in which B is true and A is false. Hence A is true in fewer possible circumstances than B is, which is just to say that A is stronger than B. But if (as in rule 2), B does imply A, then A and B are true under precisely the same sets of circumstances, and hence they are equal in strength.

SOLVED PROBLEMS

9.2 Rank the following statements in order from strongest to weakest:

(a) Either some cows are horned or some buffalo are horned.

(b) There are cows and buffalo, and all cows and buffalo are horned.

(c) There are cows, and all of them are horned.

(d) Some cows are horned.

(e) Either some cows are horned, or it is not the case that some cows are horned.

(f) Some cows are both horned and not horned.

Solution

(f), (b), (c), (d), (a), (e). This solution is obtained by applying rule 1 and the predicate calculus. One can see by using, for example, the truth tree method of Chapter 6 that each item in the solution list deductively implies all succeeding items, but that no item in the list deductively implies an earlier item. Note that statement (f) is self-contradictory, while (e) is logically necessary.

9.3 Compare the strengths of the following statements:

(a) Adair admires Adler.

(b) It is not the case that Adair doesn't admire Adler.

(c) Adler is admired by Adair.

(d) Adair admires something which is identical with Adler.

(e) Adair admires Adler, and if it rains then it rains.

Solution

These statements are logically equivalent, as the reader can check again by the techniques of Chapter 6; hence, by rule 2 they are all of equal strength. Notice that statement (e) is just the conjunction of statement (a) with the tautology 'if it rains, then it rains'. This tautology in effect says nothing, so that conjoining it with statement (a) produces a statement equivalent to (a).

Rules 1 and 2, however, are not applicable in every case. None of the strong statements of Problem 9.1, for example, deductively implies any of the others. Moreover, the differences in strength among them (if any) are too small to be intuitively apparent. Thus we have no way of ordering these strong statements with respect to strength. The same applies to the weak statements of Problem 9.1.

The logical importance of statement strength lies in its relation to inductive probability. The general rule is that inductive probability tends to vary directly with the strength of the premises and inversely with the strength of the conclusion.

SOLVED PROBLEMS

9.4 What is the effect on the following argument form of increasing the number n?

We have observed at least n daisies, and they have all had yellow centers.

∴ If we observe another daisy, it will have a yellow center.

Solution

The premise gets stronger as the number n gets larger. As a result, each increase in n also increases the argument's inductive probability. This argument form is an instance of simple induction, which will be discussed more thoroughly in Section 9.4.

9.5 Suppose that the mean height for an adult male in the United States is 5 feet 10 inches. We wish to use this fact as a premise to draw a conclusion about the height of X, an American man whom we have not yet met. Below are three different conclusions which we could draw. Which conclusion produces the strongest argument?

(a) X is exactly 5 feet 10 inches tall.

(b) X is within a foot of 5 feet 10 inches tall.

(c) X is within an inch of 5 feet 10 inches tall.

Solution

   The inductive probability (and hence the strength) of the argument will be higher the weaker we make the conclusion. Conclusion (*a*) is the strongest of the three conclusions; (*b*) is the weakest. Thus (*b*) produces the strongest argument and (*a*) the weakest.

   Strengthening the premises or weakening the conclusion of an argument may not increase the argument's inductive probability, however, if we alter their content in a way that disrupts relevance.

SOLVED PROBLEM

9.6   Evaluate the inductive probability and degree of relevance of the following argument:

   Ninety-five percent of American families have indoor plumbing, a telephone, a television, and an automobile.
   The Joneses are an American family.
   ∴ The Joneses have indoor plumbing, a telephone, a television, and an automobile.

What happens to inductive probability and relevance if we replace the conclusion with the following weaker but less relevant statement?

   (*a*)   The Chun family of Beijing owns at least one automobile.

What happens if we replace the original conclusion with *this* weaker but still highly relevant statement?

   (*b*)   The Joneses have indoor plumbing.

Solution

   The original argument has both high relevance and a fairly high inductive probability, and its conclusion is strong. Conclusion (*a*) is weaker, but when we substitute it for the original conclusion both the relevance and inductive probability of the argument decrease significantly. If we replace the conclusion by (*b*), on the other hand, we have weakened the conclusion while preserving relevance and thereby created a stronger argument. The inductive probability of this new argument is *at least* as high as that of the original, and we may reasonably suppose that it is somewhat higher.

   Since in practice we are seldom concerned with modifications to premises or conclusions unless they preserve relevance, we seldom encounter exceptions to the rule that strengthening premises or weakening the conclusion increases inductive probability.

9.2   STATISTICAL SYLLOGISM

   Inductive arguments are divisible into two types, according to whether or not they presuppose that the universe or some aspect of it is or is likely to be uniform or lawlike. Those which do not require this presupposition may be called *statistical* arguments; the premises of a statistical argument support its conclusion for purely statistical or mathematical reasons. Those which do require it we shall call *Humean* arguments, after the Scottish philosopher David Hume, who was the first to study them thoroughly and to question this presupposition.

SOLVED PROBLEMS

9.7   What type of inductive argument is the following?

> 98 percent of college freshmen can read beyond the sixth-grade level.
> Dave is a college freshman.
> ∴ Dave can read beyond the sixth-grade level.

Solution

This argument is statistical. The conclusion is quite likely, given the premises, on statistical grounds alone. (Of course, if there were some evidence that Dave's reading skills were deficient, then the conclusion would not be so likely, but that is not a special feature of this argument; all inductive arguments are vulnerable to contrary evidence. See Section 2.5.)

9.8   What type of inductive argument is the following?

> Each of the 100 college freshmen surveyed knew how to spell 'logic'.
> ∴ If we ask another college freshman, he or she will also know how to spell 'logic'.

Solution

This is a Humean argument. It is clearly nondeductive, since it is possible for its premise to be true while its conclusion is false. And it is not statistical, since its inductive probability depends on how likely it is that future observations of college freshmen resemble past ones. Any estimate of the inductive probability of this argument presupposes an appraisal of the degree of uniformity or lawlikeness of certain events—in this case, events involving the spelling abilities of college freshmen. That is the hallmark of a Humean argument.

In the remainder of this section and in Section 8.3, we focus on inductive arguments that are statistical. We shall come back to various forms of Humean inference in the second part of this chapter.

The most obvious value for the inductive probability of a statistical argument is simply the percentage figure divided by 100. Thus, the inductive probability of the argument of Problem 9.7 is .98. At least, this is true according to the so-called *logical* interpretation of inductive probability. Many theorists prefer some version of the so-called *subjective* interpretation, according to which inductive probability is a measure of a particular rational person's degree of belief in the conclusion, given the premises. According to the subjective view, the inductive probability of the argument of Problem 9.7 may deviate from .98, depending on the knowledge and circumstances of the person whose degree of belief is being measured. It is not possible to explain the details of these two kinds of interpretations here (though some further remarks are made in Chapter 10). Instead, we will simply presuppose a logical interpretation.

The form of the argument of Problem 9.7 is called *statistical syllogism* and can be represented as follows:

> n percent of F are G.
> x is F.
> ∴ x is G.

Here 'F' and 'G' are to be replaced by predicates, 'x' by a name, and 'n' by a number from 0 to 100. The inductive probability of a statistical syllogism is (by our logical interpretation) simply $n/100$. Note that in the case in which $n = 100$, the argument becomes deductive and its inductive probability is 1. For $n < 50$, it is more natural for the argument to take the form:

> n percent of F are G.
> x is F.
> ∴ x is not G.

We will regard this form too as a version of statistical syllogism. Its inductive probability is $1 - n/100$, and it becomes deductive whenever $n = 0$.

In some cases, the statistics used to draw the conclusion of a statistical syllogism are not numerically precise. This is illustrated by the arguments in the next problem.

SOLVED PROBLEM

9.9   Evaluate the inductive probability of the following statistical syllogisms:

(a)   Madame Plodsky's diagnoses are almost always right.
      Madame Plodsky says that Susan is suffering from a kidney stone.
   ∴ Susan is suffering from a kidney stone.

(b)   Most of what Dana says about his past is false.
      Dana says that he lived in Tahiti and had two wives there.
   ∴ Dana did not live in Tahiti and have two wives there.

(c)   Only a tiny fraction of commercial airline flights end in crashes.
      I will take a commercial airliner to Chicago.
   ∴ My Chicago flight will not end in a crash.

Solution

Each of these arguments has an inductive probability greater than .5, though to none of them can we assign a precise inductive probability. The terms 'almost always' and 'only a tiny fraction' in arguments (a) and (c) indicate very small and very large percentages, respectively. These arguments have reasonably high inductive probabilities. 'Most' in argument (b) simply means more than half; the inductive probability of this argument is therefore only slightly better than .5.

Reasonably high inductive probability, of course, is only one of the criteria an argument must meet in order to demonstrate the probable truth of its conclusion. It must also have true and relevant premises, and insofar as possible it must satisfy the requirement of total evidence (Section 2.5). The premises of a statistical syllogism are automatically relevant in virtue of its form. But they may not be true, and they may not be all that is known with respect to the conclusion.

Argument (a) of Problem 9.9, for example, may well have a false first premise, especially if Madame Plodsky is some sort of fortune-teller. It is an argument from authority, whose strength depends on Madame Plodsky's reliability. Although we depend on arguments from authority for much of our knowledge (i.e., we depend on the fact that much of what others tell us is true), such dependence easily becomes fallacious if the authority's reliability is in doubt or contrary evidence is supposed. If the first premise, the one asserting the authority's reliability, is omitted, then the argument is no longer a statistical syllogism. Its inductive probability drops significantly, and the remaining premise now lacks relevance to the conclusion, since in the absence of evidence that the authority is reliable, his or her pronouncements are not clearly relevant to the facts. The result is a *fallacy of appeal to authority*, a mistake discussed in Section 8.2.

Argument (b), in contrast to argument (a), reasons from the *un*reliability of a person's pronouncements. This is a form of *ad hominem* argument (argument against the person). If the premises are true and there is no suppressed evidence, argument (b) is a reasonably good argument. However, many *ad hominem* arguments, instead of addressing the veracity of the person in question, attack the person's character, reputation, or circumstances. If, for example, we replaced the first premise of argument (b) by the statement 'Dana is a terrorist', then the inductive probability of the argument would drop, the premises would lack relevance, the argument would commit an *ad hominem* fallacy (see Section 8.2). Without some premise connecting Dana's alleged terrorist activities to the reliability of his pronouncements about his past, this new argument would provide little, if any, support for its conclusion.

Argument (c) would be a very strong argument if its premises were known to be true and no evidence were suppressed. However, there is a subtle problem here. The conclusion of argument (c) concerns a future flight. Does its first premise also concern future flights, or just flights that have already occurred? That is, are we to read the first premise as

(1)   Only a tiny fraction of all commercial airline flights, past, present, and future, end in crashes.

or as

(2)   Only a tiny fraction of all past commercial airline flights have ended in crashes.

If we interpret the first premise of argument (c) as in alternative 1, then how can we know that it is true? Perhaps it is, at least if we take 'future' to mean the near or foreseeable future, but there may be some doubt. Interpreted in this way, however, the argument is clearly a statistical syllogism with relevant premises and a high inductive probability. If, on the other hand, we interpret the first premise of argument (c) as in alternative 2, then this premise is clearly true, but the argument is, strictly speaking, no longer a statistical syllogism; the flight referred to in the conclusion is not among the flights mentioned in the first premise, since it is a future flight. Under this interpretation, the first premise is weaker and (in conjunction with the second) less relevant to the conclusion than under interpretation 1, so that the argument's inductive probability is considerably lower. Indeed, since the argument now moves from premises about the past to a conclusion about the future, its reliability depends on how closely we can expect the future to resemble the past—specifically, on how consistent or uniform the pattern of airline crashes will be. This is the Humean presupposition, and hence under interpretation 2 the argument is Humean.

This is not to say that argument (c) is not a good argument. It may be a good argument under either interpretation, provided that we know the premises to be true and that there is no suppressed evidence. But under interpretation 1 we may not know the first premise to be true, and under interpretation 2, the argument's inductive probability is less than it is under interpretation 1 (how much less depends on the strength of our presupposition of uniformity).

SOLVED PROBLEM

9.10   Arrange the following arguments in order of decreasing inductive probability.

(a)   85 percent of the Snooze missiles fired so far have missed their targets.
        A Snooze missile was fired on July 4, 1997.
   ∴ This Snooze missed its target.

(b)   A Snooze missile will be fired tomorrow.
   ∴ This Snooze will miss its target.

(c)   85 percent of Snooze missiles fired so far have missed their targets.
        A Snooze will be fired tomorrow.
   ∴ This Snooze will miss its target.

(d)   No Snooze missiles have ever missed their targets.
        A Snooze was fired on July 4, 1997.
   ∴ This Snooze missed its target.

(e)   95 percent of Snooze missiles fired so far have missed their targets.
        A Snooze was fired on July 4, 1997.
   ∴ This Snooze missed its target.

(f)   No Snooze missiles have ever missed their targets.
        A Snooze will be fired tomorrow.
   ∴ This Snooze will miss its target.

Solution

(e), (a), (c), (b), (f), (d). Argument (e) is stronger than argument (a) because its first premise is stronger. Argument (a) is stronger than argument (c) because (c) extrapolates from past to future and hence presupposes the uniformity of nature, while (a) does not. Argument (c) in turn is stronger than argument (b), because (c) is based on stronger premises; (b) offers no real evidence in support of its conclusion. The conclusions of arguments (f) and (d) are *un*likely, given their premises; the inductive probability in each case is less than .5. In fact, (d) deductively implies the negation of its conclusion, so that the inductive probability of argument (d) is actually 0.

9.3 STATISTICAL GENERALIZATION

Statistical syllogism is an inference from statistics concerning a set of individuals to a (probable) conclusion about some member of that set. Statistical generalization, by contrast, moves from statistics concerning a randomly selected subset of a set of individuals to a (probable) conclusion about the composition of the set as a whole. It is the sort of reasoning used to draw general conclusions from public opinion polls and other types of random surveys.

SOLVED PROBLEM

9.11 Evaluate the inductive probability of the following argument:

Fewer than 1 percent of 1000 ball bearings randomly selected for testing from the 1997 production run of the Saginaw plant failed to meet specifications.

∴ Only a small percentage of all the ball bearings produced during the 1997 production run at the Saginaw plant fails to meet specifications.

Solution

The inductive probability of this argument is quite high. The size of the sample and the randomness of the selection strongly justify the generalization expressed by the conclusion. Moreover, the generalization itself involves a certain approximation, as indicated by the phrase 'only a small percentage'. This weakens the conclusion, adding strength to the argument.

The general form of statistical generalization is as follows:

n percent of s randomly selected F are G.
∴ About n percent of all F are G.

The number s indicates the size of the sample. F is a property which defines the population about which we are generalizing (in the case of Problem 9.11, bearings produced during the 1997 production run at the Saginaw plant). And G is the property studied by the survey (in this case, the property of failing to meet specifications).

To say that the sample was randomly selected is to say that it was selected by a method which guarantees that each of the F's had an equal chance of being sampled. This in turn implies that each s-membered subset of the F's has an equal chance of being chosen. Now it is a mathematical fact (whose proof is beyond the scope of this discussion) that if s is sufficiently large, most s-membered subsets of a given population are approximately representative of that population. In particular, for most s-membered subsets of the set of F's, the proportion of G's is about the same as it is among the F's generally. Hence if a fairly large sample of F's is randomly selected, it is likely, though not certain, that the proportion of G's which it contains will approximate the proportion of G's among all the F's.

The inductive probability of a statistical generalization is determined by purely mathematical principles. There is no need to presuppose any sort of natural uniformity. Consequently, statistical generalization is a statistical form of inference, not a Humean form.

The success of statistical generalization depends crucially on the randomness of the sampling technique. If the sample is not randomly chosen, then the sampling technique may favor samples having either an unusually high or an unusually low number of *G*'s. In such cases the sample is said to be *biased*. Attempts to apply statistical generalization with a nonrandom sampling technique commit the *fallacy of biased sample*, which is one form of the *fallacy of hasty generalization* discussed in Section 8.5. The resulting arguments are not true statistical generalizations, since true statistical generalization requires randomness. Their inductive probabilities are typically quite low.

SOLVED PROBLEM

9.12   Evaluate the inductive probability of the following argument:

> I spoke to my three friends. They took that course and they all got an A.
> ∴ Virtually everybody who took that course got an A.

Solution

This is obviously a weak argument. Here '*F*' designates the students who took the course and '*G*' those who got an A. Assuming a normal class size (number of *F*'s), a sample size of three is not large enough to justify the generalization expressed by the conclusion. Moreover, the sample is biased: the friends of the arguer do not constitute a random group of *F*'s. The argument commits the fallacy of hasty generalization.

The inductive probability of a genuine statistical generalization is primarily a function of two quantities: the sample size *s* and the strength of the conclusion. Increasing *s* strengthens the premise in a way relevant to the conclusion and thus enhances the argument's inductive probability. But to determine the argument's inductive probability we also need to take account of the strength of its conclusion. Notice that in the form given above, the conclusion is 'About *n* percent of *F* are *G*'. If it said 'Exactly *n* percent of *F* are *G*', it would be much too strong, and the argument's inductive probability would in almost all cases be close to zero. It is very unlikely that a random sample would contain *exactly* the same proportion of *G*'s as the population from which it was selected. Therefore, if we want our conclusion to be reliable, we must allow it a certain margin of error, and this is what the term 'about' (or a similar expression) signifies.

If we delineate this margin of error precisely, then there are mathematical methods for determining the argument's inductive probability numerically. The details of these methods are beyond the scope of this book, but some examples will illustrate the point.[1] Suppose we take 'about *n* percent' to mean $n\% \pm 3\%$. Then if $s = 1000$, the inductive probability of the argument turns out to be quite high, about .95 or perhaps a little higher. If we decrease *s* to 100 while keeping the conclusion the same, the inductive probability drops to something on the order of .5. For samples much smaller than 100, it becomes unlikely that the proportion of *G*'s in the sample is within 3 percent of the proportion of *G*'s in the population. In other words, the argument's inductive probability drops below .5.

If we interpret 'about' less strictly, we weaken the conclusion and hence raise the argument's inductive probability. Suppose that $s = 100$, but instead of concluding that $n\% \pm 3\%$ of *F* are *G*, we conclude that $n\% \pm 10\%$ of *F* are *G*. Now once again we have a strong argument, with an inductive probability of .95 or slightly more. If we allow an even greater margin of error in the conclusion, the inductive probability gets still closer to 1. If we are willing to accept a very wide margin of error (say, $\pm 30\%$), we can get an inductive probability of .95 with a sample as small as 20 or so. These figures remain relatively constant, regardless of the population size (number of *F*'s), provided that this number is fairly large.

---

[1]Readers who wish to understand the mathematical details of the relationships among *n*, *s*, the margin of error, and the probability of the conclusion, given the premise, of a statistical generalization should consult the material on confidence intervals in any standard work on statistics.

Inductive probability is thus enhanced both by increasing s (thereby strengthening the premise) and by increasing the margin of error in the conclusion (which weakens the conclusion). If the conclusion is too strong to be supported with reasonable inductive probability by the premise, the argument is said to commit the *fallacy of small sample*, which is another version of the fallacy of hasty generalization (Section 7.4).

Though the inductive probability of a statistical generalization varies mainly with s and the margin of error of the conclusion, n also has a small effect. If n is very large or very small (near 0 or 100), the argument's inductive probability is slightly higher (other things being equal) than it is if n is close to 50.

SOLVED PROBLEM

9.13 Arrange the following arguments in order of decreasing inductive probability.

(a) 50 percent of 100 randomly selected Americans said that they favored the president's handling of the economy.

∴ Exactly 50 percent of all Americans would say (if asked under the survey conditions) that they favor the president's handling of the economy.

(b) 50 percent of 1000 randomly selected Americans said that they favored the president's handling of the economy.

∴ 50% ± 10% of all Americans would say (if asked under the survey conditions) that they favor the president's handling of the economy.

(c) 50 out of 100 Americans said that they favor the president's handling of the economy.

∴ Exactly 50 percent of all Americans favor the president's handling of the economy.

(d) 50 percent of 100 randomly selected Americans said that they favored the president's handling of the economy.

∴ 50% ± 1% of all Americans would say (if asked under the survey conditions) that they favor the president's handling of the economy.

(e) 50 percent of 100 randomly selected Americans said that they favored the president's handling of the economy.

∴ Exactly 50 percent of all Americans favor the president's handling of the economy.

(f) 50 percent of 100 randomly selected Americans said that they favored the president's handling of the economy.

∴ 50% ± 10% of all Americans would say (if asked under the survey conditions) that they favor the president's handling of the economy.

Solution

(b), (f), (d), (a), (e), (c). Argument (b) is like argument (f), except that it employs a larger sample; so (b) is stronger than (f). Argument (f) is stronger than argument (d) because argument (d) has a stronger conclusion. For the same reason, (d) is stronger than (a). Argument (e) is still weaker than argument (a) because the conclusion of argument (e) is less relevant, being an assertion about what people actually believe, as opposed to what they would say. Argument (c), finally, has the smallest inductive probability, since it is like argument (e), except that its first premise is weaker, since it fails to indicate that the sample was random.

Inductive probabilities of statistical generalizations are usually suppressed in reports of the findings of surveys and public opinion polls. A report might say, for example, "Sixty-two percent of the voters approved the president's handling of the economy, subject to a sampling error of ±3%." What this means is that the sample was large enough (in this case about 1000) to ensure a .95 probability that the interval 62% ± 3% contains the actual proportion of voters who approve of the president's handling of the economy. Statisticians customarily take a probability of .95 as practical certainty and hence do not mention that a probability is involved here. But their conclusion that 62% ± 3% of the voters approve of the president's handling of the economy is in fact derived by statistical generalization from the premise that 62 percent of their sample approved the president's handling of the economy, a statistical generalization whose inductive probability is only .95. Thus it should be kept in mind that there is still a .05 probability that the proportion of all voters who support the president's handling of the economy lies outside the range 62% ± 3%.

Several precautions must be observed in evaluating statistical generalizations. For one thing, as noted earlier, the sample must be randomly selected. This does not mean that the sample must be known to contain the same proportion of G's as the population at large. If this were known, there would be no need for statistical generalization; we could simply *deduce* the proportion of G's in the population from the proportion of G's in the sample. It simply means that the sampling technique must ensure that the proportion of G's in the sample is *likely* to be close to the proportion of G's in the population at large.

Of course, if in making a statistical generalization someone claims that a certain sample is random but in fact it is not, then the premise of the statistical generalization is false, and the argument must be rejected. This may occur, for example, when what seems to be a random sampling method (e.g., picking names from a phone book) actually is not. Choosing names from a phone book will not, for example, provide a random sample of all homeowners, since those without phones have no chance of being chosen.

Like all inductive arguments, statistical generalization is vulnerable to suppressed evidence. If two or more random surveys get distinctly different results from true premises, then none of them alone constitutes a good argument. The requirement of total evidence demands that all of them be weighed in assessing the probability of the conclusion on which they bear.

Even slight deviations from the form of statistical generalization can seriously weaken the argument. This is illustrated by the next example.

SOLVED PROBLEM

9.14   Evaluate the following argument:

> Only 10 percent of 1000 randomly selected Americans answered "yes" to the question "Have you ever committed a felony?"
>
> ∴ About 10 percent of all Americans have committed felonies.

Solution

This argument deviates from the form of statistical generalization, because the premise concerns what the sampled Americans (F's) *said*, while the conclusion concerns what Americans *did*. That is, the property assigned to the variable G in the premise (answering "yes" to the question) is not the same as the property assigned to the variable G in the conclusion (actually having committed a felony). But for the argument to be a legitimate instance of statistical generalization, the same property must be assigned to both occurrences of the variable. The only conclusion we can legitimately draw by statistical generalization from the premise is:

> About 10 percent of all Americans would answer "yes" to the question "Have you ever committed a felony?" (if asked under the survey conditions).

With this new conclusion, the argument's inductive probability is fairly high, though we can't say exactly how high, because of the vagueness of the term 'about'. The original argument lacked clear relevance, and its inductive probability was much lower. For it is quite possible (indeed likely, given the sensitive subject matter of the survey) that some of the answers received were less than honest.

Problem 9.14 illustrates a general difficulty with polling human beings: How can we be sure that the respondents are telling the truth? In many cases (as, for example, when voters are asked which candidates they prefer) there is little motive for dishonesty, and it seems fairly safe to assume that their responses generally reflect actual opinions. But the assumption of truthfulness should not be made uncritically.

A related problem concerns human response to the way in which a question is asked. Suppose we wish to survey public opinion on a new piece of legislation by Senator S. The way we phrase our question may drastically affect the responses. If we ask, "Do you favor Senator S's government-bloating socialist bill?" we are likely to generate many more negative responses than if we asked the question more neutrally: "Do you favor Senator S's bill on government aid to the poor?" And this in turn will probably generate more negative responses than the positively worded "Do you favor Senator S's popular new bill to bring much-needed aid to the victims of poverty in America?" The final form of the argument, however, may conceal the way in which the question was asked:

SOLVED PROBLEM

9.15 Evaluate the following argument:

51 percent of 100 randomly selected registered voters said that they favored Senator S's bill.

∴ About 51 percent of all voters favor Senator S's bill.

Solution

Like the argument of Problem 9.14, this is not a statistical generalization, since it moves from what the voters *said* to what they actually *think*. But this argument has an additional source of weakness in that we are not given the exact phrasing of the question. The question itself, for all we know, may have substantially biased the responses either for or against the bill.

Biased questions are frequently a problem in polls which are poorly conducted or conducted by those with a vested interest in the outcome.

9.4 INDUCTIVE GENERALIZATION AND SIMPLE INDUCTION

Statistical generalization allows us to arrive at a conclusion concerning an entire population from a premise concerning a *random* sample of that population. The randomness of the sample ensures the probability of the conclusion on purely mathematical grounds. But often it is not possible to obtain a random sample. This is true, for example, if the relevant population concerns future objects or events. Since these objects or events do not yet exist at the time the sample is taken, they have no chance of being included in the sample. Therefore, since randomness requires that each member of the population have an equal chance of being selected, no sample which we take now can be random for a population which includes future objects or events.

SOLVED PROBLEM

9.16 Evaluate the randomness of the sample in the following generalization:

The Bats have won 10 of the 20 games they have played so far this season.

∴ The Bats will finish the season having won about half of their games.

Solution

The conclusion of the argument concerns a population (all Bats games this season) which includes future games. The sample only concerns games played so far. It is therefore not a random sample with respect to the relevant population.

The general form of the argument of Problem 9.16 may be represented as follows:

> *n* percent of *s* thus-far-observed *F* are *G*.
> ∴ About *n* percent of all *F* are *G*.

In the example given, *n* is 50, *s* is 20, *F* is 'Bats games this season', and *G* is 'are (or will be) won by the Bats'. We shall call this form *inductive generalization*.

Inductive generalization differs from statistical generalization in that its premise does not claim that the sample is random. Without a claim of randomness, the reasoning cannot be justified by mathematical principles alone. No mathematical principle can guarantee, for example, that the Bats will not suddenly improve dramatically and win all their remaining games—or finish with a long losing streak. Nor does any principle of mathematics ensure that such radical changes are not probable. The inference of Problem 9.16 therefore presupposes something substantial, namely, that the course of events (in this case, ball games) exhibits or is likely to exhibit a certain uniformity over time; that is, that future instances of wins are likely to occur with about the same frequency as past instances of wins. Inductive generalizations are therefore Humean inferences.

Inductive generalizations are weaker arguments than statistical generalizations, for the kind of uniformity they presuppose is always to some degree uncertain. But since there is no universally accepted way of calculating the inductive probabilities of Humean arguments, we cannot say exactly how much weaker. In other respects, however, evaluation of inductive generalizations employs the same principles as evaluation of statistical generalizations. Thus, for example, in both kinds of generalization, inductive probability increases as *s* gets larger.

One of the most notable forms of inductive generalization occurs when *n* = 100, so that we have:

> All the *s* thus-far-observed *F* are *G*.
> ∴ All *F* are *G*.

This form has been widely regarded as the means by which scientific laws (which can often be expressed in the form 'All *F* are *G*') are justified. Thus, for example, our knowledge that water freezes at +32 degrees Fahrenheit is said to be based on the fact that all the (very many) samples of pure water observed thus far have had a freezing point of +32 degrees Fahrenheit. Yet inductive generalization is a relatively weak form of reasoning. Some theorists reject it as too weak to establish universal laws. They argue that if *s* is small relative to the population of *F*'s, the inductive probability of the inference is near zero, and that for infinite populations and finite *s* it is strictly zero. Some have questioned whether inductive generalization really is the way in which we justify scientific laws—and, indeed, whether such laws can be justified at all. But others have disputed these contentions, and no consensus has been reached.

Despite this widespread disagreement, there are certain comparative principles on which most logicians agree. If we assume that the inductive probability of a given inductive generalization is not strictly zero, then clearly it may be increased by increasing *s*. (This is an instance of the general rule that strengthening the premise strengthens the argument.) Likewise, since reasoning is strengthened by weakening the conclusion, the smaller the population of *F*'s, the greater the inductive probability of the argument.

The most extreme way to weaken the conclusion of such an inference is to reduce the population it mentions to one individual. This yields the following form, which is called *simple induction*, *induction by enumeration*, or the *simple predictive inference*:

> *n* percent of the *s* thus-far-observed *F* are *G*.
> ∴ If one more *F* is observed, it will be *G*.

In general, simple inductions are much stronger than inductive generalizations from the same premises.

SOLVED PROBLEM

9.17   Evaluate the relative strength of the following arguments:

(a)   All the (very many) objects observed thus far exert gravitational force in proportion to their mass.

   ∴ All objects exert gravitational force in proportion to their mass.

(b)   All the (very many) objects observed thus far exert gravitational force in proportion to their mass.

   ∴ The next observed object will exert gravitational force in proportion to its mass.

Solution

   Argument (a) is a typical example of an inductive generalization; its conclusion has the form of a scientific law. Argument (b) is a simple induction whose conclusion is much weaker than the conclusion of (a). Since both arguments are based on the same premise, this means that (b) is itself considerably stronger than (a).

   Like all inductive generalizations (of which they are a special case), simple inductions get stronger as s increases, so long as $n > 50$. And like statistical syllogisms, simple inductions are highly sensitive to the value of n. They are strongest when $n = 100$ and weakest when $n = 0$. If $n < 50$, a simple induction will provide more support for the negation of its conclusion than for the conclusion itself.

   Yet unlike statistical syllogisms, simple inductions do not become deductive when $n = 100$. This is because they are Humean inferences, whose strength depends on an uncertain presupposition of the uniformity of nature. As with all Humean arguments, there is no generally accepted method for calculating the inductive probability of a simple induction.

SOLVED PROBLEM

9.18   Arrange the following arguments in order of decreasing inductive probability.

(a)   Exactly 99 percent of 500 observed meteorites contained iron.

   ∴ If another meteorite is observed, it will contain iron.

(b)   Exactly 99 percent of 500 observed meteorites contained iron.

   ∴ All meteorites contain iron.

(c)   All meteorites contain iron.

   ∴ If a meteorite is observed, it will contain iron.

(d)   All 500 meteorites observed thus far contain iron.

   ∴ If another meteorite is observed, it will contain iron.

(e)   All 500 meteorites observed thus far contain iron.

   ∴ All meteorites contain iron.

(f)   All 1000 meteorites observed thus far contain iron.

   ∴ If another meteorite is observed, it will contain iron.

(g)   All 500 meteorites observed thus far contain iron.

   ∴ All meteorites we will ever observe contain iron.

Solution

   (c), (f), (d), (a), (g), (e), (b). Argument (c) is a deductive argument. Its inductive probability is thus higher than that of the others, which are not deductive. Argument (f) is stronger than argument (d) because of its stronger first premise; it employs a larger sample.

Argument (a) is similar, but lower in inductive probability, since according to its first premise the percentage of iron-bearing meteorites is smaller. The first premise of argument (g) describes a sample like that of argument (d), but its inductive probability is still much lower than that of either (a) or (d), since its conclusion is exceedingly strong—much stronger than the conclusion of (a) or (d). The conclusion of argument (e) is stronger still; hence (e) is even weaker. The inductive probability of argument (b) is strictly zero, since according to its first premise five meteorites which do not contain iron have already been observed.

9.5 INDUCTION BY ANALOGY

Another important kind of Humean argument is *argument by analogy.* In an argument by analogy we observe that an object x has many properties, $F_1, F_2, \ldots, F_n$, in common with some other object y. We observe also that y has some further property G. Hence we consider it likely (since x and y are analogous in so many other respects) that x has G as well. The general form of the argument may be represented as follows:

$$F_1 x \ \& \ F_2 x \ \& \ldots \& \ F_n x$$
$$F_1 y \ \& \ F_2 y \ \& \ldots \& \ F_n y$$
$$Gy$$
$$\therefore \ Gx$$

SOLVED PROBLEM

9.19 Evaluate the following argument:

Specimen x is a single-stemmed plant with lanceolate leaves and five-petaled blue flowers, about 0.4 meter tall, found growing on a sunny roadside.

Specimen y is a single-stemmed plant with lanceolate leaves and five-petaled blue flowers, about 0.4 meter tall, found growing on a sunny roadside.

Specimen y is a member of the gentian family.

$\therefore$ Specimen x is a member of the gentian family.

Solution

This is a reasonably strong argument by analogy. The argument is Humean because no logical or mathematical principle can guarantee that similarities in external appearance, size, and shape correspond to taxonomic sameness. The argument thus presupposes a more or less orderly correspondence between the characteristics mentioned and taxonomic type. Its strength is in part a function of the strength of this presupposition.

Like inductive arguments generally, analogical arguments may be strengthened by strengthening their premises or by weakening their conclusions. We raise the inductive probability of the argument of Problem 9.19, for example, if we weaken the conclusion to:

Specimen x is a member of the gentian family or some closely related family.

We can also raise its inductive probability by noting more properties that x and y have in common, thus strengthening each of the first two premises. We might, for example, observe that x and y also produce similar kinds of seeds.

However, a simple count of the properties constituting the analogy is only a rough way to gauge premise strength. Some properties count more than others. We may note, for example, that both x and y have the property of being composed of matter. But this property provides only a weak and very general analogy between the two in comparison with more specific properties, like having lanceolate leaves or having five-petaled blue flowers. Thus the strength of the premises depends not only on the number of properties x and y are claimed to have in common, but also on the specificity of these properties. The more specific the resemblances are, the stronger the argument.

Another consideration in analogical reasoning is the relevance of the properties $F_1, F_2, \ldots, F_n$ to the property G (see Section 8.5). Problem 9.19 is relatively strong in part because all the properties mentioned in the first two premises are likely to be relevant to taxonomic classification (i.e., to the property G, the property of being a member of the gentian family). But where relevance is lacking and the conclusion is strong enough to be of much interest, the argument's inductive probability will be quite low.

SOLVED PROBLEM

9.20 Estimate the inductive probability of the following argument by analogy:

> Person x was born on a Monday, has dark hair, is 5 feet 8 inches tall, and speaks Finnish.
> Person y was born on a Monday, has dark hair, is 5 feet 8 inches tall, and speaks Finnish.
> Person y likes brussels sprouts.
> ∴ Person x likes brussels sprouts.

Solution

The inductive probability is low, because the properties $F_1, F_2, \ldots, F_n$ mentioned in the first two premises are almost surely irrelevant to the property G (the property of liking brussels sprouts).

Still, in advance of careful investigation, it is not always clear what is relevant and what is not. It might turn out that dark-haired people have a gene which predisposes them to have a taste for brussels sprouts, so that having dark hair, for example, is relevant after all! This, of course, is unlikely. But it does sometimes happen that previously unsuspected but genuine connections are suggested by analogical reasoning which appears at first to lack relevance.

Relevance is often difficult to determine, but in analogical arguments its role is especially problematic. Perhaps the best advice that can be given is simply that in evaluating analogical reasoning, common sense ought to prevail.

Analogical considerations can be combined with induction by enumeration to yield hybrid argument forms. For example, instead of comparing x with just one object y, we may compare it with many different objects, all of which have the properties $F_1, F_2, \ldots, F_n$, and G. This strengthens the argument by showing that G is associated with $F_1, F_2, \ldots, F_n$ in many instances, not just in one.

SOLVED PROBLEM

9.21 Arrange the following analogical arguments in order of decreasing inductive probability:

(a) A common housefly x, 8 millimeters long, is being placed in a tightly closed jar.
   A common housefly y, 8 millimeters long, was placed in a tightly closed jar.
   y died within a day.
   ∴ x will die within a day.

(b)   A common housefly x, 8 millimeters long and 14 days old, is being placed in a tightly closed jar.

A common housefly y, 8 millimeters long and 14 days old, was placed in a tightly closed jar.

y died within a day.

∴ x will die within a day.

(c)   A common housefly x, 8 millimeters long, is being placed in a tightly closed jar.

A common housefly y, 8 millimeters long, was placed in a tightly closed jar.

y died within a day.

∴ x will die within 12 hours.

(d)   A common housefly x, 8 millimeters long and 14 days old, is being placed in a tightly closed jar.

Common houseflies y, z, and w, each 8 millimeters long and 14 days old, were placed in tightly closed jars.

y, z, and w died within a day.

∴ x will die within a day.

(e)   A common housefly x, 8 millimeters long and 14 days old, is being placed in a tightly closed jar.

Common houseflies y, z, and w, each 8 millimeters long and 14 days old, were placed in tightly closed jars.

y, z, and w died within a day.

∴ x will die eventually.

(f)   A common housefly x, 8 millimeters long, is being placed in a tightly closed jar in Wisconsin.

A common housefly y, 8 millimeters long, was placed in a tightly closed jar in Wisconsin.

y died within a day.

∴ x will die within a day.

(g)   A common housefly x is being placed in a tightly closed jar.

A common housefly y was placed in a tighly closed jar.

y died within a day.

∴ x will die within 12 hours.

Solution

(e), (d), (b), (f), (a), (c), (g). Argument (e) is stronger than argument (d) because its conclusion is weaker. Argument (d) is stronger than argument (b) because of its stronger second premise; the analogy in (b) is based on a sample of just one fly, instead of three. Argument (b) is stronger than argument (f) because the state in which the experiment is performed is surely less relevant to the conclusion than the age of the fly. Yet (f) is still marginally stronger than (a), which is like (f) except that it does not mention the state in which the experiment was performed (and hence has very slightly weaker premises). Argument (c) is like argument (a), except that it has a stronger conclusion; so (c) is weaker than (a). Finally, argument (g) is slightly weaker still, since it does not mention the size of the flies and hence has weaker premises.

Analogical arguments, like all inductive arguments, are vulnerable to contrary evidence. If any evidence bearing negatively on the analogy is suppressed, then the argument violates the requirement of total evidence and should be rejected. (The conclusion should then be reconsidered in the light of the total available evidence.) Contrary evidence to analogical arguments often takes the form of a *relevant disanalogy*. (For more on faulty analogies, see Problem 8.30.)

SOLVED PROBLEM

9.22   Evaluate the following argument by analogy:

> Jim Jones was the leader of a religious movement which advocated peace, brotherhood, and a simple agrarian way of life.
>
> Mahatma Gandhi was the leader of a religious movement which advocated peace, brotherhood, and a simple agrarian way of life.
>
> Mahatma Gandhi was a saintly man.
>
> ∴   Jim Jones was a saintly man.

Solution

The argument has true premises, a fairly high inductive probability, and a reasonable degree of relevance. But it suppresses crucial contrary evidence: Jim Jones was the leader of a fanatical cult which he incited to acts of murder and mass suicide. Since the argument ignores this crucially relevant disanalogy between Jones and Gandhi, it should be rejected.

9.6   MILL'S METHODS

Often we wish to determine the cause of an observed effect. Logically, this is a two-step procedure. The first step is to formulate a list of suspected causes which, to the best of our knowledge, includes the actual cause. The second is to rule out by observation as many of these suspected causes as possible. If we narrow the list down to one item, it is reasonable to conclude that this item is probably the cause.

The justification of the first step (i.e., the evidence that the actual cause is included on our list of suspected causes) is generally inductive. The eliminative reasoning of the second step is deductive. Since both inductive and deductive reasoning are involved, the reasoning as a whole is inductive. (See Section 2.3.)

We arrive at the list of suspected causes by a process of inductive (frequently analogical) reasoning. Suppose, for example, that we wish to find the cause of a newly discovered disease. Now this disease will resemble some familiar diseases more than others. We note the familiar diseases to which it is most closely analogous and then conclude (by analogy) that its cause is probably similar to the causes of the familiar diseases which it most closely resembles. This will give us a range of suspected causes.

Suppose, for example, that the familiar diseases which the new disease most closely resembles are all viral infections. The suspected causes will then be viral. Close observation of the disease victims will establish which viruses are present in their tissues. We will conclude that the actual cause is probably one of these viruses. These viruses, then, form our list of suspected causes.

At this stage, however, our investigation is only half finished. For it is quite likely that we will find several kinds of virus in the tissues of the victims. To determine which of these actully caused the disease, we now employ a deductive process designed to eliminate from our list as many of the suspected causes as possible. The kind of eliminative process we use will depend on the kind of cause we are looking for.

Here we shall discuss four different kinds of causes and, corresponding to each, a different method of elimination. The eliminative methods were named and investigated by the nineteenth-century philosopher John Stuart Mill. Mill actually discussed five such methods, but the fifth (the method of residues) does not correspond to any specific kind of cause and will not be discussed here. Before discussing Mill's methods, however, we need to define the kinds of causes to which they apply.

The first kind of cause is a *necessary cause*, or *causally necessary condition*. A necessary cause for an effect E is a condition which is needed to produce E. If C is a necessary cause for E, then E will never occur without C, though perhaps C can occur without E. For example, the tuberculosis bacillus is a necessary cause of the disease tuberculosis. Tuberculosis never occurs without the bacillus, but the bacillus may be present in people who do not have the disease.

A given effect may have several necessary causes. Fire, for example, requires for its production three causally necessary conditions: fuel, oxygen (or some similar substance), and heat.

The second kind of cause is a *sufficient cause*, or *causally sufficient condition*. A condition C is a sufficient cause for an effect E if the presence of C invariably produces E. If C is a sufficient cause for E, then C will never occur without E, though there may be cases in which E occurs without C. For example (with respect to higher animal species), decapitation is a sufficient cause for death. Whenever decapitation occurs, death occurs. But the converse does not hold; other causes besides decapitation may result in death.

A given effect may have several sufficient causes. In addition to decapitation, as just noted, there are many sufficient causes for death: boiling in oil, crushing, vaporization, prolonged deprivation of food, water, or oxygen—to name only a few of the unpleasant alternatives.

Some conditions are both *necessary and sufficient causes* of a given effect. That is, the effect never occurs without the cause nor the cause without the effect. This is the third kind of causal relationship. For example, the presence of a massive body is causally necessary and sufficient for the presence of a gravitational field. Without mass, no gravitational field can exist. With it, there cannot fail to be a gravitational field. (This does not mean, of course, that one must experience the gravitational field. Moving in certain trajectories relative to the field will produce weightlessness, but the field is still there.)

The fourth kind of causal relation we shall discuss is *causal dependence of one variable quantity on another*. A variable quantity B is causally dependent on a second variable quantity A if a change in A always produces a corresponding change in B. For example, the apparent brightness B of a luminous object varies inversely with the square of the distance from that object, so that B is a variable quantity causally dependent on distance. We can cause an object to appear more or less bright by varying its distance from us.

An effect (such as apparent brightness) may be causally correlated with more than one quantity. If the object whose apparent brightness we are investigating is a gas flame, its apparent brightness will also depend on the amount of fuel and oxygen available to it, and on other factors as well.

SOLVED PROBLEM

9.23   Classify the kind of causality intended by the following statements:

(*a*)   Flipping the wall switch will cause the light to go on.

(*b*)   Closing the electricity supply from the main lines will cause the light to go off.

(*c*)   Making a lot of noise will cause the neighbors to complain.

(*d*)   Pulling the trigger will cause the gun to fire.

(*e*)   Raising the temperature of a gas will cause an increase in its volume.

(*f*)   Raising the temperature of the freezer above +32 degrees Fahrenheit will cause the ice cubes in the freezer to melt.

(*g*)   Killing the President will cause new presidential elections.

(*h*)   Raising the temperature in the environment will cause the death of many plants.

Solution

(*a*)   Necessary (but not sufficient: the light will not go on unless the light bulb is working).

(*b*)   Sufficient (but not necessary: the light will go off also if the wall switch is turned to the "off" position).

(*c*)   Sufficient (but not necessary: the neighbors may complain for a number of other reasons).

(*d*)   Necessary (but not sufficient: the gun won't fire unless it is loaded).

(e)   Dependent (the higher the temperature, the higher the volume).

(f)   Necessary and sufficient.

(g)   Sufficient (but not necessary).

(h)   Dependent (the higher the temperature, the greater the number of plants that will die).

Now, to reiterate, Mill's methods aim to narrow down a list of suspected causes (of one of the four kinds just described) in order to find a particular cause for an effect E. Each of the four methods listed below is appropriate to a different kind of cause:

Mill's Method of:	Rules Out Conditions Suspected of Being:
Agreement	Necessary causes of E
Difference	Sufficient causes of E
Agreement and difference	Necessary and sufficient causes of E
Concomitant variation	Quantities on which the magnitude of E is causally dependent

If by using the appropriate method we are able to narrow the list of suspected causes down to one entry, then (presuming that a cause of the type we are seeking is included in the list) this entry is a cause of the kind we are looking for. We now examine each of the four methods in detail.

The Method of Agreement

Mill's method of agreement is a deductive procedure for ruling out suspected causally necessary conditions. Recall that if a circumstance C is a causally necessary condition of an effect E, then E cannot occur without C. So to determine which of a list of suspected causally necessary conditions really is causally necessary for E, we examine a number of different cases of E. If any of the suspected necessary conditions fails to occur in any of these cases, then it can certainly be ruled out as not necessary for E. Our hope is to narrow the list down to one item.

SOLVED PROBLEM

9.24   Suppose we are looking for the necessary cause of a certain disease E, and we have by using our background knowledge and expertise formulated a list of five viral agents, V_1 through V_5, which we suspect may cause E. We examine a number of patients with E and check to see which of the suspected causes is present in each case. The results are as follows:

Case	Circumstances (suspected causes present in this case)	Effect
Patient 1	V_1, V_3, V_4	E
Patient 2	V_1, V_4, V_5	E
Patient 3	V_1, V_2	E
Patient 4	V_1, V_5	E

Solution

Only one of the five suspected causes (namely, V_1) is present in each of the four patients with the disease. This proves that none of the suspected causes, except possibly V_1, really is causally necessary for E.

Once V_2 through V_5 are eliminated as necessary causes, it follows deductively that

(1)   If the list V_1 through V_5 includes a necessary cause of E, then V_1 is that necessary cause.

This is the conclusion which Mill's method of agreement yields. If we wish to advance further to the unconditional conclusion

(2) V_1 is a necessary cause of E

then we need the premise

(3)   The list V_1 through V_5 includes a necessary cause of E.

Such a premise cannot in general be proved with certainty, but can only be established by inductive reasoning. Typically, such inductive reasoning will be analogical. In the case in question, it may look something like this:

(4)   Disease E has characteristics $F_1, F_2, \ldots, F_n$.
(5)   The known diseases similar to E have characteristics $F_1, F_2, \ldots, F_n$.
(6)   Viruses are necessary causes of the known diseases similar to E.
∴ (7)   Some virus is a necessary cause of E.

Here the characteristics $F_1, F_2, \ldots, F_n$ might be such things as infectiousness or the presence of fever. To get from statement 7 to statement 3, we need to add to statement 7 the premise that

(8)   The only viruses present in the cases of patients 1 through 4, who had E, were V_1 through V_5.

Statement 8 together with statement 7 deductively implies 3, since (by definition) any necessary cause for E must occur in every case of E. The entire argument may now be summarized in the following diagram:

$$
\begin{array}{c}
\underline{4 \;+\; 5 \;+\; 6} \\
\downarrow \\
\underline{7 \;+\; 8} \\
\downarrow \\
\underline{3 \;+\; 1} \\
\downarrow \\
2
\end{array}
$$

The basic premises in statements 4, 5, 6, and 8 are obtained by observation or previous investigation. Statement 1 is the conclusion obtained by using Mill's method of agreement. The reliability of the argument as a whole depends on the adequacy of the analogical inference (the inference from 4, 5, and 6 to 7) and on the truth of the basic premises. The premise in statement 8, for example, could prove to be false if our observations of the patients were not sufficiently thorough. That would undermine the whole argument, since in that case the real cause of E might be a virus that was present but undetected in the cases we studied. The adequacy of the analogical inference, of course, depends on the factors discussed in Section 9.5. We should be especially wary of suppressed evidence. (Does E have any unusual characteristics which suggest a nonviral cause?)

Not every application of the method of agreement works out so neatly. Suppose V_1 and V_2 both occur in all cases of E that we examine. Does that mean *both* are necessary for E? No, this does not follow. We may not have examined a large enough sample of patients to rule out one or the other. Our investigation is inconclusive, and we need to seek more data.

It may also happen that the method of agreement rules out all the suspected causes on our list. In that case, statement 3 is false, and so either 7 or 8 must be false as well. That is, either the necessary cause is not viral (as our analogical argument led us to suspect) or we failed to detect some other virus that was present in the patients. If this occurs, we need to recheck everything and probably gather more data before any firm conclusions can be drawn.

The Method of Difference

If we are seeking a sufficient cause rather than a necessary cause, the method to use is the method of difference. Recall that a sufficient cause for an effect E is an event which always produces E. If cause C ever occurs without E, then C is not sufficient for E. Often it is useful, however, to speak of sufficient causes relative to a restricted class of individuals. A small quantity of a toxic chemical, for example, may be sufficient to produce death in small animals and children but not in healthy adults. Hence, relative to the class of children and small animals it is a sufficient cause for death, but relative to a larger class which includes healthy adults it is not. Claims of causal sufficiency are often implicitly to be understood as relative to a particular class of individuals or events.

SOLVED PROBLEM

9.25 A number of people have eaten a picnic lunch which included five foods, F_1 through F_5. Many of them are suffering from food poisoning. It is assumed that among the five foods is one which is sufficient to produce the poisoning among this group of people. Now suppose that we find two individuals, one of whom has eaten all five foods and is suffering from food poisoning and the other of whom has eaten all but F_1 and is feeling fine. Thus if P is the effect of poisoning, the situation is this:

Case	Circumstances (suspected sufficient causes)	Effect
Person 1	F_1, F_2, F_3, F_4, F_5	P
Person 2	F_2, F_3, F_4, F_5	None

What is the sufficient cause of P?

Solution

Since P failed to occur in person 2 in the presence of F_2 through F_5, clearly none of these foods is sufficient for P. On the assumption, then, that a sufficient cause for P occurs among F_1 through F_5, it follows that the cause is F_1.

The weakest part of this reasoning is the assumption that a sufficient cause for P occurs among F_1 through F_5. As with the analogous premise in our discussion of necessary causes, it generally cannot be proved deductively but must be supported by an inductive argument. In this case, we might argue that in most past cases in which food poisoning has occurred, some toxic substance present in one food has been the culprit and that it has been sufficient to produce the poisoning in anyone who consumed a substantial amount of the food.

Once again, it is important to see how this sort of inductive reasoning could go wrong. Perhaps none of the foods is by itself sufficient for P, but ingestion of F_1 and F_2 together causes a chemical reaction which results in toxicity. Under these conditions we would still observe the poisoning in person 1 and no effect on person 2, but the assumption that a sufficient cause for P occurs among F_1 through F_5 would be false.

It may also happen that none of F_1 through F_5, or any combination of F_1 through F_5, is sufficient for P. A toxin may be present, say in F_1, but consumption of this toxin may produce P only in certain susceptible individuals. That is, F_1 may be sufficient for P in certain people but not in every member of the population we are concerned with. If this is so, then once again the assumption that a sufficient cause for P occurs among F_1 through F_5 is false, and so is the unqualified conclusion that F_1 is sufficient for P. These errors can occur even if we make no faulty observations, because of the fallibility of the reasoning needed to establish this assumption. Therefore, caution is needed in applying the method of difference.

In summary, Mill's method of difference is used to narrow down a list of suspected sufficient causes for an effect E. It does this by rejection of any item on the list which occurs without E. We hope to make enough observations to narrow the list down to one item. If we do, then we may conclude deductively that if there is a sufficient cause for E on the list, it is the one remaining. However, to establish that our list contains a sufficient cause, we must rely on induction from past experience.

The Joint Method of Agreement and Difference

Mill's joint method of agreement and difference is a procedure for eliminating items from a list of suspected necessary and sufficient causes. It incorporates nothing new; it merely involves the simultaneous application of the methods of agreement and difference.

If C is a necessary and sufficient cause of E, then C never occurs without E and E never occurs without C. Hence, if we find any case in which C occurs but E does not or E occurs but C does not, C can be ruled out as a necessary and sufficient cause for E (though it may still be a necessary or sufficient cause, as the case may be).

SOLVED PROBLEM

9.26 Suppose that a student in a college dormitory notices a peculiar sort of interference on her television set. She has seen similar kinds of interference before and suspects that its necessary and sufficient cause (provided the television is on) is the nearby operation of some electrical appliance. This leads her to formulate the following list of suspected causes:

S = electric shaver
H = hair dryer
D = clothes dryer
W = washing machine

She now observes which appliances are operating in nearby rooms while her television is on. The results are as follows (I is the interference):

Case	Circumstances (suspected necessary and sufficient causes)	Effect
1	H, D, W	I
2	D, W	I
3	S, H, D, W	I
4	S, H	None
5	W, S	None

Which of the suspected causes, if any, is the necessary and sufficient cause of the interference?

Solution

The only one of the suspected necessary and sufficient causes which is always present when I is present and always absent when I is absent is D. Hence, if one of the suspected necessary and sufficient causes really is necessary and sufficient for I, it must be D.

If the student goes on to conclude that D is actually necessary and sufficient for I, once again the weakest point of her reasoning will be the assumption that a necessary and sufficient cause was included on her list of suspected causes. As before, this premise can be justified only by induction from past experience with similar situations.

The Method of Concomitant Variation

Mill's method of concomitant variation differs from the other methods in that it is not concerned with the mere presence or absence of cause and effect, but with their relative magnitudes. It is a means of narrowing down a list of variable magnitudes suspected of being responsible for a specific change in the magnitude of an effect E. A variable is rejected as not responsible for a particular change if that variable remains constant throughout the change. If all but one of a list of variables remain constant while the magnitude of an effect changes, then, presuming that the variable responsible for the change appears on the list, it must be the one which has not remained constant.

SOLVED PROBLEM

9.27   A houseplant exhibits a sudden spurt of growth. We suspect that the variables relevant to its growth rate are these:

S = sunlight
W = water
F = fertilizer
T = temperature

But we observe that only one of these variables, namely, the amount of water the plant receives, has been altered recently. This observation may be schematized as follows:

Case	Circumstances (variables on which we suspect G to be relevant)	Effect
1	S, F, T, W	G
2	$S, F, T, W +$	$G +$

Here G is the growth rate and the plus signs stand for increases of magnitude. No plus sign indicates no change. Which, if any, of the variables on our list is causally relevant to the observed change in the growth rate of the plant?

Solution

Since the amount of water the plant receives is the only one of the variables on our list that has changed, only it among these variables could be responsible for the observed change in growth rate.

Notice that this method does not eliminate the possibility that changes in S, F, or T also affect G. What it shows is that these three variables were not responsible for the particular changes observed here. Hence, if some variable on our list was responsible, it must have been W. To verify our conjecture further, we may cut back on the water the plant receives. Suppose that we then find this:

3	$S, F, T, W -$	$G -$

where the minus signs indicate decreases in magnitude. Then we will be still more confident that the rate of watering is the variable responsible for the observed changes in growth rate.

As with the other three methods, the process of eliminating S, F, and T as possible causes of the observed effect is deductive, but induction from past experience with plants is required to support the premise that one of the four variables on our list caused the changes in G.

The increased confidence provided by case 3 is due to the additional support case 3 lends to this premise. For repetition of instances of the correlation between W and G enhances by simple induction the probability that W and G have varied and will continue to vary together. If we were perfectly confident that the variable responsible for the change was one of the four on our list, this additional confirmation would be superfluous, and cases 1 and 2 alone would suffice to establish that the responsible variable is W.

9.7　SCIENTIFIC THEORIES

The most sophisticated forms of inductive reasoning occur in the justification or confirmation of scientific theories. A scientific theory is an account of some natural phenomenon which in conjunction with further known facts or conjectures (called *auxiliary hypotheses*) enables us to deduce consequences which can be tested by observation. Often a theory is expressed by a *model*, a physical or mathematical structure claimed to be analogous in some respect to the phenomenon for which the theory provides an account.

For example, prior to the twentieth century there were two theories of the phenomenon of light, the corpuscular theory and the wave theory. According to the corpuscular theory (whose most notable advocate was Isaac Newton), light consists of minute particles, or corpuscles, expelled in straight trajectories by luminous objects. According to the wave theory (first propounded by the Dutch astronomer Christian Huygens), light consists of spherical waves spreading out from luminous objects like the circular ripples from a stone dropped into a lake. According to the wave theory, light waves are propagated through a fluid substance, the ether, which permeates the universe. Now, both theories were able to account for the phenomenon of color and for many of the reflective and refractive properties of light. But by the end of the nineteenth century the wave theory had temporarily won out, because of its superiority in explaining diffraction effects—patterns of light and dark bands formed when light is passed through a small aperture. Such patterns are predicted by the wave theory but difficult to explain by the corpuscular theory.

Each theory modeled light as a physical structure—moving particles, in one case; waves in a fluid medium, in the other. Both, however, were succeeded in the twentieth century by the quantum theory, in which light is modeled as a mathematical structure that has some features of both waves and particles but is not completely analogous to any familiar physical structure.

This example shows that scientific theories are justified primarily by their success in making true predictions. By 'prediction' we mean a statement about the results of certain tests or observations, not necessarily a statement about the future. Even theories about the past make predictions in this sense, since (in conjunction with appropriate auxiliary hypotheses) they imply that certain tests or observations will have certain results. A theory about the evolution of dinosaurs, for example, will have implications concerning the sorts of fossils we should expect to find in certain geological strata. These implications, then, are among its predictions. Since a theory's predictions are *deduced* from the theory together with its auxiliary hypotheses, if any of them prove false, then either the theory itself or one or more of the auxiliary hypotheses must be false. (One cannot deduce a false conclusion from a set of true premises.) If we are confident of all the auxiliary hypotheses, then we may confidently reject the part of the theory used to derive the prediction. The corpuscular theory of light, together with what seems to be the most reasonable auxiliary hypotheses about the way small particles ought to behave, implies that diffraction ought not to occur. Since it does occur, nineteenth-century physicists, confident of these auxiliary hypotheses, rejected the corpuscular theory.

In this example, a deductive process was used to refute a scientific theory. Often, however, theorists are not completely confident of the truth of the auxiliary hypotheses; hence there may be controversy about the soundness of the deduction used to reject the theory. If one or more of the auxiliary hypotheses are indeed false, then the falsity of a prediction made with the aid of those hypotheses does not entail the falsity of the theory.

Whereas the reasoning by which scientific theories are refuted is deductive, the reasoning by which they are confirmed is inductive. After the demise of the corpuscular theory of light, the wave theory became increasingly confirmed. Unlike the corpuscular theory, the wave theory (in conjunction with plausible auxiliary hypotheses about the orientation and amplitude of the waves) does predict diffraction effects. Hence, when these were observed, confidence in the wave theory increased.

However, confirmation of a prediction (or even many predictions) of a theory does not prove deductively that the theory is true. Theories together with their auxiliary hypotheses always imply many more predictions than can actually be tested. Even if all the predictions tested so far have been verified, some untested prediction may still be false. That would imply the falsity of the theory, provided that the auxiliary hypotheses are true. Hence, from a logical point of view, confidence in any scientific theory should never be absolute.

Nevertheless, it is often held that as more and more of the predictions entailed by a theory are verified, the theory itself becomes more *probable*. This principle may be formulated more precisely as follows:

> (P): If E is some initial body of evidence (including auxiliary hypotheses) and C is the additional verification of some of the theory's predictions, the probability of the theory given E & C is higher than the probability of the theory given E alone.

Principle (P) seems to be the principle underlying the inductions by which scientific theories are confirmed. But it is not self-evidently true, and it is not provable as a law of logic or probability theory. Moreover, some of its instances are evidently false, suggesting that (P) needs further restriction.

To illustrate this point, consider the situation with respect to theories of light at the time when serious attention was first paid to diffraction phenomena in the middle of the nineteenth century. What happened historically was that the corpuscular theory was rejected and the wave theory accepted. But one might have accounted for diffraction by maintaining a corpuscular theory, augmented by the hypothesis that a strange force acts on the corpuscles of light as they pass through small apertures, separating them into distinct sheaths and thus giving rise to the observed effects. Alternatively, one might have argued that the diffraction phenomena are an illusion due to peculiarities of our cameras and eyes. Or one might have rejected both the wave and corpuscular theories and argued that light is something else entirely—say, filaments or strands emitted from luminous objects. This could be made compatible with the known properties of light by adopting sufficiently ingenious auxiliary hypotheses. One could create such alternatives ad infinitum.

Each of these theories, if augmented by appropriate auxiliary hypotheses, predicts diffraction phenomena as well as the other properties of light known in the nineteenth century. Does the observation of diffraction, then, make each more probable, as unrestricted use of (P) suggests? This seems doubtful. In practice, only the wave theory was regarded as having been confirmed or rendered more probable. Theories like those mentioned in the previous paragraph were not seriously considered. The reason is that the auxiliary hypotheses required by these other theories (such as the hypothesis that a strange force affects light corpuscles traveling through small apertures) were themselves unjustified. They were not plausible independent of the theory. Auxiliary hypotheses which have no independent justification and are adopted only to make a theory fit the facts are called *ad hoc* hypotheses.

In practice, principle (P) is not applied equally to all theories, but preferentially to those theories which do not require ad hoc hypotheses. The wave theory predicted diffraction by means of auxiliary hypotheses which seemed perfectly natural. All competing theories were either extremely complex in themselves or required complex and ad hoc auxiliary hypotheses. So even though other theories could be made to imply the same predictions, only the wave theory was regarded as substantially confirmed by the observation of diffraction. (We might note, incidentally, that the wave theory itself was succeeded by the quantum theory primarily because of the discovery of new phenomena which could not be predicted by the wave theory unless it too were burdened with ad hoc hypotheses.)

Not only is (P) applied preferentially to theories which do not require ad hoc hypotheses; it is also (as suggested by the example above) applied preferentially to theories which are themselves simple. That is, other things being equal, simple theories are regarded as more highly confirmed by verification of their predictions than are complex theories. Various restrictions on (P) have been proposed by various theorists, but they are generally controversial and need not be discussed here.

Supplementary Problems

I Arrange each of the following sets of statements in order from strongest to weakest.

(1) (a) Iron is a metal.
     (b) Either iron is a metal or copper is a metal.
     (c) Either iron is a metal, or copper or zinc is a metal.
     (d) It is not true that iron is not a metal.
     (e) Iron, zinc, and copper are metals.
     (f) Something is a metal.
     (g) Some things are both metals and not metals.
     (h) Either iron is a metal or it is not a metal.
     (i) Iron and zinc are metals.

(2) (a) Most Americans are employed.
     (b) There are Americans, and all of them are employed.
     (c) Some Americans are employed.
     (d) At least 90 percent of Americans are employed.
     (e) At least 80 percent of Americans are employed.
     (f) Someone is employed.

(3) (a) About 51 percent of newborn children are boys.
     (b) Exactly 51 percent of newborn children are boys.
     (c) Some newborn children are boys.
     (d) It is not true that all newborn children are not boys.
     (e) Somewhere between one-fourth and three-fourths of all newborn children are boys.

(4) (a) Leonardo was a great scientist, inventor, and artist who lived during the Renaissance.
     (b) Leonardo did not live during the Renaissance.
     (c) Leonardo lived during the Renaissance.
     (d) Leonardo was a Renaissance artist.
     (e) Leonardo was not a Renaissance artist.
     (f) Leonardo was not a Renaissance artist and scientist.

II Arrange the following sets of argument forms in order from greatest to least inductive probability.

(1) (a) 60 per cent of observed F are G.
                x is F.
            $\therefore$ x is G.

(b) 20 per cent of F are G.

     x is F.

   ∴ x is G.

(c) 60 per cent of F are G.

     x is F.

   ∴ x is G.

(2) (a) All ten observed F are G.

   ∴ All F are G.

(b) All ten observed F are G.

   ∴ If three more F are observed, they will be G.

(c) All ten observed F are G.

   ∴ If two more F are observed, they will be G.

(d) All ten observed F are G.

   ∴ If two more F are observed, at least one of them will be G.

(e) All F are G.

   ∴ If an F is observed, it will be G.

(3) (a) 8 of 10 doctors we asked prescribed product X.

   ∴ About 80 percent of all doctors prescribe product X.

(b) 80 of 100 doctors we asked prescribed product X.

   ∴ About 80 percent of all doctors prescribe product X.

(c) 80 of 100 randomly selected doctors prescribed product X.

   ∴ About 80 percent of all doctors prescribe product X.

(d) My doctor prescribes product X.

   ∴ All doctors prescribe product X.

(e) My doctor prescribes product X.

   ∴ Some doctor(s) prescribe(s) product X.

(f) All 10 doctors we asked prescribe product X.

   ∴ All doctors prescribe product X.

(4) (a) Objects a, b, c, and d all have properties F and G.

     Objects a, b, c, and d all have property H.

     Object e has properties F and G.

   ∴ Object e has property H.

(b) Objects a, b, c, and d all have properties F, G, and H.

     Objects a, b, c, and d all have property I.

     Object e has properties F, G, and H.

   ∴ Object e has property I.

(c) Object a has property F.

     Object a has property G.

     Object b has property F.

   ∴ Object b has property G.

(d) Object a has property F.

   ∴ Object b has property F.

(e) Object a has properties F and G.

     Object a has property H.

     Object b has properties F and G.

   ∴ Object b has property H.

(f) Object a has property F.

∴ Objects b and c have property F.

(5) (a) Objects a, b, c, d, and e have property F.

∴ All objects have property F.

(b) Objects a, b, c, d, and e have property F.

∴ Objects f and g have property F.

(c) Objects a, b, and c have property F.

∴ All objects have property F.

(d) Objects a, b, c, d, and e have property F.

Objects a, b, c, d, and e have property G.

Objects f and g have property F.

∴ Objects f and g have property G.

(e) Objects a, b, c, d, and e have property F.

Objects a, b, c, d, and e have property G.

Objects f and g have property F.

∴ Object f has property G.

III Each of the following problems consists of a list of observations. For each, answer the following questions. Are the observations compatible with the assumption that exactly one cause of the type indicated (necessary, sufficient, etc.) is among the suspected causes? If so, do the observations enable us to identify it using Mill's methods? If they do, which of the unsuspected causes is it, and by what method is it identified?

(1)

Case	Circumstances (suspected necessary causes of E)	Effect
(a)	F, G, H, I	E
(b)	F, G, I	E
(c)	G, H, I	E
(d)	F, H, I	E

(2)

Case	Circumstances (suspected necessary causes of E)	Effect
(a)	F, G, H	E
(b)	G, H	E
(c)	H, I	E

(3)

Case	Circumstances (suspected necessary causes of E)	Effect
(a)	F, G	E
(b)	G, H	E
(c)	H, I	E

(4)

Case	Circumstances (suspected necessary causes of E)	Effect
(a)	F, G, H	E
(b)	F, G	E
(c)	G, H	E
(d)	F, H	None

(5)

Case	Circumstances (suspected sufficient causes of E)	Effect
(a)	F, G, H	E
(b)	F	E
(c)	H	None

(6)

Case	Circumstances (suspected necessary and sufficient causes of E)	Effect
(a)	F, G	E
(b)	G, H	E
(c)	G, H, I	E
(d)	I	None

(7)

Case	Circumstances (suspected necessary and sufficient causes of E)	Effect
(a)	F, G, H	E
(b)	G, H	E
(c)	F, G	E
(d)	F, G, H, I	None

(8)

Case	Circumstances (suspected necessary and sufficient causes of E)	Effect
(a)	F, G, H	E
(b)	F, G	E
(c)	H, I	None
(d)	H	None

(9)

Case	Circumstances (variables suspected to be relevant to E)	Effect
(a)	F, G, H	E
(b)	F, G+, H+	E+
(c)	F, G, H+	E+
(d)	F−, G, H	E

(10)

Case	Circumstances (variables suspected to be relevant to E)	Effect
(a)	F+, G+, H	E+
(b)	F−, G, H−	E−
(c)	F, G+, H+	E

Answers to Selected Supplementary Problems

I   (1)   (g), (e), (i), (a) and (d), (b), (c), (f), (h) ((a) and (d) are of equal strength)

    (4)   (a), (d), (c), (b), (e), (f)

II   (2)   (e), (d), (c), (b), (a)

    (4)   (b), (a), (e), (c), (d), (f)

III   (3)   None of the suspected causes is necessary for E (method of agreement).

    (6)   G is the only one of the suspected causes which could be necessary and sufficient for E (joint method of agreement and difference).

    (9)   H is the only one of the suspected causal variables on which E could be dependent (method of concomitant variation).

Chapter 10

The Probability Calculus

10.1 THE PROBABILITY OPERATOR

Chapter 3 illustrates how truth tables are used to calculate the truth values of complex statements from the truth values of their atomic components and thereby to determine the validity or invalidity of argument forms. It would be useful to have something analogous for inductive reasoning: a procedure that would enable us to calculate probabilities of complex statements from the probabilities of simpler ones and thereby to determine the inductive probabilities of arguments. Unfortunately, no such procedure exists. We cannot always (or even usually) calculate the probability of a statement or the inductive probability of an argument simply from the probabilities of its atomic components.

Yet significant generalizations about probabilistic relationships among statements can be made. Although these do not add up to a general method for calculating inductive probabilities, they shed a great deal of light on the nature of probability and enable us to solve some practical problems. The most important of these generalizations constitute a logical system known as the *probability calculus*.

The probability calculus is a set of formal rules governing expressions of the form '$P(A)$', meaning "the probability of A." These expressions denote numbers. We may write, for example, '$P(A) = \frac{1}{2}$' to indicate that the probability of A is $\frac{1}{2}$. The expression 'A' may stand for a variety of different kinds of objects—sets, events, beliefs, propositions, and so on—depending upon the specific application. For applications of the probability calculus to logic, 'A' generally denotes a proposition, and sometimes an event. These two sorts of entities are formally similar, except that whereas propositions are said to be true or false, events are said to occur or not occur. In particular, events, like propositions, may be combined or modified by the truth-functional operators of propositional logic. Thus, if 'A' denotes an event, '$P(\sim A)$' denotes the probability of its non-occurrence, '$P(A \lor B)$' the probability of the occurrence of either A or B, '$P(A \& B)$' the probability of the joint occurrence of A and B, and so on.

The operator 'P' can be interpreted in a variety of ways. Under the *subjective interpretation*, '$P(A)$' stands for the degree of belief a particular rational person has in proposition A at a given time. Degree of belief is gauged behaviorally by the person's willingness to accept certain bets on the truth of A.

Under the various *logical interpretations*, '$P(A)$' designates the *logical* or *a priori* probability of A. There are many notions of logical probability, but according to all of them $P(A)$ varies inversely with the information content of A. That is, if A is a weak proposition whose information content is small, $P(A)$ tends to be high, and if A is a strong proposition whose information content is great, then $P(A)$ tends to be low. (Compare Section 9.1.)

Under the *relative frequency interpretation*, A is usually taken to be an event and $P(A)$ is the frequency of occurrence of A relative to some specified reference class of events. This is the interpretation of probability most often used in mathematics and statistics.

The oldest and simplest concept of probability is the *classical interpretation*. Like the relative frequency interpretation, the classical interpretation usually takes the object A to be an event. According to the classical interpretation, probabilities can be defined only when a situation has a finite nonzero number of equally likely possible outcomes, as for example in the toss of a fair die. Here the number of equally likely outcomes is 6, one for each face of the die. The probability of A is defined as the ratio of the number of possible outcomes in which A occurs to the total number of possible outcomes:

$$P(A) = \frac{\text{Number of possible outcomes in which } A \text{ occurs}}{\text{Total number of possible outcomes}}$$

SOLVED PROBLEM

10.1 Consider a situation in which a fair die is tossed once. There are six equally likely possible outcomes: the die will show a one, or a two, or a three, . . ., or a six. Let 'A_1' denote the proposition that the die will show a one, 'A_2' denote the proposition that it will show a two, and so on. Calculate the following probabilities according to the classical interpretation:

(a) $P(A_1)$

(b) $P(A_5)$

(c) $P(\sim A_1)$

(d) $P(A_1 \lor A_3)$

(e) $P(A_1 \& A_3)$

(f) $P(A_1 \& \sim A_1)$

(g) $P(A_1 \lor \sim A_1)$

(h) $P(A_1 \lor A_2 \lor A_3 \lor A_4 \lor A_5 \lor A_6)$ [1]

Solution

(a) 1/6

(b) 1/6

(c) 5/6

(d) 2/6 = 1/3

(e) 0/6 = 0 (The die is tossed only once, so it cannot show two numbers.)

(f) 0/6 = 0

(g) 6/6 = 1

(h) 6/6 = 1

As this example illustrates, logical properties or relations among events or propositions affect the computation of the probability of complex outcomes. These properties or relations can sometimes be verified with the help of a truth table. For instance, a *contradictory* (truth-functionally inconsistent) event such as $A_1 \& \sim A_1$ can never occur, since every row of the corresponding truth table contains an F; so the probability of this event is zero (case (f)). In a similar way, we can sometimes see that some events are *mutually exclusive*, i.e., cannot jointly occur.

SOLVED PROBLEM

10.2 Show that any two events of the forms $A \& B$ and $A \& \sim B$ are mutually exclusive:

Solution

A	B	A & B	A & ~B
T	T	T	F
T	F	F	T
F	T	F	F
F	F	F	F

---

[1]Because of the associative law of propositional logic (i.e., the equivalence ASSOC), bracket placement is not crucial when three or more propositions are joined by disjunction alone or by conjunction alone. It is therefore customary to omit brackets in these cases.

Since there is no line of the table on which both A & B and A & $\sim B$ receive the value T and the table exhibits all possible situations, it is clear that there is no situation in which the two events can jointly occur.

In some cases, however, truth-functional considerations do not suffice, as an event may be impossible or two events may be mutually exclusive for non-truth-functional reasons. With reference to Problem 10.1(e), for instance, the truth table for 'A_1 & A_3' will include a line in which both A_1 and A_3 are T. Yet this line represents an impossible case: not for truth-functional reasons, but because of the nature of the die.

Two other logical relations which are important for the probability calculus are *truth-functional consequence* and *truth-functional equivalence*. The former is just the kind of validity detectable by truth tables. In other words, proposition or event A is a truth-functional consequence of proposition or event B if and only if there is no line on their common truth table in which A is false and B is true. Truth-functional equivalence holds when A and B are truth-functional consequences of each other, i.e., when their truth tables are identical. There are many examples of both kinds or relations in Chapter 3 and (in view of the completeness of the propositional calculus) in Chapter 4.

10.2   AXIOMS AND THEOREMS OF THE PROBABILITY CALCULUS

The probability calculus consists of the following three axioms (basic principles), together with their deductive consequences. These axioms are called the Kolmogorov axioms, after their inventor, the twentieth-century Russian mathematician, A. N. Kolmogorov:

AX1 $P(A) \geqslant 0$.
AX2   If A is tautologous, $P(A) = 1$.
AX3   If A and B are mutually exclusive, $P(A \vee B) = P(A) + P(B)$.

Here 'A' and 'B' may be simple or truth-functionally complex (atomic or molecular). We shall use the classical interpretation to illustrate and explain these axioms, though it should be kept in mind that they hold for the other interpretations as well.

AX1 sets the lower bound for probability values at zero. Zero, in other words, is the probability of the least probable things, those which are impossible. AX1 is true under the classical interpretation, since neither the numerator nor the denominator of the ratio which defines classical probability can ever be negative. (Moreover, the denominator is never zero, though the numerator may be.)

It is clear that tautologies ought to have the highest possible probability, since they are certainly true. Thus AX2 says in effect that we take 1 to be the maximum probability. This, too, is in accord with the classical definition, since the numerator of this definition can never exceed its denominator. (See Problem 10.1(g).)

AX3 gives the probability of a disjunction as the sum of the probabilities of its disjuncts, provided that these disjuncts are mutually exclusive. We can see that AX3 follows from the classical definition of probability by noting that a disjunctive event occurs just in case one or both of its disjuncts occurs. If the disjuncts are mutully exclusive, then in no possible outcome will they both occur, and so the number of possible outcomes in which the disjunction $A \vee B$ occurs is simply the sum of the number of possible outcomes in which A occurs and the number of possible outcomes in which B occurs. (See Problem 10.1(d).) That is, if A and B are mutually exclusive,

$$P(A \vee B) = \frac{\text{Number of possible outcomes in which } A \vee B \text{ occurs}}{\text{Total number of possible outcomes}}$$

$$= \frac{\text{Number of possible outcomes in which } A \text{ occurs}}{\text{Total number of possible outcomes}}$$

$$+ \frac{\text{Number of possible outcomes in which } B \text{ occurs}}{\text{Total number of possible outcomes}}$$

$$= P(A) + P(B)$$

We now prove some important theorems—i.e., deductive consequences of our axioms. These theorems enhance our understanding of probability itself while providing the basis for practical applications of the probability calculus. Their proofs utilize both the laws of propositional logic (Chapters 3 and 4) and some elementary arithmetic. To maximize clarity, these proofs will be presented somewhat informally.

SOLVED PROBLEMS

10.3 Prove:

$$P(\sim A) = 1 - P(A)$$

Solution

Since $A \vee \sim A$ is tautologous, by AX2 we have $P(A \vee \sim A) = 1$. And since A and $\sim A$ are mutually exclusive, by AX3, $P(A \vee \sim A) = P(A) + P(\sim A)$. Hence $1 = P(A) + P(\sim A)$, and so $P(\sim A) = 1 - P(A)$. This theorem tells how to obtain the probability of the negation of A, given the probability of A, and vice versa.

10.4 Prove:

If A is contradictory, $P(A) = 0$.

Solution

Suppose A is contradictory. Then $\sim A$ is tautologous, since negation changes every T to F in the truth table. Hence, by AX2, $P(\sim A) = 1$. So, by Problem 10.3, $P(A) = 0$.

10.5 Prove:

$$0 \leq P(A) \leq 1$$

Solution

By AX1, we know that $0 \leq P(A)$. We also know that $0 \leq P(\sim A)$. Now by Problem 10.3, $P(\sim A) = 1 - P(A)$, and so $0 \leq 1 - P(A)$, that is, $P(A) \leq 1$. This theorem summarizes the upper and lower bounds placed on probability values by the Kolmogorov axioms.

10.6 Prove:

If A and B are truth-functionally equivalent, then $P(A) = P(B)$.

Solution

Suppose A and B are truth-functionally equivalent. Then A and B have the same truth conditions. Hence A and $\sim B$ must always have opposite truth values, so that A and $\sim B$ are mutually exclusive. Moreover, $A \vee \sim B$ is a tautology. Hence by AX3 (substituting '$\sim B$' for 'B'), $P(A \vee \sim B) = P(A) + P(\sim B)$, and by AX2, $P(A \vee \sim B) = 1$. Thus $P(A) + P(\sim B) = 1$. Then, by Problem 10.3, $P(A) + 1 - P(B) = 1$, whence it follows that $P(A) = P(B)$.

This last result is very important. It allows us to substitute truth-functionally equivalent expressions for one another in probability formulas.

We next establish the general rule for computing the probability of a disjunction. This rule, which will be proved in Problem 10.7, says that the probability of a disjunction (regardless of whether or not the disjuncts are mutually exclusive) is the sum of the probabilities of the disjuncts, minus the probability of their conjunction. In the special case where the disjuncts are mutually exclusive, the probability of their conjunction is zero, and so the result of Problem 10.7 simply reduces to AX3.

To understand Problem 10.7 intuitively, consider the probability of rolling either an even number or a number less than 5 on a single toss of a fair die. Let 'E' say that the die shows an even number and 'L' that the die shows a number less than 5. Then, using the notation of Problem 10.1, we have:

$$P(E) = P(A_2 \lor A_4 \lor A_6) = \tfrac{3}{6} = \tfrac{1}{2}$$
$$P(L) = P(A_1 \lor A_2 \lor A_3 \lor A_4) = \tfrac{4}{6} = \tfrac{2}{3}$$

We wish to find $P(E \lor L)$. Notice that if we just add $\tfrac{1}{2}$ and $\tfrac{2}{3}$ together, we get the impossible value $\tfrac{7}{6}$. Clearly this is not the way to calculate $P(E \lor L)$. The error occurs because we count A_2 and A_4, the outcomes in which we get both an even number and a number less than 5, twice, whereas in fact they should be counted only once. That is, $E \lor L$ occurs in only five of the six possible outcomes of the toss, but if we count A_2 and A_4 twice, we get a total of seven outcomes, which is absurd, for only six are possible. To avoid this double counting, we subtract the probability of getting both E and L from the sum of the probability of E and the probability of L. This gives the correct figure, $\tfrac{5}{6}$, for the probability of the disjunction. That is:

$$P(E \lor L) = P(E) + P(L) - P(E \,\&\, L)$$
$$= \tfrac{1}{2} + \tfrac{2}{3} - \tfrac{1}{3}$$
$$= \tfrac{5}{6}$$

Problem 10.7 simply generalizes this idea. The proof proceeds by breaking down $P(A)$, $P(B)$, and $P(A \lor B)$ into complex equivalent probabilities in order to demonstrate the need for subtracting $P(A \,\&\, B)$ from $P(A) + P(B)$ to compensate for the double counting of $P(A \,\&\, B)$. Keeping the example just given in mind should help to clarify the proof.

SOLVED PROBLEM

10.7 Prove:

$$P(A \lor B) = P(A) + P(B) - P(A \,\&\, B)$$

Solution

We first note the following facts, which should be checked by means of truth tables:

(a) A is truth-functionally equivalent to $(A \,\&\, B) \lor (A \,\&\, {\sim}B)$.

(b) B is truth-functionally equivalent to $(A \,\&\, B) \lor ({\sim}A \,\&\, B)$.

(c) $A \lor B$ is truth-functionally equivalent to $((A \,\&\, {\sim}B) \lor ({\sim}A \,\&\, B)) \lor (A \,\&\, B)$.

(d) $A \,\&\, B$ and $A \,\&\, {\sim}B$ are mutually exclusive. (See Problem 10.2.)

(e) $A \,\&\, B$ and ${\sim}A \,\&\, B$ are mutually exclusive.

(f) $(A \,\&\, {\sim}B) \lor ({\sim}A \,\&\, B)$ and $A \,\&\, B$ are mutually exclusive.

(g) $A \,\&\, {\sim}B$ and ${\sim}A \,\&\, B$ are mutually exclusive.

By item (a) and Problem 10.6, $P(A) = P((A \,\&\, B) \lor (A \,\&\, {\sim}B))$, whence by item (d) and AX3 it follows that:

(h) $P(A) = P(A \,\&\, B) + P(A \,\&\, {\sim}B)$.

Similarly, by item (b) and Problem 10.6, $P(B) = P((A \,\&\, B) \lor ({\sim}A \,\&\, B))$, whence by item (e) and AX3 we get:

(i) $P(B) = P(A \,\&\, B) + P({\sim}A \,\&\, B)$.

Moreover, from item (c) and Problem 10.6 it follows that $P(A \lor B) = P(((A \& \sim B) \lor (\sim A \& B)) \lor (A \& B))$, whence by item (f) and AX3 we obtain $P(A \lor B) = P((A \& \sim B) \lor (\sim A \& B)) + P(A \& B)$. But then by item (g) and AX3 we see that:

(j) $P(A \lor B) = P(A \& \sim B) + P(\sim A \& B) + P(A \& B)$.

Now by adding the equations in items (h) and (i), we note that:

(k) $P(A) + P(B) = P(A \& \sim B) + P(\sim A \& B) + 2P(A \& B)$.

That is, if we simply add $P(A)$ and $P(B)$, where A and B are not mutually exclusive (i.e., where $P(A \& B) > 0$), $P(A \& B)$ will be counted twice. But as item (j) tells us, the correct value for $P(A \lor B)$ is obtained by counting $P(A \& B)$ only once. Now, subtracting the equation in item (k) from that in item (j), we obtain:

(l) $P(A \lor B) - (P(A) + P(B)) = -P(A \& B)$.

Adding $P(A) + P(B)$ to both sides of this equation yields $P(A \lor B) = P(A) + P(B) - P(A \& B)$, which was to be proved.

We next prove some additional theorems governing conjunction and disjunction.

SOLVED PROBLEMS

10.8 Prove:

$$P(A \& B) \leqslant P(A) \qquad \text{and} \qquad P(A \& B) \leqslant P(B)$$

Solution

By item (h) in the proof of Problem 10.7, $P(A) = P(A \& B) + P(A \& \sim B)$. But by AX1, $P(A \& \sim B) \geqslant 0$. Since adding a nonnegative quantity to $P(A \& B)$ gives $P(A)$, it must be that $P(A \& B) \leqslant P(A)$. Proof of the second conjunct is similar, except that item (i) from the proof of Problem 10.7 is used.

10.9 Prove:

$$P(A \lor B) \geqslant P(A) \qquad \text{and} \qquad P(A \lor B) \geqslant P(B)$$

Solution

By Problem 10.7, $P(A \lor B) = P(A) + P(B) - P(A \& B)$. And by Problem 10.8, $P(B) - P(A \& B) \geqslant 0$. Hence, $P(A \lor B) \geqslant P(A)$. Proof of the second conjunct is similar.

10.10 Prove:

If $P(A) = P(B) = 0$, then $P(A \lor B) = 0$.

Solution

Suppose $P(A) = P(B) = 0$. Then by Problem 10.8 and AX1, $P(A \& B) = 0$. Hence, by Problem 10.7, $P(A \lor B) = 0$.

10.11 Prove:

If $P(A) = 1$, then $P(A \& B) = P(B)$.

Solution

Suppose $P(A) = 1$. Now by Problem 10.9, $P(A \lor B) \geqslant P(A)$, and by Problem 10.5, $P(A \lor B) \leqslant 1$. Hence $P(A \lor B) = 1$. Furthermore, by Problem 10.7, $P(A \lor B) = P(A) + P(B) - P(A \& B)$, so that $1 = 1 + P(B) - P(A \& B)$; that is, $P(A \& B) = P(B)$.

10.12   Prove:

If A is a truth-functional consequence of B, then $P(A \& B) = P(B)$.

Solution

Suppose A is a truth-functional consequence of B. Then $A \& B$ is truth-functionally equivalent to B, since $A \& B$ is true on any line of a truth table in which B is true and false on any line of a truth table on which B is false. So by Problem 10.6, $P(A \& B) = P(B)$.

10.3   CONDITIONAL PROBABILITY

Conditional probability is the probability of one proposition (or event), given that another is true (or has occurred). This is expressed by the notation '$P(A \mid B)$', which means "the probability of A, given B." It is not to be confused with the probability of a conditional statement, $P(B \rightarrow A)$, which plays little role in probability theory. (The difference between $P(A \mid B)$ and $P(B \rightarrow A)$ is illuminated by Problem 10.30 at the end of this section.) Notice that in contrast to the notation for conditional statements, the notation for conditional probabilities lists the consequent first and the antecedent second. This unfortunate convention is fairly standard, though occasionally other conventions are employed.

The notation for conditional probability is introduced into the probability calculus by the following definition:

D1 $$P(A \mid B) =_{\text{df}} \frac{P(A \& B)}{P(B)}$$

The symbol '$=_{\text{df}}$' means "is by definition." Anywhere the expression '$P(A \mid B)$' occurs, it is to be regarded as mere shorthand for the more complex expression on the right side of D1.

The purport of this definition is easy to understand under the classical interpretation. By the classical definition of probability, we have:

$$\frac{P(A \& B)}{P(B)} = \frac{\dfrac{\text{Number of possible outcomes in which } A \& B \text{ occurs}}{\text{Total number of possible outcomes}}}{\dfrac{\text{Number of possible outcomes in which } B \text{ occurs}}{\text{Total number of possible outcomes}}}$$

$$= \frac{\text{Number of possible outcomes in which } A \& B \text{ occurs}}{\text{Number of possible outcomes in which } B \text{ occurs}}$$

Thus by the classical interpretation, $P(A \mid B)$ is the proportion of possible outcomes in which A occurs among the possible outcomes in which B occurs.

SOLVED PROBLEM

10.13   A single die is tossed once. As in Example 10.6, let 'E' state that the result of the toss is an even number and 'L' state that it is a number less than 5. What is $P(E \mid L)$?

Solution

$E \& L$ is true in two of the six possible outcomes, and L is true in four of the six. Hence

$$P(E \mid L) = \frac{P(E \& L)}{P(L)} = \frac{\frac{2}{6}}{\frac{4}{6}} = \frac{2}{4} = \frac{1}{2}$$

Conditional probabilities are central to logic. Inductive probability (Section 2.3 and Chapter 9) is the probability of an argument's conclusion, given the conjunction of its premises. It is therefore a kind of conditional probability. This sort of probability, however, is best understood not by the classical interpretation, but by the subjective interpretation or the logical interpretations, whose technical details are beyond the scope of this discussion.

Note that in the case where $P(B) = 0$, $P(A \mid B)$ has no value, since division by zero is undefined. In general, theorems employing the notation '$P(A \mid B)$' hold only when $P(B) > 0$, but it is cumbersome to repeat this qualification each time a theorem is stated. Therefore we will not repeat it, but it is to be understood implicitly.

SOLVED PROBLEMS

10.14 Prove:

$$P(A \mid A) = 1$$

Solution

$A \& A$ is truth-functionally equivalent to A. Thus, by Problem 10.6,

$$P(A \mid A) = \frac{P(A \& A)}{P(A)} = \frac{P(A)}{P(A)} = 1$$

10.15 Prove:

$$P(\sim A \mid A) = 0$$

Solution

$\sim A \& A$ is contradictory, and so by Problem 10.4, $P(\sim A \& A) = 0$. Thus, by D1,

$$P(\sim A \mid A) = \frac{P(\sim A \& A)}{P(A)} = \frac{0}{P(A)} = 0$$

10.16 Prove:

If B is tautologous, $P(A \mid B) = P(A)$.

Solution

Suppose B is tautologous. Then by AX2, $P(B) = 1$. Moreover, $A \& B$ is truth-functionally equivalent to A, so that by Problem 10.6, $P(A \& B) = P(A)$. Thus

$$P(A \mid B) = \frac{P(A \& B)}{P(B)} = \frac{P(A)}{1} = P(A)$$

In Problem 10.6 we established that truth-functionally equivalent formulas may validly be substituted for one another in a probability expression. The next two theorems establish this same result for conditional probability expressions.

SOLVED PROBLEMS

10.17 Prove:

If A and B are truth-functionally equivalent, then $P(A \mid C) = P(B \mid C)$.

Solution

Suppose A and B are truth-functionally equivalent. Then so are $A \& C$ and $B \& C$. Hence by Problem 10.6, $P(A \& C) = P(B \& C)$. But then, by D1,

$$P(A \mid C) = \frac{P(A \& C)}{P(C)} = \frac{P(B \& C)}{P(C)} = P(B \mid C)$$

10.18 Prove:

If A and B are truth-functionally equivalent, then $P(C\,|\,A) = P(C\,|\,B)$.

Solution

Suppose A and B are truth-functionally equivalent. Then, by Problem 10.6, $P(A) = P(B)$. Moreover, by the reasoning of Problem 10.17, $P(C \& A) = P(C \& B)$. Hence, by D1,

$$P(C\,|\,A) = \frac{P(C \& A)}{P(A)} = \frac{P(C \& B)}{P(B)} = P(C\,|\,B)$$

Our next result is of great importance, because it provides a way of calculating the probabilities of conjunctions.

SOLVED PROBLEMS

10.19 Prove:

$$P(A \& B) = P(A) \cdot P(B\,|\,A)$$

Solution

D1 gives $P(B\,|\,A) = P(B \& A)/P(A)$. Hence, since $B \& A$ is truth-functionally equivalent to $A \& B$, by Problem 10.6 we have $P(B\,|\,A) = P(A \& B)/P(A)$. Multiplying both sides of this equation by $P(A)$ yields the theorem.

10.20 Prove:

$$P(A) \cdot P(B\,|\,A) = P(B) \cdot P(A\,|\,B)$$

Solution

By Problem 10.19, $P(A \& B) = P(A) \cdot P(B\,|\,A)$ and also $P(B \& A) = P(B) \cdot P(A\,|\,B)$. But $A \& B$ is truth-functionally equivalent to $B \& A$. Hence, by Problem 10.6, $P(A \& B) = P(B \& A)$. This proves the theorem, which shows that the order of conjuncts is of no logical importance.

Problems 10.19 and 10.20 give us two ways of expressing the probability of any conjunction. That is, for any conjunction $A \& B$, $P(A \& B) = P(A) \cdot P(B\,|\,A) = P(B) \cdot P(A\,|\,B)$.

It sometimes happens that $P(A\,|\,B) = P(A)$, that is, that $P(A)$ is unaffected by the occurrence or nonoccurrence of B. In such a case, we say that A is *independent* of B. For instance, if A is the event of getting a one on the first toss of a single die and B is the event of getting a one on the second toss, then A is independent of B, for $P(A) = P(A\,|\,B) = \frac{1}{6}$. (The tendency to believe that tosses of dice are not independent is one version of the gambler's fallacy—see Section 8.5.) On the other hand, if A is the event of your living ten more years and B is the event of your having rabies, then clearly $P(A) > P(A\,|\,B)$, so that A is not independent of B. (Note that it is difficult to make sense of these probabilities under the classical interpretation, since there is no set of equally likely outcomes to appeal to. These probabilities would therefore be best understood by one of the other interpretations mentioned earlier.)

Independence is a concept peculiar to the probability calculus. It cannot be characterized truth-functionally as, for example, tautologousness or truth-functional equivalence can. The next theorem tells us that A is independent of B if and only if B is independent of A; that is, independence is a *symmetrical* relation.

SOLVED PROBLEM

10.21 Prove:

$$P(A \mid B) = P(A) \text{ if and only if } P(B \mid A) = P(B).$$

Solution

By D1, $P(A \mid B) = P(A)$ if and only if $P(A) = P(A \& B)/P(B)$. Now multiplying both sides of this equation by $P(B)/P(A)$ gives $P(B) = P(A \& B)/P(A)$, which (since $A \& B$ is truth-functionally equivalent to $B \& A$) is true if and only if $P(B) = P(B \& A)/P(A)$; hence by D1, $P(B) = P(B \mid A)$.

Because of the symmetry of independence, instead of saying "A is independent of B" or "B is independent of A," we may simply say "A and B are independent," without regard to the order of the two terms. Independence is important chiefly because if A and B are independent, then the calculation of $P(A \& B)$ is very simple. This is shown by the following theorem.

SOLVED PROBLEM

10.22 Prove:

If A and B are independent, then $P(A \& B) = P(A) \cdot P(B)$.

Solution

This follows immediately from Problem 10.19 and the definition of independence.

The next result also plays a central role in probability theory. It was first proved by Thomas Bayes (1702–1761), one of the founders of probability theory. Bayes' theorem enables us to calculate conditional probabilities, given converse conditional probabilities together with some nonconditional probabilities (the so-called priors). It has a number of important practical applications. We shall state it first in a simplified version (Problem 10.23) and then in a more fully articulated form (Problem 10.25). Here is the simple version:

SOLVED PROBLEM

10.23 Prove:

$$P(A \mid B) = \frac{P(A) \cdot P(B \mid A)}{P(B)}$$

Solution

This follows immediately from Problem 10.20.

This version of the theorem allows us to calculate $P(A \mid B)$ if we know the converse conditional probability $P(B \mid A)$ and the prior probabilities $P(A)$ and $P(B)$.

To state the full version of Bayes' theorem, we need two additional concepts: the concept of an exhaustive series of propositions or events, and the concept of a pairwise mutually exclusive series. A series of propositions or events $A_1, A_2, \ldots, A_n$ is *exhaustive* if $P(A_1 \lor A_2 \lor \ldots \lor A_n) = 1$. For instance, the series $A_1, A_2, \ldots, A_6$, representing the six possible outcomes of a single toss of a single die, is exhaustive, under the classical interpretation. Also, any series of the form $A \& B$, $A \& \sim B$, $\sim A \& B$, $\sim A \& \sim B$ is exhaustive, as can be seen by observing that the disjunction of these four forms is a tautology.

The second crucial concept for the general version of Bayes' theorem is that of a *pairwise mutually exclusive* series. We have already discussed mutual exclusivity for pairs of propositions or events. Now in effect we simply extend this idea to a whole series of them. Thus a series of propositions or events $A_1, A_2, \ldots, A_n$ is pairwise mutually exclusive if for each pair A_i, A_j of its members, $P(A_i \& A_j) = 0$. For instance, the exhaustive series $A_1, A_2, \ldots, A_6$ mentioned above is pairwise mutually exclusive, since only one outcome may result from a single toss. So is the other exhaustive series mentioned above, but here the reason is truth-functional: on no line of a truth table are any two of these statements both true.

Now we observe an important fact about pairwise mutually exclusive and exhaustive series:

SOLVED PROBLEM

10.24   Prove:

If $A_1 A_2, \ldots, A_n$ is a pairwise mutually exclusive and exhaustive series, then $P(B) = P(A_1 \& B) + P(A_2 \& B) + \ldots + P(A_n \& B)$.

Solution

Let $A_1, A_2, \ldots, A_n$ be a pairwise mutually exclusive and exhaustive series. Then $P(A_1 \vee A_2 \vee \ldots \vee A_n) = 1$, and so, by Problem 10.11,

(a) $P((A_1 \vee A_2 \vee \ldots \vee A_n) \& B) = P(B)$.

Now as can be seen by repeated application of the distributive law of propositional logic, $(A_1 \vee A_2 \vee \ldots \vee A_n) \& B$ is truth-functionally equivalent to $(A_1 \& B) \vee (A_2 \& B) \vee \ldots \vee (A_n \& B)$. Thus, applying Problem 10.6 to item (a) we get:

(b) $P(B) = P((A_1 \& B) \vee (A_2 \& B) \vee \ldots \vee (A_n \& B))$.

Moreover,

(c)   The series $(A_1 \& B), (A_2 \& B), \ldots, (A_n \& B)$ is pairwise mutually exclusive.

For consider any two members $(A_i \& B), (A_j \& B)$ of this series. Since the series $A_1, A_2, \ldots, A_n$ is pairwise mutually exclusive, we know that $P(A_i \& A_j) = 0$. Hence, by Problem 10.8 and AX1, $P(A_i \& A_j \& B) = 0$. So, by Problem 10.6, $P((A_i \& B) \& (A_j \& B)) = 0$.

Now from item (c) it follows that for each i such that $1 \leqslant i < n$,

(d) $P(((A_1 \& B) \vee (A_2 \& B) \vee \ldots \vee (A_i \& B)) \& (A_{i+1} \& B)) = 0$.

For again, by repeated application of the distributive law, $((A_1 \& B) \vee (A_2 \& B) \vee \ldots \vee (A_i \& B)) \& (A_{i+1} \& B)$ is truth-functionally equivalent to $((A_1 \& B) \& (A_{i+1} \& B)) \vee ((A_2 \& B) \& (A_{i+1} \& B)) \vee \ldots \vee ((A_i \& B) \& (A_{i+1} \& B))$. But by item (c), the probability of each of the disjuncts of this latter formula is zero; and repeated application of Problem 10.10 implies that any disjunction whose disjuncts all have probability zero must itself have probability zero. Hence item (d) follows by Problem 10.6.

The theorem then follows from formula (b) by repeated application of formula (d) and Problem 10.7. To see this, suppose for the sake of concreteness that $n = 3$. Then formula (b) will read:

(b') $P(B) = P((A_1 \& B) \vee (A_2 \& B) \vee (A_3 \& B))$

and we will have these two instances of formula (d):

(d') $P(((A_1 \& B) \vee (A_2 \& B)) \& (A_3 \& B)) = 0$.

(d'') $P((A_1 \& B) \& (A_2 \& B)) = 0$.

From Problem 10.7 we obtain:

(e) $P((A_1 \& B) \vee (A_2 \& B) \vee (A_3 \& B))$
       $= P((A_1 \& B) \vee (A_2 \& B)) + P(A_3 \& B) - P(((A_1 \& B) \vee (A_2 \& B)) \& (A_3 \& B))$

which reduces by (d') to:

(f) $P((A_1 \, \& \, B) \lor (A_2 \, \& \, B) \lor (A_3 \, \& \, B))$
    $= P((A_1 \, \& \, B) \lor (A_3 \, \& \, B)) + P(A_3 \, \& \, B)$

Now, again, by Problem 10.7:

(g) $P((A_1 \, \& \, B) \lor (A_2 \, \& \, B))$
    $= P(A_1 \, \& \, B) + P(A_2 \, \& \, B) - P((A_1 \, \& \, B) \, \& \, (A_2 \, \& \, B))$

which reduces by (d'') to:

(h) $P((A_1 \, \& \, B) \lor (A_2 \, \& \, B)) = P(A_1 \, \& \, B) + P(A_2 \, \& \, B)$

Combining (f) and (h), we get:

(i) $P((A_1 \, \& \, B) \lor (A_2 \, \& \, B) \lor (A_3 \, \& \, B)) = P(A_1 \, \& \, B) + P(A_2 \, \& \, B) + P(A_3 \, \& \, B)$

which, by (b'), gives:

$$P(B) = P(A_1 \, \& \, B) + P(A_1 \, \& \, B) + P(A_3 \, \& \, B)$$

which proves the theorem for $n = 3$. The reasoning for other values of n is similar.

Using the result of Problem 10.24, we now establish the general version of Bayes' theorem:

SOLVED PROBLEM

10.25 Prove:

If $A_1, A_2, \ldots, A_n$ is a pairwise mutually exclusive and exhaustive series and A_i is some member of this series, then

$$P(A_i \mid B) = \frac{P(A_i) \cdot P(B \mid A_i)}{P(A_1) \cdot P(B \mid A_1) + P(A_2) \cdot P(B \mid A_2) + \ldots + P(A_n) \cdot P(B \mid A_n)}$$

Solution

Assume $A_1, A_2, \ldots, A_n$ is a pairwise mutually exclusive and exhaustive series. Now by Problem 10.23 we know that

(a) $P(A_i \mid B) = \dfrac{P(A_i) \cdot P(B \mid A_i)}{P(B)}$

and by Problem 10.24 we have:

(b) $P(B) = P(A_1 \, \& \, B) + P(A_2 \, \& \, B) + \ldots + P(A_n \, \& \, B)$

Applying Problem 10.19 to formula (b) we get

(c) $P(B) = P(A_1) \cdot P(B \mid A_1) + P(A_2) \cdot P(B \mid A_2) + \ldots + P(A_n) \cdot P(B \mid A_n)$

Then applying formula (c) to formula (a) proves the theorem.

The general version of Bayes' theorem is often expressed in the more compact summation notation:

$$P(A_i \mid B) = \frac{P(A_i) \cdot P(B \mid A_i)}{\displaystyle\sum_{j=1}^{n} P(A_j) \cdot P(B \mid A_j)}$$

This is just a shorthand version of the equation of Problem 10.25.

One practical application of Bayes' theorem is in calculating the probability that a particular hypothesis accounts for a particular observation. This calculation requires us to formulate a series of hypotheses which is pairwise mutually exclusive and exhaustive—the series $A_1, A_2, \ldots, A_n$ mentioned in the theorem. The observation is B. Now if we wish to learn the probability that a particular member A_i of the series of hypotheses accounts for B—that is, $P(A_i \mid B)$—Bayes' theorem gives a way of finding the answer.

There is a catch, however. To make the calculation, we need to know not only the probability of the observation, given each of the hypotheses—that is, $P(B \mid A_j)$ for each j such that $1 \leq j \leq n$—but also the probability of each of the hypotheses, or $P(A_j)$ for each such j. These latter probabilities are called *prior probabilities*, or sometimes just *priors*.[2] In many cases they are difficult or impossible to determine. It can be shown, however, that as observations accumulate, the priors have less and less influence on the outcome of the calculation. This phenomenon, known as "swamping the priors," permits useful application of Bayes' theorem even in cases in which the priors are only very roughly known.

SOLVED PROBLEM

10.26 Three machines at a certain factory produce girdles of the same kind. Three-tenths of the girdles produced by machine 1 are defective, as are two-tenths of the girdles produced by machine 2 and one-tenth of the girdles produced by machine 3. Machine 1 produces four-tenths of the factory's output, machine 2 produces three-tenths, and machine 3 produces three-tenths. Alma has acquired an egregiously defective girdle from the factory. What is the probability that the girdle was produced by machine 1, given that it is defective?

Solution

Here the observation B is that Alma's girdle is defective. This can be accounted for by three pairwise mutually exclusive and exhaustive hypotheses:

A_1 = Alma's girdle was produced by machine 1.
A_2 = Alma's girdle was produced by machine 2.
A_3 = Alma's girdle was produced by machine 3.

We may take the probability of B given each of these hypotheses as just the respective proportion of defective girdles produced by each machine. Thus

$$P(B \mid A_1) = \tfrac{3}{10}$$
$$P(B \mid A_2) = \tfrac{2}{10}$$
$$P(B \mid A_3) = \tfrac{1}{10}$$

Moreover, we have definite values for the priors in this case. These are just the proportions of the total output of the factory produced by each machine:

$$P(A_1) = \tfrac{4}{10}$$
$$P(A_2) = \tfrac{3}{10}$$
$$P(A_3) = \tfrac{3}{10}$$

[2]The notion of prior probability should not be confused with that of an a priori or logical probability, which was mentioned earlier in this chapter. Prior probabilities are usually obtained from subjective estimates or (as they will be in Problem 10.26) from relative frequencies of occurrence. A priori probabilities are not based on actual or estimated frequencies of events, but on the information content of statements.

Now, by Bayes' theorem,

$$P(A_1 \mid B) = \frac{P(A_1) \cdot P(B \mid A_1)}{P(A_1) \cdot P(B \mid A_1) + P(A_2) \cdot P(B \mid A_2) + P(A_3) \cdot P(B \mid A_3)}$$

$$= \frac{\frac{4}{10} \cdot \frac{3}{10}}{(\frac{4}{10} \cdot \frac{3}{10}) + (\frac{3}{10} \cdot \frac{2}{10}) + (\frac{3}{10} \cdot \frac{1}{10})}$$

$$= \frac{12}{21}$$

This is the solution to the problem.

Our next three theorems (Problems 10.27, 10.28, and 10.29) establish the behavior of negations, disjunctions, and conjunctions as the first terms of conditional probabilities. They are conditional analogues of Problems 10.3, 10.7, and 10.19, respectively.

SOLVED PROBLEM

10.27 Prove:

$$P(\sim A \mid B) = 1 - P(A \mid B)$$

Solution

By D1,

$$P(A \lor \sim A \mid B) = \frac{P((A \lor \sim A) \& B)}{P(B)}$$

But $(A \lor \sim A) \& B$ is truth-functionally equivalent to B. Hence, by Problem 10.6, $P((A \lor \sim A) \& B) = P(B)$, so that $P(A \lor \sim A \mid B) = P(B)/P(B) = 1$. Moreover, $(A \lor \sim A) \& B$ is also truth-functionally equivalent to $(A \& B) \lor (\sim A \& B)$, so that by Problem 10.6,

$$P(A \lor \sim A \mid B) = \frac{P((A \& B) \lor (\sim A \& B))}{P(B)} = 1$$

Now $A \& B$ and $\sim A \& B$ are mutually exclusive, so that by AX3,

$$\frac{P(A \& B) + P(\sim A \& B)}{P(B)} = \frac{P(A \& B)}{P(B)} + \frac{P(\sim A \& B)}{P(B)} = 1$$

So by D1, $P(A \mid B) + P(\sim A \mid B) = 1$; that is, $P(\sim A \mid B) = 1 - P(A \mid B)$.

Since as we noted earlier, the inductive probability of an argument is a kind of conditional probability (the probability of the conclusion given the conjunction of the premises), Problem 10.27 implies the important result that the probability of $\sim A$ given a set of premises is 1 minus the probability of A given those premises. (And, likewise, the probability of A given a set of premises is 1 minus the probability of $\sim A$ given those premises.) This was our justification for remarking in Section 2.3 that any argument whose inductive probability is less than .5 is weak, since the probability of the negation of its conclusion given the same premises is therefore greater than .5.

SOLVED PROBLEMS

10.28 Prove:

$$P(A \lor B \mid C) = P(A \mid C) + P(B \mid C) - P(A \& B \mid C)$$

Solution

By D1,

$$P(A \vee B \mid C) = \frac{P((A \vee B) \& C)}{P(C)}$$

But since $(A \vee B) \& C$ is truth-functionally equivalent to $(A \& C) \vee (B \& C)$, we have, by Problem 10.6,

$$P(A \vee B \mid C) = \frac{P((A \& C) \vee (B \& C))}{P(C)}.$$

By Problem 10.7, this is equal to

$$\frac{P(A \& C) + P(B \& C) - P((A \& C) \& (B \& C))}{P(C)}$$

and since $(A \& C) \& (B \& C)$ is truth-functionally equivalent to $(A \& B) \& C$, by Problem 10.6 this in turn is equal to

$$\frac{P(A \& C)}{P(C)} + \frac{P(B \& C)}{P(C)} - \frac{P((A \& B) \& C)}{P(C)}$$

which reduces by D1 to $P(A \mid C) + P(B \mid C) - P(A \& B \mid C)$.

10.29   Prove:

$$P(A \& B \mid C) = P(A \mid C) \cdot P(B \mid A \& C)$$

Solution

By D1,

$$P(A \& B \mid C) = \frac{P((A \& B) \& C)}{P(C)}$$

which by Problem 10.6 is equal to $P((A \& C) \& B)/P(C)$. By Problem 10.18, this becomes

$$\frac{P(A \& C) \cdot P(B \mid A \& C)}{P(C)}$$

which reduces by D1 to $P(A \mid C) \cdot P(B \mid A \& C)$.

We conclude this section with a theorem that indicates the relationship between conditional probability and the probability of the material conditional.

SOLVED PROBLEM

10.30   Prove:

$$P(A \rightarrow B) = P(\sim A) + P(A) \cdot P(B \mid A)$$

Solution

$A \rightarrow B$ is truth-functionally equivalent to $\sim(A \& \sim B)$, so that by Problem 10.6:

$$P(A \rightarrow B) = P(\sim(A \& \sim B))$$

But by Problem 10.3, $P(\sim(A \& \sim B)) = 1 - P(A \& \sim B)$, and since, by Problem 10.19, $P(A \& \sim B) = P(A) \cdot P(\sim B \mid A)$, we have

$$P(A \rightarrow B) = 1 - (P(A) \cdot P(\sim B \mid A))$$

Now by Problem 10.27, $P(\sim B \mid A) = 1 - P(B \mid A)$; hence,

$$P(A \rightarrow B) = 1 - (P(A) \cdot (1 - P(B \mid A)))$$

so that:

$$P(A \rightarrow B) = 1 - (P(A) - P(A) \cdot P(B \mid A))$$

i.e.,

$$P(A \rightarrow B) = 1 - P(A) + P(A) \cdot P(B \mid A)$$

But by Problem 10.3, $1 - P(A) = P(\sim A)$, and so:

$$P(A \rightarrow B) = P(\sim A) + P(A) \cdot P(B \mid A).$$

This theorem shows that when $P(A) = 1$ (and hence $P(\sim A) = 0$), $P(A \rightarrow B) = P(B \mid A)$. But when $P(A) \neq 1$, the conditional probability generally does not equal the probability of the corresponding material conditional. Hence the two are clearly distinct, as we noted at the beginning of this section.

10.4 APPLICATION OF THE PROBABILITY CALCULUS

In this section we consider the application of the probability calculus, classically interpreted, to some simple and well-defined problems. Our first set of problems, once again, involves dice. This time, however, we will consider the toss of a pair of dice. When two fair dice are tossed, there are 36 equally likely outcomes, as listed below. (The first number of each pair indicates the outcome for the first die and the second number indicates the outcome for the second):

1-1	2-1	3-1	4-1	5-1	6-1
1-2	2-2	3-2	4-2	5-2	6-2
1-3	2-3	3-3	4-3	5-3	6-3
1-4	2-4	3-4	4-4	5-4	6-4
1-5	2-5	3-5	4-5	5-5	6-5
1-6	2-6	3-6	4-6	5-6	6-6

Now consider the following events:

A_1 = One on die 1 A_7 = One on die 2
A_2 = Two on die 1 A_8 = Two on die 2
A_3 = Three on die 1 A_9 = Three on die 2
A_4 = Four on die 1 A_{10} = Four on die 2
A_5 = Five on die 1 A_{11} = Five on die 2
A_6 = Six on die 1 A_{12} = Six on die 2

Using the classical definition of probability together with the probability calculus, we can calculate the probabilities of these events and their truth-functional combinations.

SOLVED PROBLEMS

10.31   What is the probability of A_1?

Solution

A_1 occurs in 6 of the 36 possible outcomes. Therefore, by the classical definition of probability, $P(A_1) = \frac{6}{36} = \frac{1}{6}$. Clearly, the probability of each of A_2 through A_{12} is also $\frac{1}{6}$.

10.32   Calculate $P(\sim A_1)$.

Solution

By Problem 10.3, we know that $P(\sim A_1) = 1 - P(A_1)$. We saw in Problem 10.31 that $P(A_1) = \frac{1}{6}$. Hence $P(\sim A_1) = 1 - \frac{1}{6} = \frac{5}{6}$.

10.33   Calculate $P(A_1 \,\&\, A_7)$.

Solution

We can do this directly by the classical definition, or indirectly via the probability calculus. Using the classical definition, we note that in only one of the 36 possible outcomes does $A_1 \,\&\, A_7$ occur. Hence $P(A_1 \,\&\, A_7) = \frac{1}{36}$.

To use the probability calculus, we must recognize that A_1 and A_7 are independent. That is (assuming the dice are tossed fairly, so that for example we don't start with both showing the same face and give them just a little flick, an arrangement which would tend to make them show the same face after the toss), we expect that $P(A_1 \,|\, A_7) = P(A_1)$ and $P(A_7 \,|\, A_1) = P(A_7)$. If this assumption is correct, we can apply Problem 10.19, so that $P(A_1 \,\&\, A_7) = P(A_1) \cdot P(A_7) = \frac{1}{6} \cdot \frac{1}{6} = \frac{1}{36}$. This agrees with our previous calculation.

As Problem 10.33 illustrates, we can calculate probabilities in several ways (both directly, from the classical definition, and indirectly, via the probability calculus). Checking one method of calculation by means of another helps eliminate mistakes. Problem 10.34 provides another illustration of this.

SOLVED PROBLEM

10.34   Find $P(A_1 \lor A_7)$.

Solution

To solve this problem via the calculus, we appeal to Problem 10.7:

$$P(A_1 \lor A_7) = P(A_1) + P(A_7) - P(A_1 \,\&\, A_7)$$

In Problem 10.31 we determined that $P(A_1) = P(A_7) = \frac{1}{6}$, and in Problem 10.33 we found that $P(A_1 \,\&\, A_7) = \frac{1}{36}$. Hence $P(A_1 \lor A_7) = \frac{1}{6} + \frac{1}{6} - \frac{1}{36} = \frac{11}{36}$.

But we could just as well have proceeded directly, via the classical definition. There are 11 possible outcomes in which $A_1 \lor A_7$ occurs (these are just the outcomes listed in the leftmost column, along with those listed in the first horizontal row, of our table of the 36 outcomes), Hence, once again we get $P(A_1 \lor A_7) = \frac{11}{36}$.

Our next set of problems involves the probabilities of dealing certain combinations of cards at random from a jokerless deck of 52 cards. (To say that the deal is random is to say that each card remaining in the deck is equally likely to be dealt. This provides the equally likely outcomes necessary for the classical interpretation.)

If only one card is dealt, there are 52 possible outcomes, one for each card in the deck. If two cards are dealt, the number of possible outcomes is $52 \cdot 51 = 2652$. (There are 52 possibilities for the first card and 51 for the second, since after the first card is dealt, 51 remain in the deck.) In general, the number of equally likely outcomes for a deal of n cards is $52 \cdot 51 \cdots (52 - (n - 1))$.

We adopt the following abbreviations:

$$J = \text{Jack} \qquad K = \text{King} \qquad H = \text{Heart} \qquad D = \text{Diamond}$$
$$Q = \text{Queen} \qquad A = \text{Ace} \qquad S = \text{Spade} \qquad C = \text{Club}$$

We use numerical subscripts to indicate the order of the cards dealt. Thus 'A_1' means "The first card dealt is an ace," 'D_3' means "The third card dealt is a diamond," and so on.

SOLVED PROBLEM

10.35 One card is dealt. What is the probability that it is either a queen or a heart?

Solution

We seek $P(Q_1 \lor H_1)$, which by Problem 10.7 is $P(Q_1) + P(H_1) - P(Q_1 \& H_1)$. Out of the 52 cards, 4 are queens, 12 are hearts, and 1 is the queen of hearts. Thus, by the classical definition of probability, we have $\frac{4}{52} + \frac{13}{52} - \frac{1}{52} = \frac{16}{52} = \frac{4}{13}$.

When two or more cards are dealt, we may wish to know the probability that one is of a certain type, given that certain other cards have been previously dealt. Such conditional probabilities can be computed by D1, but they are most efficiently obtained directly from consideration of the composition of the deck.

SOLVED PROBLEM

10.36 Two cards are dealt. What is the probability that the second is an ace, given that the first is?

Solution

We wish to find $P(A_2 \mid A_1)$. Given A_1, 51 cards remain in the deck, of which three are aces. Hence $P(A_2 \mid A_1) = \frac{3}{51} = \frac{1}{17}$. The same figure can be obtained with a little more work by appeal to D1. By D1 we have $P(A_2 \mid A_1) = P(A_2 \& A_1)/P(A_1)$. Now clearly $P(A_1) = \frac{4}{52} = \frac{1}{13}$. But to find $P(A_2 \& A_1)$, we must consider all the 2652 outcomes possible in a deal of two cards. Of these, only 12 give $A_2 \& A_1$. We list them below. ('AH' means "ace of hearts," 'AS' means "ace of spades," and so on.)

First card	Second card	First card	Second card
AH	AS	AD	AH
AH	AD	AD	AS
AH	AC	AD	AC
AS	AH	AC	AH
AS	AD	AC	AS
AS	AC	AC	AD

Thus

$$\frac{P(A_1 \& A_2)}{P(A_1)} = \frac{\frac{12}{2652}}{\frac{1}{13}} = \frac{1}{17}$$

Clearly, however, the first method of calculation is simpler.

Using conditional probabilities obtained by the first method of calculation explained in Problem 10.36, we can easily obtain a variety of other probabilities by applying the theorems of the probability calculus.

SOLVED PROBLEMS

10.37 Two cards are dealt. What is the probability that they are both aces?

Solution

The desired probability is $P(A_1 \& A_2)$, which, by Problem 10.19, is $P(A_1) \cdot P(A_2 \mid A_1)$. As we saw in Problem 10.36, $P(A_1) = \frac{1}{13}$; and as we also saw in that problem, $P(A_2 \mid A_1) = \frac{1}{17}$. Hence

$P(A_1 \& A_2) = \frac{1}{13} \cdot \frac{1}{17} = \frac{1}{221}$. This is the same figure obtained above for $P(A_2 \& A_1)$, since $\frac{12}{2652} = \frac{1}{221}$.

10.38 Five cards are dealt. What is the probability of a flush, i.e., five cards of a single suit?

Solution

We wish to find $P((H_1 \& H_2 \& H_3 \& H_4 \& H_5) \vee (S_1 \& S_2 \& S_3 \& S_4 \& S_5) \vee (D_1 \& D_2 \& D_3 \& D_4 \& D_5) \vee (C_1 \& C_2 \& C_3 \& C_4 \& C_5))$. Clearly these four disjuncts are mutually exclusive, and each is mutually exclusive from any disjunction of the others. Thus by Problem 10.7 the desired probability is simply the sum of the probabilities of these four disjuncts. We shall calculate $P(H_1 \& H_2 \& H_3 \& H_4 \& H_5)$. Clearly, the probabilities of the other disjuncts will be the same. Hence, to obtain our answer we simply multiply the result of our calculation by 4.

Now by four applications of Problem 10.19, $P(H_1 \& H_2 \& H_3 \& H_4 \& H_5) = P(H_1) \cdot P(H_2 \mid H_1) \cdot P(H_3 \mid H_1 \& H_2) \cdot P(H_4 \mid H_1 \& H_2 \& H_3) \cdot P(H_5 \mid H_1 \& H_2 \& H_3 \& H_4)$. Using the first method of Problem 10.36, we can see that this is $\frac{13}{52} \cdot \frac{12}{51} \cdot \frac{11}{50} \cdot \frac{10}{49} \cdot \frac{9}{48} = 0.0004952$. Multiplying by 4 gives the probability of a flush: 0.0019808.

10.39 In a game of blackjack, you have been dealt a king and an eight from a deck of 52 cards. If you ask for a third card, what is the probability that you will receive an ace, a two, or a three and thus not go over 21?

Solution

Let 'E' stand for an eight, 'W' for a two, and 'R' for a three. The probability we seek is either $P(A_3 \vee W_3 \vee R_3 \mid E_1 \& K_2)$ or $P(A_3 \vee W_3 \vee R_3 \mid K_1 \& E_2)$. Either of these will do; they are equal, and the order in which the king and the eight are dealt is not relevant to our problem. Hence we will calculate only the first.

We note that A_3 and $W_3 \vee R_3$ are mutually exclusive, as are W_3 and R_3. Hence, by two applications of Supplementary Problem I(12), below, $P(A_3 \vee W_3 \vee R_3 \mid K_1 \& E_2) = P(A_3 \mid K_1 \& E_2) + P(W_3 \mid K_1 \& E_2) + P(R_3 \mid K_1 \& E_2)$. Given $K_1 \& E_2$, the deck contains 50 cards, four of which are aces, four of which are twos, and four of which are threes. Hence the desired probability is $\frac{4}{50} + \frac{4}{50} + \frac{4}{50} = \frac{12}{50} = \frac{6}{25}$.

Supplementary Problems

I Prove the following theorems, using the axioms and theorems established above. (Keep in mind that all theorems containing expressions of the form '$P(A \mid B)$' hold only when $P(B) > 0$.)

(1) If $P(A) = 0$, then $P(A \vee B) = P(B)$.

(2) $P(A \vee B) = P(A) + P(\sim A \& B)$.

(3) If A and B are mutually exclusive, then $P(A \& B) = 0$.

(4) If $\sim A$ and $\sim B$ are mutually exclusive, then $P(A \& B) = P(A) + P(B) - 1$.

(5) If $P(A) = P(B) = 1$, then $P(A \& B) = 1$.

(6) If A is a truth-functional consequence of B, then $P(A \vee B) = P(A)$.

(7) If A is a truth-functional consequence of B, then $P(A) \leqslant P(B)$.

(8) $0 \leqslant P(A \mid B) \leqslant 1$.

(9) $P(A \mid B) = 0$ if and only if $P(B \mid A) = 0$.

(10) If A is tautologous, $P(A \mid B) = 1$.

(11) If A is contradictory, $P(A \mid B) = 0$.

(12) If A and B are mutually exclusive, $P(A \lor B \mid C) = P(A \mid C) + P(B \mid C)$.

(13) If A is a truth-functional consequence of B, then $P(A \mid B) = 1$.

(14) If B is a truth-functional consequence of C, then $P(A \& B \mid C) = P(A \mid C)$.

(15) If A is a truth-functional consequence of $B \& C$, then $P(A \& B \mid C) = P(B \mid C)$.

(16) $P(A \mid B) = P(A \& C \mid B) + P(A \& \sim C \mid B)$.

(17) If $P(A) > 0$ and $P(B) > 0$ and A and B are mutually exclusive, then A and B are not independent.

(18) $P(A \mid \sim B) = \dfrac{P(A) - P(A \& B)}{1 - P(B)}$

(19) $P(A \mid B \lor C) = \dfrac{P(A \& B) + P(A \& C) + P(A \& B \& C)}{P(B) + P(C) - P(B \& C)}$

(20) $P(A \mid B \& C) = \dfrac{P(A \& B \mid C)}{P(B \mid C)}$

II Calculate and compare the values for the following pairs of expressions:

(1) $P(A \mid A)$, $P(A \rightarrow A)$.

(2) $P(\sim A \mid A)$, $P(A \rightarrow \sim A)$.

(3) $P(A \mid B \& \sim B)$, $P((B \& \sim B) \rightarrow A)$.

(4) $P(A_1 \mid A_2)$, $P(A_2 \rightarrow A_1)$, where A_1 is the event of rolling a one and A_2 is the event of rolling a two on a single toss of a die (use the classical definition of probability).

III (1) Using the data given in Problem 10.26, calculate the probability that Alma's girdle was produced by machine 2, given that it is defective, and the probability that it was produced by machine 3, given that it is defective.

(2) Consider the following game. There are two boxes containing money: box 1 contains three $1 bills and three fifties; box 2 contains thirty $1 bills and one fifty. A blindfolded player is allowed to make a random draw of one bill from one box, but the choice of box is determined by the toss of a fair coin; if the result is heads, the person must draw from box 1; if it is tails, the draw is made from box 2. The coin is tossed and a player makes a draw. Use Bayes' theorem to calculate the probability that the draw was made from box 1, given that the player draws a $50 bill.

(3) Let Q be the hypothesis that some version of the quark theory of subatomic physics is true. Let N be the hypothesis that no proton decay is observed over a period of a year in a certain quantity of proton-rich liquid. Suppose that $P(N \mid Q) = .001$ and $P(N \mid \sim Q) = .99$. Suppose further that we estimate the prior probabilities of Q and $\sim Q$ as $P(Q) = .7$, $P(\sim Q) = .3$. Using Bayes' theorem, calculate $P(Q \mid N)$. Suppose $P(Q) = .3$, $P(\sim Q) = .7$. How strongly does this change in the priors affect our confidence in Q, given N?

IV Two fair dice are tossed once. Using the probability calculus and/or the classical definition of probability, calculate the following. (The abbreviations used are those given just before Problem 10.31.)

(1) $P(A_1 \mid A_7)$

(2) $P(A_7 \mid A_1)$

(3) $P(A_1 \mid A_2)$

(4) $P(A_1 \& A_2)$

(5) $P(A_1 \mid A_1 \lor A_2 \lor A_3)$

(6) $P(A_1 \lor A_2 \mid A_7 \lor A_8)$

    (7) $P(A_3 \lor \sim A_3)$

    (8) $P(A_1 \& \sim A_1)$

    (9) $P(\sim A_1 \& \sim A_2)$

   (10) $P(A_2 \mid A_1 \& \sim A_1)$

V Five cards are dealt at random from a deck of 52. Calculate the probabilities of the following:

    (1)   The first four cards dealt are aces.

    (2)   The fifth card dealt is an ace, given that none of the first four was.

    (3)   The fifth card dealt is a heart, given that the first four were also hearts.

    (4)   The fifth card dealt is not a heart, given that the first four were hearts.

    (5)   At least one of the first two cards dealt is an ace.

    (6)   The ten, jack, queen, king, and ace of hearts are dealt, in that order.

    (7)   Neither of the first two cards dealt is a king.

    (8)   The first card dealt is an ace, and the second one is not an ace.

    (9)   None of the five cards is a club.

   (10) The fifth card deal is an ace or a king, given that four have been dealt, two of which are aces and two of which are kings.

VI A fair coin is tossed three times. The outcomes of the tosses are independent. Find the probabilities of the following:

    (1)   All three tosses are heads.

    (2)   None of the tosses is heads.

    (3)   The first toss is heads and the second and third are tails.

    (4)   One of the tosses is heads; the other two are tails.

    (5)   At least one of the tosses is heads.

    (6)   The first toss is heads, given that all three tosses are heads.

    (7)   The second toss is heads, given that the first is heads.

    (8)   The first toss is heads, given that the second is heads.

    (9)   The first toss is heads, given that exactly one of the three is heads.

   (10) Exactly two heads occur, given that the first toss is heads.

Answers to Selected Supplementary Problems

I   (5)   Suppose $P(A) = P(B) = 1$. Then, by Problem 10.11, $P(A \& B) = P(B) = 1$.

   (10) Suppose A is tautologous. Then, by AX2, $P(A) = 1$; and by Problem 10.11, $P(A \& B) = P(B)$. But by D1,

$$P(A \mid B) = \frac{P(A \& B)}{P(B)} = \frac{P(B)}{P(B)} = 1$$

   (15) Suppose A is a truth-functional consequence of $B \& C$. Then clearly $(A \& B) \& C$ is a truth-functional consequence of $B \& C$. Moreover, $B \& C$ is a truth-functional consequence of

$(A \& B) \& C$, so that $(A \& B) \& C$ and $B \& C$ are truth-functionally equivalent. Thus, by Problem 10.6, $P((A \& B) \& C) = P(B \& C)$. Now, by D1,

$$P(A \& B \mid C) = \frac{P((A \& B) \& C)}{P(C)} = \frac{P(B \& C)}{P(C)}$$

which, again by D1, is just $P(B \mid C)$.

(20)　By D1 and Problem 10.6,

$$P(A \mid B \& C) = \frac{P(A \& (B \& C))}{P(B \& C)} = \frac{P((A \& B) \& C)/P(C)}{P(B \& C)/P(C)} = \frac{P(A \& B \mid C)}{P(B \mid C)}$$

III　(3)　Q and $\sim Q$ are mutually exclusive and exhaustive. Therefore, by Bayes' theorem,

$$P(Q \mid N) = \frac{P(Q) \cdot P(N \mid Q)}{P(Q) \cdot P(N \mid Q) + P(\sim Q) \cdot P(N \mid \sim Q)}$$

$$= \frac{.7 \cdot .001}{(.7 \cdot .001) + (.3 \cdot .99)} = .00235$$

For $P(Q) = .3$, $P(\sim Q) = .7$, the answer is .00043. The change in the prior probabilities is drastic, but the change in $P(Q \mid N)$ is rather small; Q is still very unlikely, given the observations N. Within a wide range of values for the priors, Q remains very unlikely, given N.

IV　(5)　Note first that A_1 is truth-functionally equivalent to $A_1 \& (A_1 \lor A_2 \lor A_3)$. So, by D1 and Problem 10.6,

$$P(A_1 \mid A_1 \lor A_2 \lor A_3) = \frac{P(A_1 \& (A_1 \lor A_2 \lor A_3))}{P(A_1 \lor A_2 \lor A_3)} = \frac{P(A_1)}{P(A_1 \lor A_2 \lor A_3)}$$

Since $A_1 \lor A_2$ and A_3 are mutually exclusive, as are A_1 and A_2, we have by two applications of AX3:

$$P(A_1 \lor A_2 \lor A_3) = P(A_1) + P(A_2) + P(A_3)$$

But the probabilities of A_1, A_2, and A_3 are each $\frac{1}{6}$. Thus the desired value is:

$$\frac{\frac{1}{6}}{\frac{1}{6} + \frac{1}{6} + \frac{1}{6}} = \frac{1}{3}$$

(10)　This probability is undefined, since $P(A_1 \& \sim A_1) = 0$.

V　(5)　Using 'A_1' for 'The first card is an ace' and 'A_2' for 'The second card is an ace', the probability we seek is $P(A_1 \lor A_2)$, which is, by Problem 10.6,

$$P((A_1 \& A_2) \lor (A_1 \& \sim A_2) \lor (\sim A_1 \& A_2))$$

Since the disjuncts here are mutually exclusive, two applications of AX3, give simply

$$P(A_1 \& A_2) + P(A_1 \& \sim A_2) + P(\sim A_1 \& A_2)$$

which by Problem 10.19 becomes:

$$P(A_1) \cdot P(A_2 \mid A_1) + P(A_1) \cdot P(\sim A_2 \mid A_1) + P(\sim A_1) \cdot P(A_2 \mid \sim A_1)$$

i.e.,

$$\frac{4}{52} \cdot \frac{3}{51} + \frac{4}{52} \cdot \frac{48}{51} + \frac{48}{52} \cdot \frac{4}{51} = \frac{12}{2652} + \frac{192}{2652} + \frac{192}{2652} = \frac{396}{2652} = \frac{33}{221}$$

(10) If two aces and two kings have been dealt, there are two aces and two kings among the 48 remaining cards. That is, there are four chances out of 48 to draw an ace or a king. The probability of drawing an ace or a king, given that two aces and two kings have been dealt, is thus $\frac{4}{48} = \frac{1}{12}$.

VI (5) Let the probability of getting a head on the nth toss be H_n. We wish to determine $P(H_1 \vee H_2 \vee H_3)$. By Problem 10.3, this is equal to $1 - P(\sim(H_1 \vee H_2 \vee H_3))$, which, since $\sim(H_1 \vee H_2 \vee H_3)$ is truth-functionally equivalent to $\sim H_1 \& \sim H_2 \& \sim H_3$, is just:

$$1 - P(\sim H_1 \& \sim H_2 \& \sim H_3)$$

Because the tosses are independent, two applications of Problem 10.22 to $P(\sim H_1 \& \sim H_2 \& \sim H_3)$ give:

$$P(\sim H_1) \cdot P(\sim H_2) \cdot P(\sim H_3) = \tfrac{1}{2} \cdot \tfrac{1}{2} \cdot \tfrac{1}{2} = \tfrac{1}{8}$$

Thus $P(H_1 \vee H_2 \vee H_3) = 1 - \tfrac{1}{8} = \tfrac{7}{8}$.

(10) Using the notation of the answer to Supplementary Problem VI(5), the probability we wish to determine is $P((H_2 \& \sim H_3) \vee (\sim H_2 \& H_3) \mid H_1)$. Since the tosses are independent, this equals $P((H_2 \& \sim H_3) \vee (\sim H_2 \& H_3))$. The disjuncts here are mutually exclusive; so, by AX3, this becomes

$$P(H_2 \& \sim H_3) + P(\sim H_2 \& H_3)$$

and since the tosses are independent, this is, by Problem 10.22,

$$P(H_2) \cdot P(\sim H_3) + P(\sim H_2) \cdot P(H_3) = \tfrac{1}{2} \cdot \tfrac{1}{2} + \tfrac{1}{2} \cdot \tfrac{1}{2} = \tfrac{1}{2} .$$

Chapter 11

Further Developments in Formal Logic

In this chapter we introduce a smattering of topics to illustrate some of the directions in which the study of formal logic may be pursued. Readers who wish to explore any of these topics in detail should consult the works mentioned in the footnotes at various points in the chapter.

11.1 EXPRESSIVE LIMITATIONS OF PREDICATE LOGIC

The system of predicate logic presented in Chapter 7 is only the beginning of formal deductive logic. Though complete in one sense, it is incomplete in another. As noted at the beginning of Chapter 7, it is complete in the sense that its inference rules generate proofs for all argument forms valid in virtue of the semantics of the identity predicate, the truth-functional operators, and the kinds of quantifiers expressible within the system.[1] This is known as *semantic completeness*. But it is incomplete in the sense that there are valid argument forms whose validity depends on the semantics of other sorts of expressions. Predicate logic does not generate proofs for these forms.

SOLVED PROBLEM

11.1 Consider the following form, where 'T' is interpreted as "is taller than" and 'a', 'b', and 'c' are names of individuals:

$Tab, Tbc \vdash Tac$

From the assumptions that a is taller than b and b is taller than c, it plainly follows that a is taller than c. Yet the predicate calculus counts this form as invalid. If we construct a truth tree with the premises and the negation of the conclusion, we get simply this:

Tab

Tbc

$\sim Tac$

No further moves are possible, and the single path does not close, indicating that the inference is invalid. What has gone wrong?

Solution

Just as the truth tables or truth trees of propositional logic detect validity only insofar as it is due to the semantics of the truth-functional operators, so the methods of predicate logic detect validity only insofar as it is due to the semantics of these operators, the quantifiers, and the identity predicate. But since none of these expressions occurs in this inference, its validity is due not to their semantics, but to the semantics of the predicate 'is taller than'.

The fact is that the meanings of the predicate letters of predicate logic vary from problem to problem; unlike quantifiers, truth-functional operators, and the identity predicate, they do not have fixed meanings. Consequently, predicate logic provides no rules to account for the semantics of specific predicates—with the sole exception of the identity predicate. It is therefore insensitive to validity generated by their distinctive semantics.

---

[1]More general kinds of quantifiers will be discussed in the next section.

We could, however, decide to give the predicate letter 'T' the same logical status given to the symbol '$=$' in Chapter 6. That is, we could stipulate for this letter a fixed interpretation, so that it always means "is taller than". And we could adopt a new set of inference rules to go along with this interpretation as we did in Chapter 7 for '$=$'. The result would be an expanded version of predicate logic, which we might call *T logic*. A plausible deductive version of T logic (the *T calculus*) could be obtained by adding the following two rules to the predicate calculus with identity:

Rule T1:   We may introduce the following formula at any line of a proof:

$$\forall x \forall y \forall z ((Txy \mathbin{\&} Tyz) \rightarrow Txz)$$

Rule T2:   We may introduce the following formula at any line of a proof:

$$\forall x \forall y (Txy \rightarrow {\sim} Tyx)$$

Rule T1 is a general rule to justify the sort of reasoning we recognized as valid in Problem 11.1. It says that the relation 'is taller than' is *transitive*; that is, if one thing is taller than a second and the second is taller than a third, then the first thing is taller than the third. Rule T2 says that this relation is *asymmetric*; if one thing is taller than a second, then the second is not taller than the first. Both rules permit us to introduce obvious general truths involving the predicate 'T' at any line of a proof. Like statements introduced by the rule $=$I, these general truths do not count as assumptions. Statements which function this way in proofs are called *axioms*. We have encountered axioms before. The rule $=$I allows us to introduce axioms of the form '$a = a$' at any line of a proof. And we examined some axioms for the probability calculus in a semiformal way in Chapter 10. T logic is a purely formal system with two axioms. In T logic we can construct proofs for a variety of new argument forms and theorems.

SOLVED PROBLEM

11.2   Construct a proof for the form of Problem 11.1 in T logic.

Solution

1	Tab	A
2	Tbc	A
3	$\forall x \forall y \forall z ((Txy \mathbin{\&} Tyz) \rightarrow Txz)$	T1
4	$\forall y \forall z ((Tay \mathbin{\&} Tyz) \rightarrow Taz)$	3 $\forall$E
5	$\forall z ((Tab \mathbin{\&} Tbz) \rightarrow Taz)$	5 $\forall$E
6	$(Tab \mathbin{\&} Tbc) \rightarrow Tac$	5 $\forall$E
7	$Tab \mathbin{\&} Tbc$	1, 2 $\&$I
8	Tac	6, 7 $\rightarrow$E

In view of this proof, one may object that the argument form of Problem 11.1 was not invalid from the point of view of predicate logic, but only incomplete, requiring for its completion the suppressed assumption '$\forall x \forall y \forall z ((Txy \mathbin{\&} Tyz) \rightarrow Txz)$'. But that objection would miss the point. Although adding this assumption does produce a form provable in predicate logic, it also changes the example. The form we wished to consider in Problem 11.1 was precisely the form given there; and *that* form is valid (its conclusion cannot be false if its premises are true), even though it is not provable in standard predicate logic. The objection, however, is not wholly incorrect, since the suggested assumption or something that implies it must be incorporated into any logical system in which this form is provable. The point is that in such a system it functions not as an assumption (whether suppressed or explicit), but as a truth of the logic itself.

T logic has other desirable consequences. For example, it seems to be a logical or conceptual truth that the relation 'is taller than' is *irreflexive*, i.e., that nothing is taller than itself. Yet this is not a theorem of the predicate calculus. It is, however, a theorem of T logic:

SOLVED PROBLEM

11.3   Prove the following theorem in T logic:

$$\forall x \sim Txx$$

Solution

1	Taa	H (for $\sim$I)
2	$\forall x \forall y (Txy \rightarrow \sim Tyx)$	T2
3	$\forall y (Tay \rightarrow \sim Tya)$	2 $\forall$E
4	$Taa \rightarrow \sim Taa$	3 $\forall$E
5	$\sim Taa$	1, 4 $\rightarrow$E
6	$Taa \,\&\, \sim Taa$	1, 5 &I
7	$\sim Taa$	1–6 $\sim$I
8	$\forall x \sim Txx$	7 $\forall$I

There are, of course, a vast number of predicates besides 'is taller than' that give rise to new valid forms and new theorems. We could go on expanding the predicate calculus by adding axioms for new predicates almost indefinitely. One field into which the predicate calculus has been developed by the addition of a new predicate is *set theory*. Set theory is the study of *sets*, collections of objects considered without regard to order or description. It is considered by many to be the foundation of mathematics, since virtually all mathematical theorems can be proved from its axioms. A formalized version of set theory may be obtained from the predicate calculus with identity by adding several axioms for the predicate 'is a member of', which is usually represented by the symbol '∈' or by the Greek letter epsilon: 'ε'. All other notions of set theory, and indeed virtually all the notions of mathematics, can be defined in terms of this predicate, but that is a story too long to tell here.[2]

There are other ways of expanding the predicate calculus besides fixing the interpretations of certain predicates and adopting new axioms or inference rules for them. One way, discussed in the next section, is to add new kinds of variables. Another is to add new expressions belonging to wholly different grammatical categories. In Sections 11.3, 11.4, and 11.7 we discuss two new categories of expressions: function symbols and non-truth-functional propositional operators.

11.2   HIGHER-ORDER LOGICS

The quantifiers of predicate logic as described in Chapters 6 and 7 bind variables standing for individuals; when instantiating these variables, we replace them by names. But it is also possible to have quantifiers binding variables that stand for properties or relations; when these are instantiated, they are replaced by predicates or open formulas. To see how such quantifiers could be useful, consider the following problem:

SOLVED PROBLEM

11.4   Using 'M' for "is Mexican," 'a' for "Anita," and 'c' for "Carlos," formalize the following argument:

Carlos and Anita are both Mexican.

∴ Carlos and Anita have something in common.

---

[2]For an introduction to set theory and the foundations of mathematics see Y. N. Moschovakis, *Notes on Set Theory*, New York, Springer-Verlag, 1994. See also *Schaum's Outline of Set Theory*, by Seymour Lipschutz.

Solution

Formalization of the premise is easy:

$$Mc \ \& \ Ma$$

But the conclusion presents problems. We could introduce the three-place predicate 'has in common with', so that '$Hxyz$', for example, would mean "x has y in common with z," and then write the conclusion as:

$$\exists x Hcxa$$

But the result is a form which is not provable in predicate logic, and the argument seems obviously valid. Fortunately, there is a better solution. The conclusion says that the two individuals have *something* in common. The term 'something' indicates existential quantification. Now the relevant thing they have in common is, of course, the property of being Mexican, and this property is expressed by the predicate letter 'M'. This suggests that the quantifier applies in this case not to individuals, but to a property. What is replaced by the variable it binds is a predicate, not a name. Following this suggestion, we can formalize the argument as follows:

$$Mc \ \& \ Ma \vdash \exists P(Pc \ \& \ Pa)$$

The conclusion says, "There exists a property P, such that Carlos has P and Anita has P." We use a capital 'P' rather than lowercase, to indicate that this is a special kind of variable which ranges over properties, not over individuals.

Now the conclusion of this argument is not a wff by the formation rules of Chapter 6, but it is a wff of a generalization of predicate logic called *second-order logic*. And, having formalized the argument this way, we see at once that it is valid by a generalized form of existential introduction.

The name 'second-order logic' originates with a conception of logic as a hierarchical structure describing individuals, properties, properties of properties, and so on. Any language whose variables range only over individuals (people, trees, electrons, nations, or planets, for example) is a *first-order language*. Thus the system of logic presented in Chapter 6 and Chapter 7 is often called the *first-order predicate logic*, because its variables stand only for individuals. Now individuals have a variety of properties (such as being Mexican, being human, and being green), and any language whose variables range just over individuals and their properties is called a *second-order language*. Moreover, properties themselves may have properties; the property of being Mexican, for example, has the property of being shared by more than two people. So there are also *third-order languages* (whose variables range over individuals, their properties, and properties of their properties). Indeed, there are order n languages for any integer n. An entire array of such languages is called a *type system* or *theory of types*.

Theories of types are syntactically complex, since a different style of variables is used for each different level of objects. We shall not consider a complete type theory here, but shall instead discuss only some aspects of second-order logic. Moreover, although it is possible to have second-order quantifiers over relations (so that the variables replace n-place predicates for $n > 2$), we shall consider only quantification over properties (so that the variables are replaceable only by one-place predicates or open formulas on a single variable).

SOLVED PROBLEM

11.5 Formalize the following statements in second-order logic, using 'a' for "Albert" and 'b' for "Bob":

(*a*) There is at least one property which Albert has and at least one which he lacks.

(*b*) For any object and any property, either that object has that property or it does not.

(*c*)   All objects are such that there is no property which they both have and lack.

(*d*)   Albert and Bob have exactly the same properties.

(*e*)   For all objects x and y, x and y have exactly the same properties if and only if they are identical.

Solution

(*a*) $\exists P Pa \,\&\, \exists P \sim Pa$

(*b*) $\forall x \forall P (Px \lor \sim Px)$

(*c*) $\forall x \sim \exists P (Px \,\&\, \sim Px)$

(*d*) $\forall P (Pa \leftrightarrow Pb)$

(*e*) $\forall x \forall y (\forall P (Px \leftrightarrow Py) \leftrightarrow x = y)$

The syntax of second-order logic, as Problems 11.4 and 11.5 suggest, is just that of predicate logic, with the addition of quantified variables over predicates. (We may reserve some letters of the alphabet to represent predicate variables, say '*P*' through '*Z*'.) Thus, to the four formation rules of Section 6.3, we need add only this:

> If ϕ is a wff containing a one-place predicate Π, then any formula of the form $\forall \Psi \phi^\Psi_\Pi$ or $\exists \Psi \phi^\Psi_\Pi$ is a wff, where ϕ^Ψ_Π is the result of replacing one or more of the occurrences of Π in ϕ by some predicate variable Ψ not already in ϕ.

The semantics of this language is like that of predicate logic (see Section 6.4), except of course that the truth conditions of wffs involve second-order quantification. The simplest way to formulate these conditions is patterned after the treatment of first-order quantification. If M is a model and Π a predicate letter, define a Π-*variant* of M to be any model that results from M by freely interpreting Π as a subset of the universe of M. Thus, if M does not assign any interpretation to Π, a Π-variant will extend M by providing an interpretation for the additional symbol, Π; otherwise it will simply represent one alternative way of interpreting Π on the universe of M, keeping everything else exactly as in M. (Every model that is already defined for Π counts as a Π-variant of itself.) The truth conditions for quantified sentences can then be defined as follows:

(1)   A universal quantification $\forall \Psi \phi$ is true in a model M if the wff ϕ^Π_Ψ is true in every Π-variant of M, where Π is the first predicate letter in the alphabetic order not occurring in ϕ and ϕ^Π_Ψ is the result of replacing all occurrences of Ψ in ϕ by Π; if ϕ^Π_Ψ is false in some Π-variant of M, then $\forall \Psi \phi$ is false in M.

(2)   An existential quantification $\exists \Psi \phi$ is true in M if the wff ϕ^Π_Ψ is true in some Π-variant of M, where Π and ϕ^Π_Ψ are as in (1); if ϕ^Π_Ψ is false in every Π-variant of M, then $\exists \Psi \phi$ is false in M.

SOLVED PROBLEMS

11.6   Evaluate the following wff in the model M given below:

$\exists P (Pa \,\&\, \sim Pb)$

Universe: the class of all people

a : Bill Clinton

b : Barbra Streisand

Solution

With reference to condition (2), here the wff ϕ is '$(Pa \,\&\, \sim Pb)$', the variable Ψ is '*P*', the predicate letter Π is '*A*', and the formula ϕ^Π_Ψ is '$(Aa \,\&\, \sim Ab)$'. It is easy to see that this sentence is true in any model where '*A*' is interpreted as a class containing Clinton but not Barbra Streisand—e.g., the class of all male people. So the sentence is true by condition (2).

11.7   Evaluate the following wff in the same model of Problem 11.6:

$$\exists x \forall P(Pa \rightarrow Px)$$

Solution

Here we must see whether there exists any b-variant of M in which the wff '$\forall P(Pa \rightarrow Pb)$' is true (see conditon (5) of Section 6.4). Consider the b-variant M' which interprets 'b' as Bill Clinton, i.e., as the same object designated by 'a'. Then every A-variant of M' is sure to make the wff '$Aa \rightarrow Ab$' true. Thus, '$\forall P(Pa \rightarrow Pb)$' is indeed true in M' (by condition (1) above). Therefore '$\exists x \forall P(Pa \rightarrow Px)$' is true in M.

As for the notion of a proof, second-order logic includes all the inference rules of the predicate calculus, but the four quantifier rules of Sections 7.2 and 7.3 are generalized to permit $\forall$E, $\forall$I, $\exists$E, and $\exists$I for quantifiers binding predicate variables. The details are fairly straightforward, and we shall not bother to state them precisely here. We must note, however, that in the application of these inference rules, not only one-place predicates but also open formulas on a single variable are allowed as replacements for predicate variables. This is because, like one-place predicates, open formulas on a single variable represent properties, though these properties are logically complex. Problem 11.9 will illustrate the use of a quantifier rule with an open formula.

SOLVED PROBLEMS

11.8   Prove that formula (b) of the solution to Problem 11.5 is a theorem of second-order logic.

Solution

1	$Fa \lor \sim Fa$	TI 4.44
2	$\forall P(Pa \lor \sim Pa)$	1 $\forall$I
3	$\exists x \forall P(Px \,\&\, \sim Px)$	2 $\exists$I

11.9   Prove that it is a theorem of second-order logic that there is some property which every object has.

Solution

The theorem to be proved is:

$$\vdash \exists P \forall x Px$$

The immediate problem is to decide what sort of property 'P' might represent. One property which every object has is the logically complex property of being a fish or not a fish, which is representable by the open formula:

$$Fx \lor \sim Fx$$

As noted above, the second-order quantifier rules allow us to replace such open formulas with predicate variables. Thus we may reason as follows:

1	$Fa \lor \sim Fa$	TI 3.45
2	$\forall x(Fx \lor \sim Fx)$	1 $\forall$I
3	$\exists P \forall x Px$	2 $\exists$I

In step 3, we replace the open formula '$(Fx \lor \sim Fx)$' by the expression 'Px' and add an existential quantifier on 'P'. This corresponds to saying that since any x has the complex property of being a fish or not a fish, then there is some property which any x has.

Some philosophers and logicians have felt that there is something "fishy" about complex properties of the sort employed in this proof. Thus it is not universally agreed that the proof is valid. It has been argued on philosophical grounds that whereas the property of being a fish is indeed a property, there is no such thing as the logically complex property of being a fish or not being a fish. Adherents of this point of view would ban instantiation of second-order variables by open sentences and allow instantiation only by predicate letters. This point of view, however, trivializes second-order logic and will not be considered here.

If we want to have the identity predicate in second-order logic, we need at least one new inference rule to take account of how it interacts with second-order quantification. One way to do this is to take formula (e) of Problem 11.7, which is known as *Leibniz' law*, as an axiom or as a definition of identity. The rules =E and =I then become redundant (in the sense that anything provable by means of these rules may be proved from the axiom alone) and may be dropped.

SOLVED PROBLEM

11.10 Using formula (e) of Problem 11.5 as an axiom, prove in second-order logic the theorem:

$$\vdash \forall x\, x = x$$

Solution

1	$\forall x \forall y (\forall P(Px \leftrightarrow Py) \leftrightarrow x = y)$	Axiom
2	$\forall y (\forall P(Pa \leftrightarrow Py) \leftrightarrow a = y)$	1 $\forall$E
3	$\forall P(Pa \leftrightarrow Pa) \leftrightarrow a = a$	2 $\forall$E
4	$Fa \rightarrow Fa$	TI Supp. Prob. IV (1), Chap. 4
5	$Fa \leftrightarrow Fa$	4, 4 $\leftrightarrow$I
6	$\forall P(Pa \leftrightarrow Pa)$	5 $\forall$I
7	$\forall P(Pa \leftrightarrow Pa) \rightarrow a = a$	3 $\leftrightarrow$E
8	$a = a$	6, 7 $\rightarrow$E
9	$\forall x\, x = x$	8 $\forall$I

Are the four quantifier rules plus Leibniz' law a complete set of rules for second-order logic with identity (in the sense that they generate all forms whose validity is a function of the semantics of the logical symbols of our second-order language)? In part, the answer to this question depends on our concept of a property, for which forms count as valid depends on what sorts of properties there are. If we identify properties with sets, as our semantic rules indicate, then it can be shown that our second-order rules are not complete and that it is impossible in principle to state a complete set of rules.[3] Some authors have developed alternative semantics for second-order logic, where properties are not identified with sets. These matters, however, are beyond the scope of this book.

One last point is worth making. Some properties seem to be properties of themselves. For example, the property of being nonhuman is itself nonhuman. We call such properties *self-predicable*. Others are clearly not properties of themselves; the property of being a person, for example, is not itself a person. Such properties are said to be *non-self-predicable*. It would seem natural to express the statement that the property of being nonhuman is nonhuman by a formula of the form '*NN*', where '*N*' designates this property, and similarly to express the statement that the property of being a person is not a person by the formula '~*PP*'. This, however, is prohibited by the formation rules. The predicates of second-order logic may be attached only to names of individuals, not to other predicates. Historically, the reason for this restriction was the appearance of certain *antinomies* (unexpected inconsistencies) produced when predicates are permitted to apply to themselves without restriction.

---

[3]The semantic incompleteness of second-order logic is established by the argument of Gödel's theorem (see the end of Section 11.4). For an elegant presentation, see Richard Jeffrey, *Formal Logic: Its Scope and Limits*, 2d edn, New York, McGraw-Hill, 1981, Chap. 7.

The simplest of these was first noticed by Bertrand Russell in 1901 and later reformulated in semantic terms by K. Grelling. One version of it concerns the property of being non-self-predicable: is it self-predicable or not? Suppose it is not. Then the property of being non-self-predicable is itself non-self-predicable, and so it is self-predicable. That contradicts our hypothesis, so the property of being non-self-predicable must be self-predicable. But then it does not have itself as a property, and so it *is* non-self-predicable! If we allow ourselves even to consider this property, we are driven to contradiction. (For a formalization of this argument in first-order logic, see Chapter 7, answer to Supplementary Problem (25).)

To avoid such contradictions, Russell developed the notion of a type hierarchy (mentioned above) in which predicates of higher orders apply only to predicates or objects of lower orders; no predicate may be applied to itself. Second-order logic constitutes the first two levels of such a hierarchy.

11.3 PREDICATE LOGIC WITH FUNCTION SYMBOLS

We can expand the predicate calculus by adding other new categories of symbols. Function symbols (also known as *operation symbols*) are a good example. A *function symbol* is a symbol which, when attached to one or more names or other denoting terms, produces an expression which denotes a single object and functions grammatically as a new name. Thus function symbols differ from predicates, which when applied to one or more names produce sentences. One English expression which may be treated as a function symbol (if the subject under discussion is people) is 'the father of' (understood in the biological sense). Applying this English phrase to a name produces a new English phrase which denotes a single new individual. Thus, applying 'the father of' to the name 'Isaac ', we get the phrase 'the father of Isaac', which designates the individual Abraham (assuming that the biblical Isaac is intended). This new phrase, which we shall call a *functional expression*, acts grammatically as a name; that is, it may occur in any grammatical context in which a name is appropriate. We may, for example, attach it to a one-place predicate to create a sentence, as in 'The father of Isaac was a prophet'.

We now consider how this sentence might be expressed in an extension of the predicate calculus. We shall use the lowercase letters 'f', 'g', and 'h' for function symbols. The names or other functional expressions to which they are applied are written after them and enclosed in brackets. Thus, using 'P' for 'was a prophet', 'i' for 'Isaac', and 'f' for 'the father of', we may write 'the father of Isaac' as '$f(i)$' and 'The father of Isaac was a prophet' as '$Pf(i)$'.

Function symbols may be applied repeatedly. For example, to say that Isaac's paternal grandfather (the father of Isaac's father) was a prophet, we can write '$Pf(f(i))$'. Moreover, the names contained in function symbols (as well as the functional expressions themselves) are accessible to quantification. Thus we may write 'Someone's paternal grandfather was a prophet' as '$\exists x P(f(f(x)))$'.

An expression may be treated as a functional symbol only if each expression formed by applying it to a name denotes a unique individual. Thus 'the son of' may not be treated as a function symbol, since when applied to the name of anyone who has more than one son, it denotes not one individual, but several, and it denotes nobody when applied to the name of anyone who has no sons. By contrast, 'the father of' denotes a unique individual when applied to the name of any person, since each person has one and only one (biological) father.

Notice, however, that if 'a' is the name of an inanimate object, then also the expression 'the father of a' denotes nothing. That is why we were careful to restrict our comments above to discussions of people. When using function symbols it is important to specify the *domain*, i.e., the set of individuals to which the function symbol applies. 'The father of' may not be treated as a function symbol for domains which include fatherless objects.[4]

Semantically, function symbols may be thought of as denoting operations called *functions*. An *n-place function* is an operation which assigns a unique object to each *n-tuple* (i.e., linear ordering, or

---

[4]This is the case so long as we require each functional expression to denote something. There are logics of *partial functions* which dispense with this requirement, but we will not consider these logics here.

list, of n objects) of members of the domain. The father-of function, for example, is a one-place (monadic) function which assigns to each person a single individual who is that person's father. The addition function of arithmetic is a two-place (binary) function which assigns to each 2-tuple (ordered pair) of numbers a third number which is their sum. A member of an n-tuple to which the function is applied is called an *argument* of the function. (Be careful not to confuse this meaning of 'argument' with the idea of a set of premises and their conclusion.) The object assigned by an n-place function to a given n-tuple of arguments is called the *value* of the function for that n-tuple. For example, the value of the father-of function for the argument Isaac is Abraham. The set of all values for a given domain is the *range* of the function.

SOLVED PROBLEM

11.11   Using 'f' for 'the father of', 'm' for 'the mother of', 'a' for 'Abel', 'c' for 'Cain', and 'S' for 'is a sinner', formalize each of the following English sentences in the language of the predicate calculus with identity and function symbols. (Assume a domain of people.)

(*a*)   Cain's father is a sinner.

(*b*)   If a person's father is a sinner, then that person is a sinner.

(*c*)   If a person's paternal grandfather is a sinner, then that person is a sinner.

(*d*)   Abel's mother and father are sinners.

(*e*)   Abel's father is also Cain's father.

(*f*)   Abel's father is someone else's father.

Solution

(*a*) $Sf(c)$

(*b*) $\forall x(Sf(x) \rightarrow Sx)$

(*c*) $\forall x(Sf(f(x)) \rightarrow Sx)$

(*d*) $Sm(a) \mathbin{\&} Sf(a)$

(*e*) $f(a) = f(c)$

(*f*) $\exists y(f(a) = f(y) \mathbin{\&} \sim y = a)$

As parts (*e*) and (*f*) of Problem 11.11 suggest, two different individuals may have the same father. Thus a function may take the same value for two distinct arguments. The father of Abel is the same person as the father of Cain—namely, Adam. Still, for a given argument, the father-of function has one and only one value.

The father-of function takes only one argument at a time. It is thus a *one-place*, or *monadic*, function. But, as noted above, there are also functions which take two or more arguments at a time. Consider, for example, the domain of points on a flat (Euclidean) plane. For each pair of points, there is exactly one point which lies midway between them. (For a pair consisting of a point and itself, the midpoint is that point.) Thus the expression 'the midpoint between' denotes a function which takes two arguments, i.e., a *two-place*, or *binary*, function on this domain. Applied to a pair of points, this function gives a point as its value.

SOLVED PROBLEM

11.12   Using 'P' for 'is a point' and 'm' for 'the midpoint between', formalize the following statements:

(*a*)   The midpoint between a point and itself is that point.

(b)   For any points x and y, the midpoint between x and y is identical to the midpoint between y and x.

(c)   Any point is the midpoint between some pair of points.

(d)   For any points x and y, there is some point z such that y is the midpoint between x and z.

(e)   For any points x and y, the midpoint between x and the midpoint between y and y is the midpoint between x and y.

Solution

When a function has more than one argument place, we write the names of the arguments in brackets after the function symbol and separate them with commas; thus:

(a) $\forall x \, m(x, x) = x$

(b) $\forall x \forall y \, m(x, y) = m(y, x)$

(c) $\forall x \exists y \exists z \, x = m(y, z)$

(d) $\forall x \forall y \exists z \, y = m(x, z)$

(e) $\forall x \forall y \, m(x, m(y, y)) = m(x, y)$

In part (e), the second argument of the first occurrence of 'm' is the value of the function m applied to y and y. Incidentally, all the propositions in this problem are true of points on the Euclidean plane.

To accommodate function symbols in the predicate calculus with identity, we need make only minor modifications in the formation rules and inference rules. The new formation rules allow functional expressions to occur anywhere names can occur, and the rules =I, =E, $\forall$E, and $\exists$I may now be performed with functional expressions in addition to names. (However, if we want any of our function symbols to have fixed interpretations, then we must add special rules or axioms for them. This is illustrated in the next section.)

SOLVED PROBLEMS

11.13 Prove in the predicate calculus with identity and function symbols that statement (c) of Problem 11.11, 'If a person's paternal grandfather is a sinner, then that person is a sinner', may be deduced from statement (b), 'If a person's father is a sinner, then that person is a sinner'. (Use the notation of Problem 11.11.)

Solution

We are to prove the following:

$$\forall x(Sf(x) \to Sx) \vdash \forall x(Sf(f(x)) \to Sx)$$

Here is the proof:

1	$\forall x(Sf(x) \to Sx)$	A
2	$Sf(f(a)) \to Sf(a)$	1 $\forall$E
3	$Sf(a) \to Sa$	1 $\forall$E
4	$Sf(f(a)) \to Sa$	2, 3 HS
5	$\forall x(Sf(f(x)) \to Sx)$	4 $\forall$I

At step 2 we perform $\forall$E by replacing the variable 'x' with the functional expression '$f(a)$'. At step 3 we perform $\forall$E again, using the name 'a'.

11.14 Let 'f' be a one-place function symbol. Prove as a theorem of the predicate calculus with identity and function symbols that for any argument the function f has one and only one value.

Solution

The theorem to be proved is:

$$\vdash \forall x \exists y (y = f(x) \,\&\, \forall z(z = f(x) \to z = y))$$

Here is the proof:

1	$f(a) = f(a)$	$=$I
2	$b = f(a) \to b = f(a)$	TI Supp. Prob. IV(1), Chap. 4
3	$\forall z(z = f(a) \to z = f(a))$	2 $\forall$I
4	$f(a) = f(a) \,\&\, \forall z(z = f(a) \to z = f(a))$	1, 3 &I
5	$\exists y(y = f(a) \,\&\, \forall z(z = f(a) \to z = y))$	4 $\exists$I
6	$\forall x \exists y(y = f(x) \,\&\, \forall z(z = f(x) \to z = y))$	5 $\forall$I

The trick in setting up the proof is to realize that the object y mentioned in the theorem must be $f(x)$ in order to satisfy the condition '$y = f(x)$'. Thus we can see that we need to prove something of the form '$f(a) = f(a) \,\&\, \forall z(z = f(a) \to z = f(a))$', which we do at line 4, in order to apply $\exists$I and $\forall$I to obtain the theorem.

11.4 FORMAL ARITHMETIC

Many of the familiar expressions of arithmetic are function symbols, though they are not all written before the names to which they apply, as in the previous section. (The names used in arithmetic are the numerals, '0', '1', '2', etc.) The symbols '+' and '·', for example, denote the addition and multiplication functions, each of which takes two arguments. These symbols are written, however, in an *infix* position (between the names of their arguments, often with brackets surrounding the result) rather than in the *prefix* position, which is usual in logic. Thus instead of writing 'the sum of 1 and 2' as '+(1, 2)', we write it as '1 + 2' or '(1 + 2)'. By contrast, the function symbol '−', when used to denote the one-place negative function (which applies to a single number) rather than subtraction (which gives the difference between two numbers), is written in the prefix position, though the brackets are often omitted.

In this section we examine six axioms and one inference rule for the arithmetic of the *nonnegative integers* (zero and the positive whole numbers). The resulting formal system generates proofs for many arithmetical truths and enables us to formalize and evaluate a great deal of arithmetical reasoning.

Arithmetic can be regarded as an extension of the predicate calculus with identity. It is obtained by adding one name and three function symbols to the vocabulary and special axioms and/or inference rules for them. The name is '0', which designates the number zero. The first two function symbols are the familiar '+' and '·', which we shall write in the customary infix position with surrounding brackets. The third is 's', which means 'the successor of'. The successor function is a monadic function which, when applied to any integer, gives as its value the next integer. Thus the successor of 0 is 1, the successor of 1 is 2, and so on. When using 's', we shall follow bracketing conventions similar to those used in mathematics for '−'. That is, where 'x' denotes a number, we shall write 'sx' rather than '$s(x)$' for 'the successor of x'. Omitting the brackets makes long formulas more readable.

Numerals other than '0' are treated as abbreviations for functional expressions built up from 's' and '0'. Thus '1' abbreviates '$s0$', '2' abbreviates '$ss0$', and so on. These abbreviations are not officially part of our language, but may be employed in writing formulas to avoid prolixity. As in the predicate calculus with identity, the symbol '=' is a predicate, not a function symbol. Like the identity symbol, the four arithmetical symbols have fixed interpretations. We will adopt six special axioms and one new rule of inference to express these interpretations.[5] If we keep in mind that the intended domain is the nonnegative integers, the axioms are fairly easy to understand:

---

[5]The resulting system is similar to that presented in Elliot Mendelson, *Introduction to Mathematical Logic*, 2d edn, New York, Van Nostrand, 1979, Chap. 3. Mendelson's treatment is excellent and far more detailed than ours.

A1 $\forall x \sim 0 = sx$
A2 $\forall x \forall y (sx = sy \rightarrow x = y)$
A3 $\forall x \, (x + 0) = x$
A4 $\forall x \forall y \, (x + sy) = s(x + y)$
A5 $\forall x \, (x \cdot 0) = 0$
A6 $\forall x \forall y \, (x \cdot sy) = ((x \cdot y) + x)$

Their meanings are as follows (where x and y are any nonnegative integers):

A1   0 is not the successor of any nonnegative integer.
A2   If x and y have the same successor, then they are identical, i.e., no number is the
              successor of more than one number.
A3 x plus 0 equals x.
A4 x plus the successor of y equals the successor of $x + y$. (For instance, $2 + s3 = s(2 + 3) = 6$.)
A5 x times 0 equals 0.
A6 x times the successor of y equals $((x \cdot y) + x)$. (For instance, $2 \cdot s3 = (2 \cdot 3) + 2 = 8$.)

As axioms, any of these formulas may be introduced at any line of a proof.

SOLVED PROBLEMS

11.15   Prove as a theorem of formal arithmetic:

$$\vdash 2 + 2 = 4$$

Solution

In successor notation, what we are to prove is:

$$\vdash (ss0 + ss0) = ssss0$$

The proof of this simple statement is surprisingly complex:

1	$\forall x \forall y \, (x + sy) = s(x + y)$	A4
2	$\forall y \, (ss0 + sy) = s(ss0 + y)$	1 $\forall$E
3	$(ss0 + ss0) = s(ss0 + s0)$	2 $\forall$E
4	$(ss0 + s0) = s(ss0 + 0)$	2 $\forall$E
5	$(ss0 + ss0) = ss(ss0 + 0)$	3, 4 =E
6	$\forall x \, (x + 0) = x$	A3
7	$(ss0 + 0) = ss0$	6 $\forall$E
8	$(ss0 + ss0) = ssss0$	5, 7 =E

We instantiate the variable 'x' with '$ss0$' at step 2, 'y' with '$s0$' at step 3, and 'y' again with '0' at step 4. Using line 4, we then replace '$(ss0 + s0)$' in line 3 with '$s(ss0 + 0)$' by =E to obtain line 5. A similar =E move occurs at step 8.

11.16   Prove as a theorem of formal arithmetic:

$$\vdash 1 \cdot 2 = 2$$

Solution

In successor notation, the theorem is:

$$\vdash (s0 \cdot ss0) = ss0$$

Here the proof is even more complex:

1	$\forall x \forall y \, (x \cdot sy) = ((x \cdot y) + x)$	A6
2	$\forall y \, (s0 \cdot sy) = ((s0 \cdot y) + s0)$	1 $\forall$E
3	$(s0 \cdot ss0) = ((s0 \cdot s0) + s0)$	2 $\forall$E

4	$\forall x \forall y \, (x + sy) = s(x + y)$	A4
5	$\forall y \, ((s0 \cdot s0) + sy) = s((s0 \cdot s0) + y)$	4 $\forall$E
6	$((s0 \cdot s0) + s0) = s((s0 \cdot s0) + 0)$	5 $\forall$E
7	$(s0 \cdot ss0) = s((s0 \cdot s0) + 0)$	3, 6 =I
8	$\forall x \, (x + 0) = x$	A3
9	$((s0 \cdot s0) + 0) = (s0 \cdot s0)$	8 $\forall$E
10	$(s0 \cdot ss0) = s(s0 \cdot s0)$	7, 9 =I
11	$(s0 \cdot s0) = ((s0 \cdot 0) + s0)$	2 $\forall$E
12	$(s0 \cdot ss0) = s((s0 \cdot 0) + s0)$	10, 11 =I
13	$\forall x \, (x \cdot 0) = 0$	A5
14	$(s0 \cdot 0) = 0$	13 $\forall$E
15	$(s0 \cdot ss0) = s(0 + s0)$	12, 14 =I
16	$\forall y \, (0 + sy) = s(0 + y)$	4 $\forall$E
17	$(0 + s0) = s(0 + 0)$	16 $\forall$E
18	$(s0 \cdot ss0) = ss(0 + 0)$	15, 17 =I
19	$(0 + 0) = 0$	8 $\forall$E
20	$(s0 \cdot ss0) = ss0$	18, 19 =I

This proof repays careful study, since it provides good illustrations of the uses of =I and $\forall$E with function symbols. It also demonstrates how repeated application of instances A4 and A6 can be used to manipulate successor expressions until the desired result is attained. Roughly, the strategy is this: Use instances of A6 to reduce formulas containing '·' followed by long successor expressions to formulas containing '·' followed by shorter successor expressions and a sum. Then use instances of A4 to reduce formulas containing '+' followed by long successor expressions to formulas containing '+' followed by shorter successor expressions. Eventually, by continually shortening successor expressions following '+' and '·', we reduce them to '0', at which point A3 or A5 can be used to eliminate these occurrences of '+' and '·' and simplify the formula to achieve the desired result.

Proofs of this sort have been criticized on the grounds that statements such as '$2 \cdot 1 = 2$' are more certain than the principles used to prove them. That may well be true. But though such proofs hardly increase our confidence in their conclusions, they do systematically illuminate the connections between simple arithmetical truths and the more general principles from which they can be deduced. Moreover, the real value of formal arithmetic lies not in the proofs of such simple truths, but in the careful examination of more advanced forms of mathematical reasoning. Dissection of mathematical reasoning under the microscope of formal logic has in fact facilitated a number of fundamental mathematical discoveries (e.g., Gödel's theorem, which is briefly discussed at the end of this section) and significantly enhanced the unity of mathematical knowledge.

Notice that much of the work in Problems 11.15 and 11.16 involved the removal of universal quantifiers by $\forall$E. In mathematics, initial universal quantifiers are generally omitted, so that A5, for example, would be written simply as '$(x \cdot 0) = 0$'. Moreover, proofs in mathematics often skip many of the steps (such as $\forall$E) which are required for complete formal rigor. Given the demanding complexity of formal proof exemplified in the last two problems, such shortcuts are obviously both desirable and necessary.

The axioms alone, although they imply the particular facts codified in addition and multiplication tables and some other relatively simple truths, do not enable us to prove very interesting arithmetical generalizations. For that we need a more powerful principle of reasoning, the rule of *mathematical induction*. The name is a bit misleading; mathematical induction is actually a form of deductive reasoning. In fact, it can be seen as a generalization of the rule of conditional elimination. The idea is this: We wish to prove that a certain general fact holds for all the nonnegative integers, i.e., that they all have a certain (possibly very complex) property. To do this, it suffices to show two things:

(1)   Zero has this property.

(2)   For any number x, if x has this property, then so does the successor of x.

For condition (2) is equivalent (under the standard interpretation of arithmetic) to the following infinite sequence of conditional statements:

If 0 has this property, then so does 1.
If 1 has this property, then so does 2.
If 2 has this property, then so does 3.

Now condition 1 together with the first of these conditionals implies by →E that 1 has the property in question. And the fact that 1 has this property, together with the second conditional, implies by →E that 2 has it, and so on. Thus, by what are in effect infinitely many steps of →E, we reach the conclusion that all the nonnegative integers have this property, i.e., that the desired generalization is true of them all.

Of course, we cannot actually write out a proof containing infinitely many steps of →E. Yet the reasoning outlined above is clearly valid. So we adopt a new rule of inference which licenses this reasoning.[6] Intuitively, the rule says that if we have established conditions 1 and 2, we can directly infer:

(3)   Every nonnegative integer has this property.

In formal terms, the rule is this:

Mathematical Induction (I):   Given a wff ϕ containing the name letter '0' and a wff of the form $\forall\alpha(\phi\% \to \phi^{s}\%)$, we may infer $\forall\alpha\phi\%$, where $\phi\%$ is the result of replacing one or more occurrences of '0' in ϕ by some variable α not already in ϕ, and $\phi^{s}\%$ is the result of replacing those same occurrences of '0' in ϕ by $s\alpha$.

Here $\phi\%$ is an open formula on α expressing the property about which we wish to generalize, ϕ is condition 1, $\forall\alpha(\phi\% \to \phi^{s}\%)$ is condition 2, and $\forall\alpha\phi\%$ is the conclusion (3).

SOLVED PROBLEM

11.17   Prove the following theorem of formal arithmetic:

$$\vdash \forall x\,(0 + x) = x$$

Solution

Notice that this is not the same as A3, since the positions of the occurrences of 'x' and '0' flanking the plus sign are reversed. Despite its similarity to A3, this theorem cannot be proved directly from the axioms, since we have no rule which allows us to go directly from something of the form '$a + 0$' to something of the form '$0 + a$'. Its proof requires rule I. We wish to show that all numbers have a certain complex property: the property of being such that, when added to zero, they yield themselves as the result. The proof is as follows:

---

[6]Many authors treat mathematical induction as an additional axiom, rather than as a rule of inference. The difference is merely one of style; it does not change what we can prove. We treat it as an inference rule to emphasize its similarities with other rules, particularly →E, and also because it generally functions as an inference rule in mathematical proofs.

1	$\forall x\,(x + 0) = x$	A3
2	$(0 + 0) = 0$	$1\,\forall$E
3	$\quad(0 + a) = a$	H (for $\rightarrow$I)
4	$\quad\forall x\forall y\,(x + sy) = s(x + y)$	A4
5	$\quad\forall y\,(0 + sy) = s(0 + y)$	$4\,\forall$E
6	$\quad(0 + sa) = s(0 + a)$	$5\,\forall$E
7	$\quad(0 + sa) = sa$	$3, 6$ =E
8	$(0 + a) = a \rightarrow (0 + sa) = sa$	3–$7 \rightarrow$I
9	$\forall x((0 + x) = x \rightarrow (0 + sx) = sx)$	$8\,\forall$I
10	$\forall x\,(0 + x) = x$	$2, 9$ I

We prove that 0 has the relevant property at step 2. Then at step 9 we deduce that if any number has it, then so does that number's successor. The theorem follows by mathematical induction at step 10. With reference to the formal statement of the rule I, the formula ϕ containing '0' is that shown at step 2. Replacing the last two occurrences of 0 in this formula by some variable α (in this case, 'x') yields $\phi\%$, that is, '$(0 + x) = x$'. The formula $\forall\alpha(\phi\% \rightarrow \phi^{s}\%)$ is represented in step 9, and the conclusion $\forall\alpha\phi\%$ appears in line 10.

Problem 11.17 illustrates a pattern common to most elementary proofs by mathematical induction. Zero is shown to have the relevant property, and then a hypothetical derivation followed by a step of $\rightarrow$I and a step of $\forall$I establishes that if any number has it, so does that number's successor. The derivation of the statement that zero has the property is called the *basis case* of the proof. Thus, in Problem 11.17, the basis case consists of steps 1 and 2. The hypothetical derivation together with the steps of $\rightarrow$I and $\forall$I (steps 3 through 9 in Problem 11.17) is known as the *inductive portion* of the proof. This inductive portion begins with a hypothesis called the *inductive hypothesis* (step 3). This same pattern is evident again in the next problem.

SOLVED PROBLEM

11.18 Prove the following theorem of formal arithmetic:

$$\vdash \forall y\forall x\,(sx + y) = s(x + y)$$

Solution

Notice that this theorem is related to A4 in much the same way that Problem 11.17 is related to A3. Here, however, there are two universally quantified variables, rather than one. When two or more initial universal quantifiers are present, the induction can be attempted in several different ways. In this case, we can try to prove either that all numbers x have the property expressed by the open formula '$\forall y\,(sx + y) = s(x + y)$' or that all numbers y have the property expressed by the open formula '$\forall x\,(sx + y) = s(x + y)$'. In the first case we say that the induction is *on* 'x'; in the second, that it is *on* 'y'. Sometimes a theorem can be proved by induction on one variable but not by induction on another. Even if it can be proved by induction on several different variables, the proof is often much easier by induction on one than by induction on others. In this case proof by induction on 'y' is most efficient:

1	$\forall x\,(x + 0) = x$	A3
2	$(a + 0) = a$	$1\,\forall$E
3	$sa = sa$	=I
4	$s(a + 0) = sa$	$2, 3$ =E
5	$(sa + 0) = sa$	$1\,\forall$E
6	$(sa + 0) = s(a + 0)$	$4, 5$ =E

7	$\forall x\,(sx + 0) = s(x + 0)$	6 $\forall$I
8	$\forall x\,(sx + b) = s(x + b)$	H (for $\rightarrow$I)
9	$(sa + b) = s(a + b)$	8 $\forall$E
10	$s(sa + b) = s(sa + b)$	=I
11	$s(sa + b) = ss(a + b)$	9, 10 =E
12	$\forall x \forall y\,(x + sy) = s(x + y)$	A4
13	$\forall y\,(sa + sy) = s(sa + y)$	12 $\forall$E
14	$(sa + sb) = s(sa + b)$	13 $\forall$E
15	$(sa + sb) = ss(a + b)$	11, 14 =E
16	$\forall y\,(a + sy) = s(a + y)$	12 $\forall$E
17	$(a + sb) = s(a + b)$	16 $\forall$E
18	$s(a + sb) = s(a + sb)$	=I
19	$s(a + sb) = ss(a + b)$	17, 18 =E
20	$(sa + sb) = s(a + sb)$	15, 19 =E
21	$\forall x\,(sx + sb) = s(x + sb)$	20 $\forall$I
22	$\forall x(sx + b) = s(x + b) \rightarrow \forall x\,(sx + sb) = s(x + sb)$	8–21 $\rightarrow$I
23	$\forall y(\forall x\,(sx + y) = s(x + y) \rightarrow \forall x\,(sx + sy) = s(x + sy))$	22 $\forall$I
24	$\forall y \forall x\,(sx + y) = s(x + y)$	7, 23 I

The basis case is steps 1 to 7, and the inductive portion is steps 8 to 23. We now use the theorems of Problems 11.17 and 11.18 to prove one of the fundamental laws of arithmetic, the commutative law for addition.

SOLVED PROBLEM

11.19 Prove the following theorem of formal arithmetic:

$$\vdash \forall y \forall x\,(x + y) = (y + x)$$

Solution

Again we use mathematical induction on 'y'.

1	$\forall x\,(x + 0) = x$	A3
2	$(a + 0) = a$	1 $\forall$E
3	$\forall x\,(0 + x) = x$	TI 11.17
4	$(0 + a) = a$	3 $\forall$E
5	$(a + 0) = (0 + a)$	2, 4 =I
6	$\forall x\,(x + b) = (b + x)$	H (for $\rightarrow$I)
7	$(a + b) = (b + a)$	7 $\forall$E
8	$s(a + b) = s(a + b)$	=I
9	$s(a + b) = s(b + a)$	7, 8 =E
10	$\forall x \forall y\,(x + sy) = s(x + y)$	A4
11	$\forall y\,(a + sy) = s(a + y)$	10 $\forall$E
12	$(a + sb) = s(a + b)$	11 $\forall$E
13	$\forall y \forall x\,(sx + y) = s(x + y)$	TI 11.18
14	$\forall x\,(sx + a) = s(x + a)$	13 $\forall$E
15	$(sb + a) = s(b + a)$	14 $\forall$E
16	$(a + sb) = s(b + a)$	9, 12 =E
17	$(a + sb) = (sb + a)$	15, 16 =E
18	$\forall x\,(x + sb) = (sb + x)$	17 $\forall$I
19	$\forall x\,(x + b) = (b + x) \rightarrow \forall x\,(x + sb) = (sb + x)$	6–18 $\rightarrow$I
20	$\forall y(\forall x\,(x + y) = (y + x) \rightarrow \forall x\,(x + sy) = (sy + x))$	19 $\forall$I
21	$\forall y \forall x\,(x + y) = (y + x)$	5, 20 I

The basis case is steps 1 to 5. The inductive portion is steps 6 to 20.

The formal arithmetic outlined in this section is strong enough to provide proofs for all of the familiar laws governing the nonnegative integers. It is possible, using the symbols of this system, to define additional predicates such as 'is less than' and 'is greater than' (see Supplementary Problem IV, below) and additional function symbols such as 'the square of', 'the smallest power of 10 greater than', and so on. Subtraction and division are not function symbols for the nonnegative integers, since these operations may result in numbers which are not nonnegative integers. They cannot therefore be represented by function symbols; but, as we shall see in the next section for the case of subtraction, they can be handled in other ways.

Nevertheless, the formal arithmetic presented here is semantically incomplete; some truths about the nonnegative integers which are formulable in this notation cannot be proved as theorems. This incompleteness cannot be remedied. A celebrated metalogical proof due to Kurt Gödel has established that for any system containing the axioms and rules given here (and indeed, any system containing certain even weaker axioms and rules), some truths expressible in that system cannot be proved within that system.[7]

11.5   FORMAL DEFINITIONS

The formalization of relatively simple ideas often produces forbiddingly complex formulas. Such formulas can be made more comprehensible by systematic rewriting or abbreviation. That is the role of formal definition. The simplest sort of formal definition is direct replacement of one expression by another, generally shorter expression. This was illustrated in the previous section, where we took the familiar numerals as abbreviations for successor expressions.

More complicated, but also more useful, are *contextual definitions*, or *definitions in use*. In a contextual definition, a symbol is defined not by substituting it for other symbols, but by showing how entire formulas in which it occurs can be systematically translated into formulas in which it does not occur. Consider, for example, the definition of the familiar subtraction symbol '−' in the language of formal arithmetic:

D1: $(x - y) = z =_{df} (y + z) = x$

(The symbol '$=_{df}$' means "is by definition.") Thus D1 indicates that '$(x - y) = z$' is to be regarded as an alternative way of writing '$(y + z) = x$'. Notice that '−' does not simply abbreviate some other symbol or set of symbols. Rather, the formula in which '−' occurs is translated into a different formula in which 'x', 'y', and 'z' occur in a different order.

The definition is to be understood in a completely general way, so that the variables 'x', 'y', and 'z' may stand for any name, variable, or functional expression. It applies to subformulas as well as whole formulas. Thus D1 allows us to rewrite '$(2 + 1) = 3$' as '$(3 - 2) = 1$', '$(s0 + s0) = ss0$' as '$(ss0 - s0) = s0$', '$\forall x (x + 0) = x$' as '$\forall x (x - x) = 0$', and so on.

Definitions may be used like inference rules in proofs, allowing replacement of either version of the defined formula by the other. They are not counted as inference rules, however, since they do not enable us to infer anything new, but only to rewrite previously stated formulas in an alternative notation.

SOLVED PROBLEMS

11.20   Prove the following theorem of formal arithmetic:

$\vdash \forall x (x - 0) = x$

---

[7]For a very readable introduction to Gödel's incompleteness theorem, see Ernest Nagel and James R. Newman, *Gödel's Proof*, New York, New York University Press, 1958.

Solution

The theorem is simply an alternative way of writing the formula '$\forall x\,(0 + x) = x$'. But this formula was proved as a theorem in Problem 11.17. Hence the proof is straightforward:

1   $\forall x\,(0 + x) = x$ TI 11.17
2   $\forall x\,(x - 0) = x$ 1 D1

11.21   Prove the following theorem of formal arithmetic:

$$\vdash \forall x \forall y \forall z((x - y) = z \rightarrow (x - z) = y)$$

Solution

1	$(a - b) = c$	H
2	$(b + c) = a$	1 D1
3	$\forall x \forall y\,(x + y) = (y + x)$	TI 11.19
4	$\forall y\,(b + y) = (y + b)$	3 $\forall$E
5	$(b + c) = (c + b)$	4 $\forall$E
6	$(c + b) = a$	2, 5 =E
7	$(a - c) = b$	6 DI
8	$(a - b) = c \rightarrow (a - c) = b$	1–7 $\rightarrow$I
9	$\forall z((a - b) = z \rightarrow (a - z) = b)$	8 $\forall$I
10	$\forall y \forall z((a - y) = z \rightarrow (a - z) = y)$	9 $\forall$I
11	$\forall x \forall y \forall z((x - y) = z \rightarrow (x - z) = y)$	10 $\forall$I

The theorem is a triply quantified conditional. We use a $\rightarrow$I strategy involving D1 to obtain the appropriate conditional at 8, and then introduce the quantifiers by $\forall$I.

Contextual definitions have their limitations. D1 enables us to introduce '$-$' (flanked by names, variables, or functional expressions) only on the left side of an identity predicate. It does not allow us to use '$-$' in other contexts. It does not, for example, permit us to write '$1 = (3 - 2)$', where '$-$' appears to the right of '$=$'. For that we need the parallel definition:

D2: $z = (x - y) =_{\mathrm{df}} (y + z) = x$

But even with D2 there are contexts in which we may not write '$-$', such as in the expression '$s(0 - 0) = s0$', where '$-$' occurs within a functional expression. Introducing '$-$' into functional expressions presents special problems, since whenever x is less than y, $(x - y)$ is not a nonnegative integer and hence is not in the domain. Definitions to handle such cases must therefore include certain restrictions. We shall not discuss these technicalities here.[8]

11.6   DEFINITE DESCRIPTIONS

One of the most notable applications of contextual definition occurs in Bertrand Russell's theory of definite descriptions. *Definite descriptions* are expressions which purportedly denote a single object by enumerating properties which uniquely identify it. In English they are usually phrases beginning with the definite article 'the'. The following expressions are definite descriptions:

The current president of the United States

The summit of Mount Everest

The steel tower in Paris

Isaac's father (i.e., the father of Isaac)

---

[8]For detailed discussions, see Patrick Suppes, *Introduction to Logic*, New York, Van Nostrand, 1957; or Benson Mates, *Elementary Logic*, 2d edn, New York, Oxford University Press, 1972, pp. 197–204.

Functional expressions may be used to formalize some definite descriptions (e.g., 'the father of Isaac'—see Section 11.3), but they cannot be used to formalize others. The phrase 'the big tower in', for example, does not denote a function when applied to the domain of cities (or virtually any other domain), since some cities have no big tower and others have more than one. Russell's theory handles definite descriptions in either case.

Russell uses the symbol '$\imath$' for 'the'. Like a quantifier, this symbol must be followed by a variable. For example, if 'F' is the predicate 'is the father of' (as opposed to the functional phrase 'the father of') and 'i' is the name 'Isaac', the expression 'Isaac's father' is written as '$\imath xFxi$'. This means: "the (unique) x such that x is the father of Isaac." Definite descriptions, like functional expressions, act grammatically like names. So to say 'Isaac's father was a prophet', we write '$P\imath xFxi$', with the definite description following the predicate.

SOLVED PROBLEM

11.22   Using 'S' for 'is made of steel', 'T' for 'is a tower', 'P' for 'is in Paris', and 'M' for 'is magnificent', formalize the sentence 'The steel tower in Paris is magnificent'.

Solution

The definite description 'The steel tower in Paris' is formalized as '$\imath x((Sx \& Tx) \& Px)$'. Read literally, this means, "the x such that x is made of steel and x is a tower and x is in Paris." To formalize the entire statement, we simply attach the predicate 'M' to the description: '$M\imath x((Sx \& Tx) \& Px)$'.

All formalized definite descriptions have the form '$\imath xFx$', where 'x' may be replaced by any variable and 'Fx' by any open sentence on that variable. Thus, the simplest statements containing definite descriptions all have the form '$G\imath xFx$', where G may be replaced by any one-place predicate. This may be read as "The F is G." The statements 'The father of Isaac was a prophet' and 'The steel tower in Paris is magnificent' both have this form. (In the latter case, 'Fx' is replaced by the open formula '$(Sx \& Tx) \& Px$'.)

According to Russell's theory, such expressions make the following rather complex assertion:

There is exactly one F, and it is G.

Now this can be formalized without using the symbol '$\imath$' in predicate logic with identity:

$\exists x((Fx \& \forall y(Fy \rightarrow x = y)) \& Gx)$

(compare part (l) of Problem 6.40). Thus the symbol '$\imath$' may be defined contextually for occurrences following one-place predicates as follows:

D: $G\imath xFx =_{df} \exists x((Fx \& \forall y(Fy \rightarrow x = y)) \& Gx)$

where 'G' is any one-place predicate. By this definition, the formula of Problem 11.22 is shorthand for the rather intimidating formula:

$\exists x((((Sx \& Tx) \& Px) \& \forall y(((Sy \& Ty) \& Py) \rightarrow x = y)) \& Mx)$

This says "There is an x such that x is made of steel and x is a tower and x is in Paris, and for all y if y is made of steel and y is a tower and y is in Paris, y is the same as x; and x is magnificent," or, more succinctly, "There is exactly one steel tower, x, in Paris, and x is magnificent."

The formula is a legitimate wff of the predicate calculus, regardless of whether there is one steel tower in Paris, or many or none. (In the latter two cases it is simply false.) Since the definition eliminates '$\imath$' in favor of the terminology of the predicate calculus, no new axioms or inference rules are needed. All the implications of formulas containing definite descriptions can be handled by the rules of the predicate calculus with identity. The following problem, for example, illustrates the validity of existential introduction over definite descriptions:

SOLVED PROBLEM

11.23   Prove:

$$G\iota xFx \vdash \exists xGx$$

Solution

1	$G\iota xFx$	A
2	$\exists x((Fx \,\&\, \forall y(Fy \rightarrow x = y)) \,\&\, Gx)$	1 D
3	$(Fa \,\&\, \forall y(Fy \rightarrow a = y)) \,\&\, Ga$	H (for $\exists$E)
4	Ga	3 &E
5	$\exists xGx$	4 $\exists$I
6	$\exists xGx$	2, 3–5 $\exists$E

A number of interesting complexities and ambiguities arise when definite descriptions occur in contexts more complicated than the subject position of a one-place predicate. For these cases, the definition D must be generalized to allow '$G\iota xFx$' to stand for any wff containing a single occurrence of 'ιxFx'. But we shall not discuss these complexities here.[9]

11.7   MODAL LOGIC

Formal logic can be expanded in still other directions. One extension which has undergone rapid development recently is *modal logic*. Modal logic—the name comes from the Latin '*modus*' (mode or mood)—is the logic of expressions such as 'could', 'would', 'should', 'may', 'might', 'ought', and so on, which express the grammatical "mood" of a sentence. The most thoroughly investigated of the modal expressions are 'It is possible that' and 'It is necessary that'. In modal logic these are represented respectively by the symbols '$\diamond$' and '$\square$'.[10] Syntactically they function just like negation, prefixing a wff to create a new wff. But semantically they are quite different, since they are not truth-functional; we cannot always determine the truth value of a sentence of the form '$\diamond P$' or '$\square P$' simply from the truth value of 'P'.

Suppose, for example, that a certain stream is polluted. Let 'P' represent the statement that this is so. Then 'P' is true and '$\sim P$' is false. But the statements

$\square P$ (It is necessary that this stream is polluted.)

and

$\square \sim P$ (It is necessary that this stream is not polluted.)

are both false. Neither condition is necessary. The condition of the stream is a contingent fact; it is not bound to be one way or the other. Thus the operator '$\square$' may produce a false sentence when prefixed either to a false sentence or to a true sentence.

Similarly, the statements

$\diamond P$ (It is possible that this stream is polluted.)

---

[9]A brief account may be found in Jeffrey, *op. cit.*, pp. 118–122. A more thorough but more technical treatment is given in Donald Kalish, Richard Montague, and Gary Mar, *Logic: Techniques of Formal Reasoning*, 2d edn, New York, Harcourt Brace Jovanovich, 1980, Chaps. VI and VIII.

[10]Sometimes the letter 'M', the first letter of the German term '*möglicherweise*' (possibly), is used for 'it is possible that', and 'L' (signifying logical truth) or 'N' for 'it is necessary that'.

and

$\Diamond \sim P$ (It is possible that this stream is not polluted.)

are both true. Both conditions are possible. Hence, again we cannot determine the truth value of the modal statement solely from the truth value of its nonmodal component.

There are two exceptions to this rule. If 'P' is true, than '$\Diamond P$' is certainly true, since if something is actually the case it is clearly possible. And if 'P' is false, then '$\Box P$' is false, since what is not the case is certainly not necessary. In general, however, the truth value of a modal statement depends not only on the actual truth values of its components, but also on the truth values that these components might have. The truth value of a modal statement, then, can be thought of as a function of the truth values of its components in various *possible worlds*. (The notion of a "possible world" is a source of some philosophical controversy. Here we shall use it to refer to any conceivable state of affairs, any one of the many ways things could have been besides the way they actually are.)

Statements of the form '$\Diamond P$' are true if and only if 'P' is true in at least one possible world. Since the actual world (the universe) is a possible world, if 'P' is true in the actual world, that makes '$\Diamond P$' true. But '$\Diamond P$' may be true even if 'P' is false in the actual world, provided only that 'P' is true in some possible world.

Similarly, statements of the form '$\Box P$' are true if and only if 'P' is true in all possible worlds. Thus if 'P' is false in the actual world, '$\Box P$' is false. But if 'P' is true in the actual world, '$\Box P$' may be either true or false, depending upon whether or not 'P' is true in all the other possible worlds as well.

SOLVED PROBLEM

11.24 Formalize the following statements, using the operators '$\Box$' and '$\Diamond$' and the sentence letter 'P' for 'Pete is a postman'.

(a) It is impossible that Pete is a postman.

(b) Pete might not be a postman.

(c) It is not true that Pete must be a postman.

(d) It is necessary that Pete is not a postman.

(e) Necessarily, if Pete is a postman, then Pete is a postman.

(f) It is necessarily possible that Pete is a postman.

(g) It is possible that Pete is a postman and possible that he is not.

(h) It is impossible that Pete both is and is not a postman.

(i) If it is necessary that Pete is a postman, then it is necessary that it is necessary that Pete is a postman.

(j) It is necessary that it is necessary that if Pete is a postman then Pete is a postman.

Solution

(a) $\sim \Diamond P$

(b) $\Diamond \sim P$

(c) $\sim \Box P$

(d) $\Box \sim P$

(e) $\Box(P \rightarrow P)$

(f) $\Box \Diamond P$

(g) $\Diamond P \,\&\, \Diamond \sim P$

(h) $\sim \Diamond(P \,\&\, \sim P)$

(i) $\Box P \rightarrow \Box \Box P$

(j) $\Box \Box(P \rightarrow P)$

Notice the difference in meaning when the relative positions of a modal operator and a negation sign are changed, as in the pair (*a*) and (*b*) or the pair (*c*) and (*d*).

Actually, the terms 'possible' and 'necessary' themselves have various shades of meaning. There are conditions which are *logically* possible (they violate no law of logic) but not *physically* possible (they violate the laws of physics). It is, for example, logically possible but not physically possible to accelerate a body to a speed faster than the speed of light. And there are conditions which are physically possible but not *practically* possible (at least with current technologies). For example, it is physically possible to supply all of the earth's energy needs with solar power; but this is not practically possible now. And there are other senses of possibility as well.

The different concepts of possibility have different logics, and for some the appropriate logic is still a matter of controversy. We shall be concerned exclusively with the logic of logical possibility. Thus we shall count something as possible if we can describe it without inconsistency. But what, exactly, is inconsistency? We have thus far characterized this notion only vaguely as falsehood due to semantics alone. That's clear enough when we are working with a formal language which has a well-defined semantics (e.g., propositional or predicate logic), but none too clear for natural languages.

Alternatively, we might say that a thing is logically possible if its description violates no law of logic. But what is a law of logic? There is near universal agreement that the theorems of the predicate calculus should count as logical laws. But what of the theorems of T logic (Section 11.1), second-order logic (Section 11.2), or formal arithmetic (Section 11.4)? Just how far does the domain of logic extend? Moreover, if logic includes formal arithmetic (which, as we noted, is incomplete), should we count those arithmetical truths which cannot be proved in our system of formal arithmetic as logical laws?

There is yet a deeper issue here. In Section 2.3 we defined a valid argument as one such that it is logically impossible (not logically possible) for its conclusion to be false while its premises are true. If we now define logical possibility in terms of logical laws, how can we define the notion of a logical law without presupposing some concept of validity and thus making our definitions viciously circular? Will the concept of a possible world help, or is that too ultimately circular or insufficiently clear?

These difficulties have led some philosophers to reject the notions of possibility and necessity altogether, along with our conception of validity and a whole array of related ideas,[11] but that response is drastic. Still, it must be admitted that these concepts are fraught with unresolved difficulties. Fortunately, the formal development of much of modal logic does not depend upon solutions to these philosophical difficulties, so that much progress has been made in spite of them.

This section presents a system of propositional modal logic known as S5. We shall not discuss quantified modal logic. S5, the most powerful of five modal calculi invented by the logician C. I. Lewis, is widely regarded as the system which best codifies the concept of logical possibility, though it may well be too powerful for other concepts of possibility.

Our version of S5 will be based on the propositional calculus of Chapter 4, together with four axiom schemas and a new rule of inference. An *axiom schema*, as opposed to an axiom, is a formula each substitution instance of which counts as an axiom. (Recall from Section 4.4 that a substitution instance of a wff is the result of replacing zero or more of its sentence letters by wffs, each occurrence of the same sentence letter being replaced by the same wff.) An axiom schema thus represents infinitely many axioms. The axiom schemas are as follows:

AS1 $\diamond P \leftrightarrow \sim\square\sim P$

AS2 $\square(P \rightarrow Q) \rightarrow (\square P \rightarrow \square Q)$

AS3 $\square P \rightarrow P$

AS4 $\diamond P \rightarrow \square \diamond P$

---

[11]Most notably, W. V. O. Quine, in "Two Dogmas of Empiricism," in *From a Logical Point of View*, Cambridge, Mass., Harvard University Press, 1953. For a discussion of these and other criticisms of modal logic, see Susan Haack, *Philosophy of Logics*, Cambridge, Cambridge University Press, 1978, Chap. 10.

Any substitution instance of these schemas counts as an axiom. Thus, for example, the following substitution instances of AS1 are all axioms:

$$\Diamond Q \leftrightarrow \sim\Box\sim Q$$
$$\Diamond\Box S \leftrightarrow \sim\Box\sim\Box S$$
$$\Diamond(P \,\&\, R) \leftrightarrow \sim\Box\sim(P \,\&\, R)$$

AS1 itself says that it is possible that P if and only if it is not necessary that not-P.[12] It asserts, in other words, that 'P' is true in some possible world if and only if it is not the case that in all possible worlds '$\sim P$' is true. This is clearly correct, since 'P' is false in any possible world in which '$\sim P$' is true.

AS2 says that if it is necessary that if P then Q, then if it is necessary that P, it is necessary that Q. Thinking in terms of possible worlds, this means that if '$P \to Q$' is true in all possible worlds, then if 'P' is true in all possible worlds, so is 'Q'.

AS3 asserts that if it is necessary that P, then P. That is, if 'P' is true in all possible worlds, then 'P' is true in the actual world (or in any world at which we consider the axiom's truth).

AS4 is the only one of the axioms whose truth may not be obvious upon reflection. The intuition behind it is this: What counts as logically possible is not a matter of contingent fact but is fixed necessarily by the laws of logic. Thus if it is possible that P, then it is necessary that it is possible that P.[13]

In the language of possible worlds, AS4 means: If 'P' is true in some possible world, then it is true in all possible worlds that 'P' is true in some possible world.

In addition to the four axioms, S5 has a special inference rule, the rule of necessitation:

<u>Necessitation (N):</u>   If ϕ has been proved as a theorem, then we may infer $\Box\phi$.

Theorems, in other words, are to be regarded as necessary truths. But N allows us to infer $\Box\phi$ from ϕ only if ϕ is a theorem, i.e., has been proved without making any assumptions. The following use of N, for example, is not legitimate:

1	P	A
2	$\Box P$	N (incorrect)

However, any formula in a derivation containing no assumptions (provided that the formula is not part of a hypothetical derivation) is a theorem, since it is proved on no assumptions by the part of the derivation which precedes it.

We now prove some theorems of S5. The equivalences of propositional logic (Section 4.6), which are valid for modal logic as well as propositional logic, figure prominently in the proofs. The first theorem establishes the obvious truth that if P then it is possible that P.

SOLVED PROBLEMS

11.25 Prove the theorem:

$$\vdash P \to \Diamond P$$

---

[12]In many treatments of modal logic, AS1 is regarded as a definition; the language of such systems contains only one modal operator, '$\Box$'; and '$\Diamond$' is introduced as mere abbreviation for '$\sim\Box\sim$'. However, treating AS1 as an axiom schema yields equivalent results. Still other systems have '$\Diamond$' as the sole modal operator and treat '$\Box$' as an abbreviation for '$\sim\Diamond\sim$'. (Problem 11.26 provides a clue as to why this works.) These, too, yield equivalent results.

[13]This reasoning is not plausible for some nonlogical forms of possibility; hence AS4 is rejected in some versions of modal logic. Omitting AS4 from our system yields a weak modal logic called T. Other versions of modal logic have been constructed by replacing AS4 by various weaker axioms. (See the comments following Problem 11.30, below.)

Solution

1	$\Box \sim P \rightarrow \sim P$	AS3
2	$\sim\sim P \rightarrow \sim\Box\sim P$	1 TRANS
3	$P \rightarrow \sim\Box\sim P$	2 DN
4	$\Diamond P \leftrightarrow \sim\Box\sim P$	AS1
5	$\sim\Box\sim P \rightarrow \Diamond P$	4 ↔E
6	$P \rightarrow \Diamond P$	3, 5 HS

Notice that a substitution instance of AS3 is used at 1. The proof can be shortened by treating AS1 as an equivalence in the manner of Section 4.6. Then, instead of steps 4, 5, and 6, we simply have:

4	$P \rightarrow \Diamond P$	3 AS1

11.26 Prove the theorem:

$$\vdash \Box P \leftrightarrow \sim \Diamond \sim P$$

Solution

1	$\Diamond \sim P \leftrightarrow \sim\Box\sim\sim P$	AS1
2	$\Diamond \sim P \leftrightarrow \sim\Box P$	1 DN
3	$\Diamond \sim P \rightarrow \sim\Box P$	2 ↔E
4	$\sim\Box P \rightarrow \Diamond \sim P$	2 ↔E
5	$\sim\sim\Box P \rightarrow \sim\Diamond\sim P$	3 TRANS
6	$\sim\Diamond\sim P \rightarrow \sim\sim\Box P$	4 TRANS
7	$\sim\sim\Box P \leftrightarrow \sim\Diamond\sim P$	5, 6 ↔I
8	$\Box P \leftrightarrow \sim\Diamond\sim P$	7 DN

Compare this theorem with AS1. Again, this result can be proved more efficiently if AS1 is used as an equivalence. The proof is left to the reader as an exercise.

11.27 Prove the theorem:

$$\vdash \sim\Diamond(P \,\&\, \sim P)$$

Solution

1	$\sim(P \,\&\, \sim P)$	TI 4.40
2	$\Box\sim(P \,\&\, \sim P)$	1 N
3	$\Box\sim(P \,\&\, \sim P) \leftrightarrow \sim\Diamond\sim\sim(P \,\&\, \sim P)$	TI 11.26
4	$\Box\sim(P \,\&\, \sim P) \rightarrow \sim\Diamond\sim\sim(P \,\&\, \sim P)$	3 ↔E
5	$\sim\Diamond\sim\sim(P \,\&\, \sim P)$	2, 4 →E
6	$\sim\Diamond(P \,\&\, \sim P)$	5 DN

11.28 Prove the theorem:

$$\vdash \Box\Box P \rightarrow P$$

Solution

1	$\Box P \rightarrow P$	AS3
2	$\Box(\Box P \rightarrow P)$	1 N
3	$\Box(\Box P \rightarrow P) \rightarrow (\Box\Box P \rightarrow \Box P)$	AS2
4	$\Box\Box P \rightarrow \Box P$	2, 3 →E
5	$\Box\Box P \rightarrow P$	1, 4 HS

The next theorem is somewhat surprising. It shows that if it is possible that a proposition is necessary, then it is indeed necessary.

SOLVED PROBLEM

11.29　Prove the theorem:

$$\vdash \Diamond\Box P \to \Box P$$

Solution

1	$\Diamond\Box P \leftrightarrow \sim\Box\sim\Box P$	AS1
2	$\Diamond\Box P \to \sim\Box\sim\Box P$	2 $\leftrightarrow$E
3	$\Box P \leftrightarrow \sim\Diamond\sim P$	TI 11.26
4	$\Box P \to \sim\Diamond\sim P$	3 $\leftrightarrow$E
5	$\sim\sim\Diamond\sim P \to \sim\Box P$	4 TRANS
6	$\Diamond\sim P \to \sim\Box P$	5 DN
7	$\Box(\Diamond\sim P \to \sim\Box P)$	6 N
8	$\Box(\Diamond\sim P \to \sim\Box P) \to (\Box\Diamond\sim P \to \Box\sim\Box P)$	AS2
9	$\Box\Diamond\sim P \to \Box\sim\Box P$	7, 8 $\to$I
10	$\sim\Box\sim\Box P \to \sim\Box\Diamond\sim P$	9 TRANS
11	$\Diamond\Box P \to \sim\Box\Diamond\sim P$	2, 10 HS
12	$\Diamond\sim P \to \Box\Diamond\sim P$	AS4
13	$\sim\Box\Diamond\sim P \to \sim\Diamond\sim P$	12 TRANS
14	$\Diamond\Box P \to \sim\Diamond\sim P$	11, 13 HS
15	$\sim\Diamond\sim P \to \Box P$	3 $\leftrightarrow$E
16	$\Diamond\Box P \to \Box P$	14, 15 HS

To see just how surprising this theorem is, consider the fact that many theologians have held that God exists necessarily. If we interpret 'P' as 'God exists', then the theorem implies that if it is even possible that it is necessary that God exists, then necessarily God exists. Thus, given the assumption that it is possible that it is necessary that God exists, by conditional elimination it is necessary that God exists, and hence by AS3 God exists! The reasoning here seems impeccable, provided that possibility and necessity are understood in the logical sense, so that the laws of S5 hold. However, the assumption that it is possible that it is necessary that God exists is controversial. Logicians generally do not regard this as a sound argument.

SOLVED PROBLEM

11.30　Prove the theorem:

$$\vdash \Box P \to \Box\Box P$$

Solution

1	$\Box P \to \Diamond\Box P$	TI 11.25
2	$\Diamond\Box P \to \Box\Diamond\Box P$	AS4
3	$\Box P \to \Box\Diamond\Box P$	1, 2 HS
4	$\Diamond\Box P \to \Box P$	TI 11.29
5	$\Box(\Diamond\Box P \to \Box P)$	4 N
6	$\Box(\Diamond\Box P \to \Box P) \to (\Box\Diamond\Box P \to \Box\Box P)$	AS2
7	$\Box\Diamond\Box P \to \Box\Box P$	5, 6 $\to$I
8	$\Box P \to \Box\Box P$	3, 7 HS

The theorem of Problem 11.30 asserts that if a proposition is necessary, then it is necessarily necessary. It is sometimes taken as an axiom schema in place of AS4. The resulting system, known as S4, is weaker than S5 (in particular, AS4 and Problem 11.29 are not theorems) but more plausible for some applications.[14]

---

[14]For more thorough treatments of modal logic, see Brian F. Chellas, *Modal Logic: An Introduction*, Cambridge, Cambridge University Press, 1980; or G. E. Hughes and M. J. Cresswell, *A New Introduction to Modal Logic*, London, Routledge, 1996.

We noted in Chapter 3 that the material conditional is not the only kind of conditional; in fact, it is the very weakest kind. In modal logic we can define the *strict conditional*, which is one of the strongest. Like the other modal operators, the strict conditional is non-truth-functional. We shall represent it by the symbol '$\dashv$'.[15] Statements of the form '$P \dashv Q$' are often regarded as mere shorthand for '$\Box(P \rightarrow Q)$'. Thus P strictly implies Q if and only if the material conditional '$P \rightarrow Q$' is necessary. Where '$\Box$' represents logical necessity, this means that '$P \dashv Q$' is true if and only if the inference with 'P' as premise and 'Q' as conclusion is valid. Using the strict conditional notation, AS2 can be written as:

$$(P \dashv Q) \rightarrow (\Box P \rightarrow \Box Q)$$

Although in the version of modal logic which we have presented here the operators '$\Box$' and '$\Diamond$' mean "it is necessary that" and "it is possible that," respectively, in other versions of modal logic these operators are given different interpretations. There are, for example, *deontic logics* (logics dealing with moral concepts), in which '$\Box$' means "it ought to be the case that" or "it should be the case that" and '$\Diamond$' means "it is morally permissible that." There are also *epistemic logics* (logics of knowledge and belief), in which these operators or operators similar to them are interpreted as representing various knowledge-related concepts. There are *tense logics*, in which they represent various temporal notions. And there are metalogical interpretations, in which they represent the notions of provability and consistency in some formal system (such as the formal arithmetic of Section 11.4).[16] All these interpretations require formal systems different from and generally weaker than S5.

Another form of logic closely related to the logic of possibility and necessity is the logic of *counterfactual* or *subjunctive* conditionals. These are the sorts of conditionals expressed in English by the subjunctive mood, as, for example, in the sentence 'If I had not been ill, I would have run the race'. Counterfactual conditionals are intermediate in strength between material and strict conditionals, and their logic is different from the logic of either.[17]

Supplementary Problems

I   Suppose that we expand the predicate calculus without identity to a second-order logic and then introduce the identity predicate by taking Leibniz' law as the definition:

   LL: $x = y =_{df} \forall P(Px \leftrightarrow Py)$

Prove the following laws of identity as theorems of this second-order logic:

(1) $a = b \rightarrow b = a$

(2) $(Fa \,\&\, a = b) \rightarrow Fb$

(3) $(a = b \,\&\, b = c) \rightarrow a = c$

(4) $\exists P(Pa \,\&\, {\sim}Pb) \rightarrow {\sim}a = b$

(5) $a = b \leftrightarrow \forall P(Pa \rightarrow Pb)$

---

[15]In systems which use '$\supset$' for the material conditional, the strict conditional is sometimes represented by the symbol '$\rightarrow$', which we use here for the material conditional.

[16]A good introduction to deontic logic is R. Hilpinen, ed., *Deontic Logic: Introductory and Systematic Readings*, Dordrecht, Netherlands, D. Reidel, 1981. The classic work on epistemic logic is J. Hintikka, *Knowledge and Belief: An Introduction to the Logic of the Two Notions*, Ithaca, N.Y., Cornell University Press, 1962. For an introduction to tense logic, see R. P. McArthur, *Tense Logic*, Dordrecht, Netherlands, D. Reidel, 1976. For the metalogical interpretation of modality, see G. Boolos, *The Logic of Provability*, Cambridge, Cambridge University Press, 1993.

[17]A classic work on the logic of counterfactual conditionals is D. K. Lewis, *Counterfactuals*, Cambridge, Mass., Harvard University Press, 1973.

II Prove the following theorems of the predicate calculus with identity and function symbols:

(1) $\forall x \exists y\, y = f(x)$

(2) $\forall x \forall y\, x = y \rightarrow \forall x\, f(x) = x$

(3) $\forall x \forall y (\sim f(x) = f(y) \rightarrow \sim x = y)$

(4) $\forall x Fg(x) \rightarrow \forall x Fg(g(x))$

(5) $\forall x \forall y ((x = f(y)\ \&\ y = g(x)) \rightarrow x = f(g(x)))$

III Prove the following theorems of formal arithmetic:

(1) $\sim 1 = 2$

(2) $(1 + 1) = 2$

(3) $(1 \cdot 1) = 1$

(4) $\forall x \exists y\, y = sx$

(5) $\forall x \sim x = sx$

(6) $\forall x\, (0 \cdot x) = 0$

(7) $\forall y \forall x\, (x \cdot y) = (y \cdot x)$

IV The predicates 'is less than' ('$<$'), 'is greater than' ('$>$'), 'is less than or equal to' ('$\leqslant$'), and 'is greater than or equal to' ('$\geqslant$') are definable in formal arithmetic as follows:

D3 $x < y =_{df} \exists z (\sim 0 = z\ \&\ (x + z = y))$

D4 $x \leqslant y =_{df} x < y \lor x = y$

D5 $x > y =_{df} y < x$

D6 $x \geqslant y =_{df} y \leqslant x$

Using these definitions together with D1, prove the following additional theorems of formal arithmetic:

(1) $\forall x \forall y (x < y \leftrightarrow y > x)$

(2) $\forall x \forall y (x < y \rightarrow x \leqslant y)$

(3) $\forall x \forall y (x > y \rightarrow x \geqslant y)$

(4) $\forall x\, x \leqslant x$

(5) $\forall x \exists y\, y > x$

(6) $\forall x \forall y\, x \leqslant x + y$

(7) $\forall x\, 0 < sx$

V Using Russell's theory of descriptions, formalize the following arguments and prove their validity, using 'a' for 'Andy', 'C' for 'is a chair', 'L' for 'loves', and 'B' for 'is blue':

(1) The chair that Andy loves is blue. Therefore Andy loves a chair.

(2) The chair that Andy loves is blue. Therefore, there is at most one chair that Andy loves.

(3) The chair that Andy loves is blue. Therefore, every chair that Andy loves is blue.

(4) It is not the case that the chair that Andy loves is blue. Andy loves a blue chair. Therefore, Andy loves at least two chairs.

(5) It is not the case that the chair that Andy loves is blue. Therefore, if there is exactly one chair that Andy loves, then Andy loves a chair which is not blue.

VI Prove the following theorems of S5 modal logic:

(1) $P \rightarrow \Diamond\Diamond P$

(2) $\sim\Diamond P \leftrightarrow \Box\sim P$

(3) $\Diamond\sim P \leftrightarrow \sim\Box P$

(4) $\sim\Diamond\sim\Diamond(P \vee \sim P)$

(5) $\Box(P \,\&\, Q) \leftrightarrow (\Box P \,\&\, \Box Q)$

(6) $\sim\Diamond(P \vee Q) \leftrightarrow (\sim\Diamond P \,\&\, \sim\Diamond Q)$

(7) $\Diamond(P \vee Q) \leftrightarrow (\Diamond P \vee \Diamond Q)$

(8) $(\Box P \vee \Box Q) \rightarrow \Box(P \vee Q)$

Answers to Selected Supplementary Problems

I (5)

1	$a = b$	H (for $\rightarrow$I)
2	$\forall P(Pa \leftrightarrow Pb)$	1 LL
3	$Fa \leftrightarrow Fb$	1 $\forall$E
4	$Fa \rightarrow Fb$	1 $\leftrightarrow$E
5	$\forall P(Pa \rightarrow Pb)$	4 $\forall$I
6	$a = b \rightarrow \forall P(Pa \rightarrow Pb)$	1, 5 $\rightarrow$I
7	$\forall P(Pa \rightarrow Pb)$	H (for $\rightarrow$I)
8	Fb	H (for $\rightarrow$I)
9	$\sim Fa$	H (for $\sim$I)
10	$\sim Fa \rightarrow \sim Fb$	7 $\forall$I
11	$\sim Fb$	9, 10 $\rightarrow$E
12	$Fb \,\&\, \sim Fb$	8, 11 &I
13	$\sim\sim Fa$	9–12 $\sim$I
14	Fa	13 $\sim$E
15	$Fb \rightarrow Fa$	8–14 $\rightarrow$I
16	$Fa \rightarrow Fb$	7 $\forall$I
17	$Fa \leftrightarrow Fb$	15, 16 $\leftrightarrow$I
18	$\forall P(Pa \leftrightarrow Pb)$	17 $\forall$I
19	$a = b$	18 LL
20	$\forall P(Pa \rightarrow Pb) \rightarrow a = b$	7, 19 $\rightarrow$I
21	$a = b \leftrightarrow \forall P(Pa \rightarrow Pb)$	6, 20 $\leftrightarrow$I

Comment: This theorem is important, because it shows that the condition that a and b have the same properties (i.e., $a = b$) is equivalent to the simpler condition that b have all the properties that a has. Thus Leibniz' law itself may be formulated more simply but equivalently as:

$$a = b =_{df} \forall P(Pa \rightarrow Pb)$$

Notice how 'P' is instantiated by the complex predicate '$\sim F$' at step 10.

II (5)

1	$a = f(b) \,\&\, b = g(a)$	H (for $\rightarrow$I)
2	$a = f(b)$	1 &E
3	$b = g(a)$	1 &E
4	$a = f(g(a))$	2, 3 =E
5	$(a = f(b) \,\&\, b = g(a)) \rightarrow a = f(g(a))$	1–4 $\rightarrow$I
6	$\forall y((a = f(y) \,\&\, y = g(a)) \rightarrow a = f(g(a)))$	5 $\forall$I
7	$\forall x\forall y((x = f(y) \,\&\, y = g(x)) \rightarrow x = f(g(x)))$	6 $\forall$I

III (5) 1 $\forall x \sim 0 = sx$ — A1
2 $\sim 0 = s0$ — 1 $\forall$E
3 $\sim a = sa$ — H (for $\rightarrow$I)
4 $\forall x \forall y (sx = sy \rightarrow x = y)$ — A2
5 $\forall y (sa = sy \rightarrow a = y)$ — 4 $\forall$E
6 $sa = ssa \rightarrow a = sa$ — 5 $\forall$E
7 $\sim sa = ssa$ — 3, 6 MT
8 $\sim a = sa \rightarrow \sim sa = ssa$ — 3–7 $\rightarrow$I
9 $\forall x (\sim x = sx \rightarrow \sim sx = ssx)$ — 8 $\forall$I
10 $\forall x \sim x = sx$ — 2, 9 I

IV (5) 1 $\forall x \sim 0 = sx$ — A1
2 $\sim 0 = s0$ — 1 $\forall$E
3 $\forall x \forall y \, (x + sy) = s(x + y)$ — A4
4 $\forall y \, (a + sy) = s(a + y)$ — 3 $\forall$E
5 $(a + s0) = s(a + 0)$ — 4 $\forall$E
6 $\forall x \, (x + 0) = x$ — A3
7 $(a + 0) = a$ — 6 $\forall$E
8 $(a + s0) = sa$ — 5, 7 =E
9 $\sim 0 = s0 \,\&\, (a + s0) = sa$ — 2, 8 &I
10 $\exists z (\sim 0 = z \,\&\, (a + z) = sa)$ — 9 $\exists$I
11 $a < sa$ — 10 D3
12 $sa > a$ — 11 D5
13 $\exists y \, y > a$ — 12 $\exists$I
14 $\forall x \exists y \, y > x$ — 13 $\forall$I

V (5) The argument form is:

$\sim B\imath x(Cx \,\&\, Lax) \vdash \exists x((Cx \,\&\, Lax) \,\&\, \forall y((Cy \,\&\, Lay) \rightarrow x = y)) \rightarrow \exists x((Cx \,\&\, Lax) \,\&\, \sim Bx)$

Here is the proof:

1 $\sim B\imath x(Cx \,\&\, Lax)$ — A
2 $\sim \exists x(((Cx \,\&\, Lax) \,\&\, \forall y((Cy \,\&\, Lay) \rightarrow x = y)) \,\&\, Bx)$ — 1 D
3 $\forall x \sim (((Cx \,\&\, Lax) \,\&\, \forall y((Cy \,\&\, Lay) \rightarrow x = y)) \,\&\, Bx)$ — 2 QE
4 $\exists x((Cx \,\&\, Lax) \,\&\, \forall y((Cy \,\&\, Lay) \rightarrow x = y))$ — H (for $\rightarrow$I)
5 $(Cb \,\&\, Lab) \,\&\, \forall y((Cy \,\&\, Lay) \rightarrow b = y)$ — H (for $\exists$E)
6 $\sim (((Cb \,\&\, Lab) \,\&\, \forall y((Cy \,\&\, Lay) \rightarrow b = y)) \,\&\, Bb)$ — 3 $\forall$E
7 $\sim ((Cb \,\&\, Lab) \,\&\, \forall y((Cy \,\&\, Lay) \rightarrow b = y)) \lor \sim Bb$ — 6 DM
8 $\sim\sim((Cb \,\&\, Lab) \,\&\, \forall y((Cy \,\&\, Lay) \rightarrow b = y))$ — 5 DN
9 $\sim Bb$ — 7, 8 DS
10 $Cb \,\&\, Lab$ — 5 &E
11 $(Cb \,\&\, Lab) \,\&\, \sim Bb$ — 9, 10 &I
12 $\exists x((Cx \,\&\, Lax) \,\&\, \sim Bx)$ — 11 $\exists$I
13 $\exists x((Cx \,\&\, Lax) \,\&\, \sim Bx)$ — 4, 5–12 $\exists$E
14 $\exists x((Cx \,\&\, Lax) \,\&\, \forall y((Cy \,\&\, Lay) \rightarrow x = y)) \rightarrow \exists x((Cx \,\&\, Lax) \,\&\, \sim Bx)$ — 4–13 $\rightarrow$I

VI (5) 1 $P \,\&\, Q$ — H (for $\rightarrow$I)
2 P — 2 &E
3 $(P \,\&\, Q) \rightarrow P$ — 1–2 $\rightarrow$I
4 $\Box((P \,\&\, Q) \rightarrow P)$ — 3 N
5 $\Box((P \,\&\, Q) \rightarrow P) \rightarrow (\Box(P \,\&\, Q) \rightarrow \Box P)$ — AS2

6	$\Box(P \& Q) \to \Box P$	$4, 5 \to E$
7	$P \& Q$	H (for $\to$I)
8	Q	7 &E
9	$(P \& Q) \to Q$	7–$8 \to$I
10	$\Box((P \& Q) \to Q)$	9 N
11	$\Box((P \& Q) \to Q) \to (\Box(P \& Q) \to \Box Q)$	AS2
12	$\Box(P \& Q) \to \Box Q$	$10, 11 \to$E
13	$\Box(P \& Q)$	H (for $\to$I)
14	$\Box P$	$6, 13 \to$E
15	$\Box Q$	$12, 13 \to$E
16	$\Box P \& \Box Q$	$14, 15$ &I
17	$\Box(P \& Q) \to (\Box P \& \Box Q)$	13–$16 \to$I
18	$P \to (Q \to (P \& Q))$	TI Supp. Prob. IV(2), Chap. 4
19	$\Box(P \to (Q \to (P \& Q)))$	18 N
20	$\Box(P \to (Q \to (P \& Q))) \to (\Box P \to \Box(Q \to (P \& Q)))$	AS2
21	$\Box P \to \Box(Q \to (P \& Q))$	$19, 20 \to$E
22	$\Box(Q \to (P \& Q)) \to (\Box Q \to \Box(P \& Q))$	AS2
23	$\Box P \to (\Box Q \to \Box(P \& Q))$	$21, 22$ HS
24	$\Box P \& \Box Q$	H (for $\to$I)
25	$\Box P$	24 &E
26	$\Box Q$	24 &E
27	$\Box Q \to \Box(P \& Q)$	$23, 25 \to$E
28	$\Box(P \& Q)$	$26, 27 \to$E
29	$(\Box P \& \Box Q) \to \Box(P \& Q)$	24–$28 \to$I
30	$\Box(P \to Q) \leftrightarrow (\Box P \& \Box Q)$	$17, 29 \leftrightarrow$I

The overall strategy here is $\leftrightarrow$I. The first 17 lines of the proof establish the first of the two required conditionals. Lines 18 to 29 establish the second. On lines 1 to 12 we prove two conditionals, '$\Box(P \& Q) \to \Box P$' and '$\Box(P \& Q) \to \Box Q$', which are needed in the proof of the first conditional.

(10)	1	$\Diamond(P \& Q)$	H (for $\to$I)
	2	$\sim\Box\sim(P \& Q)$	1 AS1
	3	$\sim\Box(\sim P \lor \sim Q)$	2 DM
	4	$(\Box\sim P \lor \Box\sim Q) \to \Box(\sim P \lor \sim Q)$	TI Supp. Prob. IV(8)
	5	$\sim(\Box\sim P \lor \Box\sim Q)$	$3, 4$ MT
	6	$\sim\Box\sim P \& \sim\Box\sim Q$	5 DM
	7	$\Diamond P \& \sim\Box\sim Q$	6 AS1
	8	$\Diamond P \& \Diamond Q$	7 AS1
	9	$\Diamond(P \& Q) \to (\Diamond P \& \Diamond Q)$	1–$8 \to$I

This proof makes frequent use of equivalences. DM is used at steps 3 and 6, and AS1 is used as an equivalence at steps 2, 7, and 8.

Glossary

_A_-form categorical statement. A statement of the form 'All S are P'.

a priori probability. Inherent probability; probability considered apart from the evidence. The a priori probability of a statement is inversely related to its strength.

α-variant. Any model that results from a given model upon freely interpreting the name letter α.

absorption (ABS). The valid inference reference rule $P \rightarrow Q \vdash P \rightarrow (P \& Q)$.

abusive _ad hominem_ argument. A species of _ad hominem_ argument which attempts to refute a claim by attacking personal characteristics of its proponents.

accent, fallacy of. A fallacy of ambiguity caused by misleading emphasis.

ad baculum argument. (See _appeal to force_.)

ad hoc hypothesis. An auxiliary hypothesis adopted without independent justification in order to enable a scientific theory to explain certain anomalous facts.

ad hominem argument. An argument which attempts to refute a claim by discrediting its proponents.

ad ignorantiam argument. (See _appeal to ignorance._)

ad misericordiam argument. (See _appeal to pity._)

ad populum argument. (See _appeal to the people._)

ad verecundiam argument. (See _appeal to authority._)

affirmative categorical statement. An _A_- or _I_-form categorical statement.

affirming the consequent. The invalid argument form $P \rightarrow Q, Q \vdash P$.

algorithm. A rigorously specified test which could in principle be carried out by a computer and which always yields an answer after a finite number of finite operations.

ambiguity. Multiplicity of meaning.

amphiboly. Ambiguity resulting from sentence structure.

antecedent. The statement P in a conditional of the form 'If P, then Q'.

antinomy. An unexpected inconsistency.

appeal to authority. An argument that a claim is true because some person or authority says so.

appeal to force. An argument which attempts to establish its conclusion by threat or intimidation.

appeal to ignorance. Reasoning from the premise that a claim has not been disproved to the conclusion that it is true.

appeal to pity. An argument that a certain action should be excused or a special favor be granted on grounds of extenuating circumstances.

appeal to the people. Reasoning from the premise that an idea is widely held to the conclusion that it is correct.

argument. A sequence of statements, of which one is intended as a conclusion and the others, called premises, are intended to prove or at least provide some evidence for the conclusion.

argument of an _n_-place function. One of an _n_-tuple of objects to which an _n_-place function is applied.

association (ASSOC).   The equivalences $(P \vee (Q \vee R)) \leftrightarrow ((P \vee Q) \vee R)$ and $(P \mathbin{\&} (Q \mathbin{\&} R)) \leftrightarrow ((P \mathbin{\&} Q) \mathbin{\&} R)$.

assumption.   A premise that is not also a conclusion from previous premises.

asymmetric relation.   A relation R such that for all x and y, if Rxy, then not Ryx.

atomic formula.   The simplest sort of wff of a formal language; an atomic formula of the language of predicate logic is a predicate letter followed by zero or more name letters.

axiom.   A formula which does not count as an assumption and yet may be introduced anywhere in a proof; intuitively, an axiom should express an obvious general truth.

axiom schema.   A formula each substitution instance of which counts as an axiom.

basic premise.   (See *assumption*.)

begging the question.   (See *circular reasoning*.)

biconditional.   A statement of the form 'P if and only if Q'.

biconditional elimination ($\leftrightarrow$E).   The valid inference rule $P \leftrightarrow Q \vdash P \rightarrow Q$ or $P \leftrightarrow Q \vdash Q \rightarrow P$.

biconditional introduction ($\leftrightarrow$I).   The valid inference rule $P \rightarrow Q, Q \rightarrow P \vdash P \leftrightarrow Q$.

bivalence, principle of.   The principle that says that true and false are the only truth values and that in every possible situation each statement has one of them.

categorical statement.   A statement which is either an A-, E-, I-, or O-form categorical statement, or a negation thereof.

categorical syllogism.   A two-premise argument consisting entirely of categorical statements and containing exactly three class terms, one of which occurs in both premises, and none of which occurs more than once in a single premise.

circular reasoning.   The fallacy of assuming a conclusion we are attempting to prove.

circumstantial *ad hominem*.   A kind of *ad hominem* argument which attempts to refute a claim by arguing that its proponents are inconsistent in their endorsement of that claim.

class.   A collection (possibly empty) of objects considered without regard to order or description. Also called a set.

class term.   A term denoting a class (set) of objects.

classical interpretation.   The notion of probability by which the probability of an event A relative to a situation is the number of equipossible outcomes in which that event occurs divided by the total number of equipossible outcomes.

commutation (COM).   The equivalences $(P \vee Q) \leftrightarrow (Q \vee P)$ and $(P \mathbin{\&} Q) \leftrightarrow (Q \mathbin{\&} P)$.

complement of a set.   The set of all things which are not members of the set in question.

completeness, semantic.   With respect to a formal system and a semantics for that system, the property that each argument form expressible in that system which is valid by that semantics is provable in the system.

complex argument.   An argument consisting of more than one step of reasoning.

composition, fallacy of.   Invalidly inferring that a thing must have a property because one, or some, or all of its parts do.

conclusion indicator.   An expression attached to a sentence to indicate that it states a conclusion.

conditional.   A statement of the form 'If P, then Q'. Conditionals are also expressible in English by such locutions as 'only if' and 'provided that'.

conditional elimination ($\rightarrow$E).   The valid inference rule $P \rightarrow Q, P \vdash Q$. Also called *modus ponens*.

conditional introduction ($\rightarrow$I).   The valid inference rule which allows us to deduce a conditional after deriving its consequent from the hypothesis of its antecedent.

conditional probability.   The probability of one proposition or event, given another.

conjunct.   Either of a pair of statements joined by the operator 'and'.

conjunction.   The truth functional operation expressed by the term 'and'; the conjunction of two statements is true if and only if both statements are true. Derivatively, a conjunction is any statement whose main operator is '&'.

conjunction elimination (&E).   The valid inference rule $P \& Q \vdash P$ or $P \& Q \vdash Q$.

conjunction introduction (&I).   The valid inference rule $P, Q \vdash P \& Q$.

consequence, truth-functional.   Statement or statement form A is a truth-functional consequence of statement or statement form B if there is no line on their common truth table on which B is true and A is false.

consequent.   The statement Q in a conditional of the form 'If P, then Q'.

constructive dilemma (CD).   The valid inference rule $P \vee Q, P \rightarrow R, Q \rightarrow S \vdash R \vee S$.

contextual definition.   A type of formal definition in which the use of a symbol is explained by showing how entire formulas in which it occurs can be systematically translated into formulas in which it does not occur.

contradiction.   A truth-functionally inconsistent statement. Also, the valid inference rule CON: $P, \sim P \vdash Q$.

contradictories.   Two statements, each of which validly implies the negation of the other.

contrapositive.   The result of replacing the subject term of a categorical statement with the complement of its predicate term and replacing its predicate term with the complement of its subject term.

convergent argument.   An argument containing several steps of reasoning which support the same intermediate or final conclusion.

converse.   The result of exchanging the subject and predicate terms of a categorical statement.

copula.   Any variant of the verb 'to be' which links the subject and predicate terms of a categorical statement.

counterexample.   A possible situation in which the conclusion of an argument or argument form is false while the assumptions are true. A counterexample shows the argument or argument form to be invalid.

counterfactual conditional.   The sort of conditional expressed in English by the subjunctive mood. Counterfactuals are intermediate in strength between material and strict conditionals.

decidability.   A formal system is decidable if there exists an algorithm for determining for any argument form expressible in that system whether or not that form is valid. Propositional logic is decidable; predicate logic is not.

deductive argument.   An argument such that it is logically impossible for its conclusion to be false while its assumptions are all true.

definite description.   An expression which purportedly denotes a single object by enumerating properties which uniquely identify it. Definite descriptions in English typically begin with the word 'the'.

De Morgan's laws (DM).   The equivalences $\sim(P \& Q) \leftrightarrow (\sim P \vee \sim Q)$ and $\sim(P \vee Q) \leftrightarrow (\sim P \& \sim Q)$.

denying the antecedent.   The invalid inference rule $P \rightarrow Q, \sim P \vdash \sim Q$.

deontic logics.   Logics dealing with moral concepts.

derivation.   (See *Proof.*)

derived rule of inference.   A rule of inference which is not one of the basic or defining rules of a formal system but which can be proved in that system.

disjunct.   Either of a pair of statements joined by the operator 'or'.

disjunction.   Any statement whose main operator is 'or' in either the inclusive or exclusive sense.

disjunction elimination ($\vee$E).   The valid inference rule $P \vee Q, P \rightarrow R, Q \rightarrow R \vdash R$.

disjunction, exclusive.   A binary truth-functional operation which when applied to a pair of statements yields a third statement which is true if and only if exactly one of the pair is true.

disjunction, inclusive.   A binary truth-functional operation which when applied to a pair of statements yields a third statement which is true if and only if at least one of the pair is true.

disjunction introduction ($\vee$I).   The valid inference rule $P \vdash P \vee Q$ or $Q \vdash P \vee Q$.

disjunctive syllogism (DS).   The valid inference rule $P \vee Q, \sim P \vdash Q$.

distribution (DIST).   The equivalences $(P \mathbin{\&} (Q \vee R)) \leftrightarrow ((P \mathbin{\&} Q) \vee (P \mathbin{\&} R))$ and $(P \vee (Q \mathbin{\&} R)) \leftrightarrow ((P \vee Q) \mathbin{\&} (P \vee R))$.

division, fallacy of.   Invalidly inferring that a thing's parts must have a property because the thing itself does.

domain of a function.   The set of objects to which a function may be applied; the set of possible arguments for the function.

domain of interpretation.   (See *universe of interpretation*.)

double negation (DN).   The equivalence $P \leftrightarrow \sim\sim P$.

***E*-form categorical statement.**   A statement of the form 'No S are P'.

entailment.   The relationship which holds between statement or set of statements S and a statement A if it is logically impossible for A to be false while S is true, and if at the same time S is relevant to A.

epistemic logics.   Logics dealing with the concepts of knowledge and belief.

equivalence.   A biconditional which is a theorem.

equivalence, truth-functional.   The relationship that holds between two statements or statement forms if they have the same truth table.

equivocation.   (See *ambiguity*.)

existential elimination ($\exists$E).   A rule of inference which, provided that certain restrictions are met (see Section 7.3), allows us to infer a conclusion after deriving that conclusion from a hypothesized instance of an existentially quantified formula.

existential introduction ($\exists$I).   The valid inference rule that allows us to infer an existentially quantified wff from any of its instances.

existential quantifier.   The symbol '$\exists$', which means "for at least one." English terms expressing the same meaning are also sometimes called existential quantifiers.

exportation (EXP).   The equivalence $((P \mathbin{\&} Q) \rightarrow R) \leftrightarrow (P \rightarrow (Q \rightarrow R))$.

fallacy.   Any mistake that affects the cogency of an argument.

fallacy of relevance.   Reasoning which is mistaken because the premises are not sufficiently relevant to the conclusion.

false-cause fallacy.   Concluding that one event causes another on the basis of insufficient evidence.

false dichotomy, fallacy of.   Reasoning which is faulty because of a false disjunctive premise.

faulty analogy, fallacy of.   Analogical reasoning whose inductive probability is low because of the inadequacy of the analogy on which it is based.

first-order logic.   Quantificational logic whose quantifiers range only over individual objects.

formal fallacy.   Mistaken reasoning resulting from misapplication of a valid rule of inference or application of an invalid rule.

formal language.   A rigorously defined language.

formal logic.   The study of argument forms.

formal system.   A formal language together with a set of inference rules and/or axioms for that language. (The propositional calculus, for example, is a formal system.)

formation rules.   A set of rules which define the well-formed formulas (wffs) of some formal language.

formula.   Any finite sequence of elements of the vocabulary of some formal language.

free logic.   A modification of predicate logic in which proper names are not assumed to designate existing objects.

function.   An operation which for some n assigns unique objects to n-tuples of objects.

function symbol.   A symbol which when applied to some definite number n of names or other denoting terms produces a complex term which denotes a single object.

functional expression.   An expression formed by applying an n-place function symbol to n denoting terms.

gambler's fallacy.   An argument of the form:

> x has not occurred recently.
> $\therefore$ x is likely to happen soon.

where 'x' designates an event whose occurrences are independent.

guilt by association, fallacy of.   A species of *ad hominem* argument in which a claim is attacked by pointing out that one or more of its proponents are associated with allegedly unsavory characters.

hasty generalization.   Fallaciously inferring a statement about an entire class of things on the basis of information about some of its members.

higher-order logics.   Logics which use special variables for quantification over properties or relations in addition to the variables used for quantification over individual objects.

Humean argument.   An inductive argument which presupposes the uniformity of nature.

hypothesis.   An assumption introduced into a proof in order to show that certain consequences follow, but which must be discharged before the proof is completed.

hypothetical derivation.   A derivation which begins with a hypothesis and ends when that hypothesis is discharged.

hypothetical syllogism (HS).   The valid inference rule $P \to Q, Q \to R \vdash P \to R$.

***I*-form categorical statement.**   A statement of the form 'Some S are P'.

identity elimination (=E).   The rule that allows us to infer the result of replacing one or more occurrences of a name letter α by a name letter β in any wff containing α, using that wff together with $\alpha = \beta$ or $\beta = \alpha$ as premises.

identity introduction (=I).   The rule that allows us to write $\alpha = \alpha$, for any name letter α, at any line of a proof.

identity predicate.   The symbol '=', which means "is identical to."

ignoratio elenchi.   The fallacy of drawing a conclusion different from (and perhaps opposed to) what the argument actually warrants. Also called *missing the point*.

immediate inference.   An inference from a single categorical statement as premise to a categorical statement as conclusion.

inconsistent set of statements.   A set of statements whose semantics alone prevents their simultaneous truth; a set of statements whose simultaneous truth is logically impossible.

inconsistent statement.   A statement whose semantics alone prevents its truth; a statement whose truth is logically impossible.

independence.   The relationship which holds between two propositions (or events) when the probability of either is unaffected by the truth (or occurrence) of the other.

indirect proof. (See *negation introduction*.)

induction by analogy. An argument of the form:

$$F_1 x \,\&\, F_2 x \,\&\, \cdots \,\&\, F_n x$$
$$F_1 y \,\&\, F_2 y \,\&\, \cdots \,\&\, F_n y$$
$$Gy$$
$$\therefore\ Gx$$

induction by enumeration. (See *simple induction*.)

inductive argument. A nondeductive argument; an argument such that it is logically possible for its conclusion to be false while all its assumptions are true.

inductive fallacy. Overestimation of an argument's inductive probability.

inductive generalization. An argument of the form:

n percent of s thus-far-observed F are G.
$\therefore$ About n percent of all F are G.

inductive probability. The probability of a conclusion, given a set of premises.

inference indicator. A premise indicator or conclusion indicator.

inference rules. The rules of a formal system that determine which steps of reasoning are admissible in proofs.

informal logic. The study of particular arguments in their natural-language contexts.

instance of a quantified wff. The result of removing the initial quantifier and replacing the variable it binds by a name letter.

intermediate conclusion. (See *nonbasic premise*.)

interpretation structure. (See *model*.)

invalid form. An argument form at least one instance of which is not valid.

irreflexive relation. A relation R such that there is no x such that Rxx.

Leibniz' law. The second-order principle: $\forall x \forall y (\forall P(Px \leftrightarrow Py) \leftrightarrow x = y)$.

logical operator. An element of the vocabulary of some formal language which has a fixed interpretation with respect to that language. (For example, '&' is a logical operator of the language of propositional logic.) Derivatively, a logical operator is a natural-language expression of such a term. (Thus, 'and' is a logical operator of English.)

logical probability. (See *a priori probability*.)

logically impossible. Impossible purely in virtue of semantic considerations.

logically necessary. Necessary purely in virtue of semantic considerations.

logically possible. Not ruled out by semantic considerations alone.

main operator. The operator in a wff whose scope is the entire wff.

major term. The predicate term of the conclusion of a categorical syllogism.

material conditional. The kind of conditional which is true if and only if it is not the case that its antecedent is true and its consequent is false. The material conditional is formalized in this book by the symbol '$\rightarrow$'.

material implication (MI). The equivalence $(P \rightarrow Q) \leftrightarrow (\sim P \lor Q)$.

mathematical induction. The principle that if 0 has a certain property and, for any nonnegative integer N, if N has that property, then so does $N + 1$, then all the nonnegative integers have it. (In spite of the name, this principle licenses *deductive* inferences.)

metalogic. Logical reasoning about logical theories.

middle term.   The class term which occurs in both premises of a categorical syllogism.

Mill's methods.   Forms of reasoning used to rule out suspected causes of an observed effect.

minor term.   The subject term of the conclusion of a categorical syllogism.

missing the point.   (See *ignoratio elenchi*.)

modal logic.   The logic of modal expressions, particularly 'it is possible that' and 'it is necessary that'.

model.   The assignment of an interpretation to the nonlogical symbols of the language of predicate logic, relative to a specified universe or domain of interpretation.

modus ponens.   (See *conditional elimination*.)

modus tollens (MT).   The valid inference rule $P \rightarrow Q, \sim Q \vdash \sim P$.

mutually exclusive.   Two propositions are mutually exclusive if they cannot both be true.

***n*-tuple.**   A linear ordering (list) of n objects.

necessary cause.   A condition needed to produce a certain effect.

necessitation (N).   The inference rule of modal logic that allows us to infer $\Box$A from any theorem A.

negation.   The truth-functional operation expressed by the term 'not'; the negation of a statement is true if and only if the statement itself is false. Derivatively, a negation is any statement whose main operator is '$\sim$'.

negation elimination ($\sim$E).   The valid inference rule $\sim\sim P \vdash P$.

negation introduction ($\sim$I).   The valid inference rule that allows us to infer a conclusion of the form $\sim P$ after having derived a contradiction from the hypothesis P. Also called indirect proof or *reductio ad absurdum*.

negative categorical statement.   An E- or O-form categorical statement.

non sequitur.   Any fallacy of relevance.

nonbasic premise.   A premise of a complex argument which functions as the conclusion of one step of reasoning and a premise of a succeeding step.

nonlogical symbols.   Those symbols of a formal language whose interpretation varies from context to context.

***O*-form categorical statement.**   A statement of the form 'Some S are not P', where 'not' represents complementation.

obverse.   The result of changing the quality of a categorical statement and replacing the predicate term by its complement.

open formula.   A formula which results from removing one or more initial quantifiers from a quantified wff.

operator.   (See *logical operator*.)

Π-variant.   Any model that results from a given model upon freely interpreting the predicate letter Π.

particular categorical statement.   An I- or O-form categorical statement.

petitio principii.   (See *circular reasoning*.)

possible world.   A conceivable state of affairs; unless otherwise specified, 'possible' or 'conceivable' in this context usually means "logically possible."

***post hoc* fallacy.**   The fallacy of reasoning from the premise that A precedes B to the conclusion that A causes B.

predicate calculus.   The system of inference rules designed to yield a proof of the valid argument forms of predicate logic. Also called first-order predicate calculus, among other designations.

predicate logic.   The study of the logical concept of validity insofar as this is due to the truth-functional operators, the quantifiers, and the identity predicate.

predicate term.   The second of the two class terms in a categorical statement.

premise indicator.   An expression attached to a sentence to indicate that it states a premise.

principle of charity.   The principle that in formulating implicit statements for argument analysis, one should remain as faithful as possible to what is known of the arguer's thought.

probability calculus.   Axioms AX1 through AX3 (Section 10.2) and their deductive consequences.

proof.   A sequence of wffs of some formal system, each member of which is either an assumption (possibly hypothetical) or an axiom, or is derived from previous wffs by the inference rules of that system.

proposition.   (See *statement.*)

propositional calculus.   The system of inference rules designed to yield a proof of the valid argument forms of propositional logic. Also called the sentential calculus, or the statement calculus.

propositional logic.   The study of the logical concept of validity insofar as this is due to the truth-functional operators.

quality.   The classification of a categorical statement as affirmative or negative.

quantifier.   A logical operator which binds the variable of an open formula to create a wff. Quantifiers are expressed in English by such words as 'any', 'every', 'each', 'some', 'all', and 'no'.

quantifier exchange rules (QE).   A set of inference rules of the predicate calculus based on the equivalences listed in Table 7-1.

quantity.   The classification of a categorical statement as universal or particular.

range of a function.   The set of possible values for a function.

red herring.   An extraneous matter introduced to divert attention from the issue posed by an argument.

reductio ad absurdum.   (See *negation introduction.*)

reflexive relation.   A relation R such that for all x, Rxx.

refutation tree.   An exhaustive search for ways in which the premises and the negation of the conclusion of an argument form can be true. Used to test the validity of the argument form.

relational predicate.   A predicate which combines with two or more names to make a sentence.

relevance logic.   The study of entailment.

repeat (RE).   The valid inference rule $P \vdash P$.

requirement of total evidence.   The principle that if an argument is inductive, its premises must contain all known evidence that is relevant to the conclusion.

scope.   The scope of an occurrence of a logical operator in a wff is the smallest subwff which contains that operator; intuitively, that occurrence of the operator together with that part of the wff to which it applies.

second-order logic.   Quantificational logic whose quantifiers range only over individual objects and their properties (which may not be construed as sets).

semantic fallacy.   Mistaken reasoning resulting from vagueness or multiplicity of meaning.

semantics.   The study of meaning; the semantics of an expression in the contribution it makes to the truth or falsity of statements which contain it.

set.   (See *class.*)

simple induction.   An argument of the form:

> n percent of the s thus-far-observed F are G.
>
> $\therefore$ If one more F is observed, it will be G.

slippery-slope argument.   An argument of the form:

$A_1 \to A_2$.
$A_2 \to A_3$.
$\vdots$
$A_n \to A_{n+1}$.
It should not be the case that A_{n+1}.
$\therefore$ It should not be the case that A_1.

sound argument.   A valid argument whose assumptions are all true.

standard form.   A format for writing arguments in which premises are listed before conclusions and the conclusions are prefixed by '$\therefore$'.

statement.   A thought of the kind expressible by a declarative sentence.

statistical argument.   An inductive argument which does not presuppose the uniformity of nature.

statistical generalization.   An argument of the form:

n percent of s randomly selected F are G.
$\therefore$ About n percent of all F are G.

statistical syllogism.   An argument of the form:

n percent of F are G.
x is F.
$\therefore$ x is G.

straw man fallacy.   An attempt to refute a person's claim which confuses that claim with a less plausible claim not advanced by that person.

strength of a statement.   The information content of a statement.

strict conditional.   A kind of conditional which is true if and only if its antecedent necessarily implies its consequent.

strong reasoning.   High inductive probability.

subject term.   The first of the two class terms in a categorical statement.

subjective probability.   A person's degree of belief in a proposition, as gauged by that person's willingness to accept certain wagers regarding that proposition.

subwff.   A part of a wff which is itself a wff.

sufficient cause.   A condition which always produces a certain effect.

suppressed evidence, fallacy of.   A violation of the requirement of total evidence so serious as to constitute a substantial mistake.

symmetrical relation.   A relation R such that for all x and y, if Rxy, then Ryx.

syntax.   Grammar; the syntax of a formal language is encoded in its formation rules. Formal inference rules are syntactic, since they refer only to grammatical form, not to truth conditions; whereas truth tables, Venn diagrams, and refutation trees are semantic in nature.

tautology.   A wff of propositional logic whose truth table contains only T's under its main operator. Derivatively, any statement whose formalization is such a wff. (The term 'tautology' is also sometimes used, though not in this book, to designate any logically necessary truth.) Also, either of the equivalences called TAUT: $P \leftrightarrow (P \mathbin{\&} P)$ or $P \leftrightarrow (P \lor P)$.

theorem.   A wff of some formal system which is the conclusion of some proof of that system that does not contain any nonhypothetical assumptions.

theorem introduction (TI).   The inference rule that permits inserting a theorem at any line of a proof.

transitive relation.   A relation R such that for all x, y, and z, if Rxy and Ryz, then Rxz.

truth-functionally contingent wff.   A wff of the propositional calculus whose truth table contains a mix of T's and F's under its main operator.

truth-functionally inconsistent wff.   A wff of the propositional calculus whose truth table contains only F's under its main operator. Any statement whose formalization is such a wff is also said to be truth-functionally inconsistent.

truth table.   A table listing the truth value of a wff under all possible assignments of truth values to its atomic subwffs.

truth value.   The truth or falsity of a statement or wff, often designated by 'T' or 'F' respectively.

***tu quoque* fallacy.**   A species of *ad hominem* fallacy in which an attempt is made to refute a belief by arguing that its proponents hold it hypocritically.

type system.   A formal system with a number of different styles of variables, each intended to be interpreted over one of a series of hierarchically arranged domains.

universal categorical statement.   An A- or E-form categorical statement.

universal elimination ($\forall$E).   The valid inference rule that allows us to infer from a universally quantified wff any instance of that wff.

universal generalization.   (See *universal introduction*.)

universal instantiation.   (See *universal elimination*.)

universal introduction ($\forall$I).   A valid inference rule which, provided that certain restrictions are met (see Section 7.2), allows us to infer a universally quantified statement from a proof of one of its instances.

universal quantifier.   The symbol '$\forall$', which means "for all." English terms which express the same meaning are sometimes also called universal quantifiers.

universe (of interpretation).   A nonempty class of objects relative to which a model specifies an interpretation for the nonlogical symbols of predicate logic.

valid argument.   (See *deductive argument*.)

valid form.   An argument form every instance of which is valid.

value of a function.   The object assigned by an n-place function to a given n-tuple of arguments.

Venn diagram.   A representation of relationships among class terms used to display the semantics of categorical statements and to test some arguments in which they occur for validity. (See Chapter 5.)

vested interest fallacy.   A kind of *ad hominem* fallacy in which an attempt is made to refute a claim by arguing that its proponents are motivated by desire for personal gain or to avoid personal loss.

weak reasoning.   Low inductive probability.

wff.   A well-formed formula of some formal language, as defined by the formation rules of that language.

Index

The letter *n* following a page number refers to a footnote

Truth table

Disjunction

P ∧ Q
T T T
T F F
F F T
F F F

Exclusive Disjunction

P ⌄ Q
T F T
T T F
F T T
F F F

Inclusive Disjunction

P ∨ Q
T T T
T T F
F T T
F F F

v = or (vel)

∧ = and

Deductive = conclusion deduced from what you already know

Inductive = Generalising from set of instances

Infinite regress [A] , [B]

↓

has to be something between holding the 2 together.